Fundamentals of Psychiatry for Health Care Professionals

Roberto Cavallaro • Cristina Colombo
Editors

Fundamentals of Psychiatry for Health Care Professionals

Springer

Editors
Roberto Cavallaro
Vita-Salute San Raffaele University
General Psychiatry Unit IRCCS San Raffaele Scientific Institute
Milan, Italy

Cristina Colombo
Vita-Salute San Raffaele University
Mood Disorders Unit IRCCS San Raffaele Scientific Institute
Milan, Italy

ISBN 978-3-031-07714-2 ISBN 978-3-031-07715-9 (eBook)
https://doi.org/10.1007/978-3-031-07715-9

This Springer imprint is published by the registered company Springer Nature Switzerland AG
The registered company address is: Gewerbestrasse 11, 6330 Cham, Switzerland

Preface

The World Health Organization (WHO) defines normality as a state of complete physical, mental, and social well-being, but this definition is limited, because it defines physical and mental health simply as the absence of a physical or mental disease.

The text revision of the Diagnostic and Statistical Manual of Mental Disorders offers no definition of normality or mental health, although a definition of mental disorder is presented: Disturbances of an individual's behaviour or of his psychological functioning that are not culturally expected and that lead to psychological distress, behavioural disability, or impaired overall functioning (DSM-5).

A mental disorder is defined as a syndrome characterized by clinically significant disturbance in an individual's cognition, emotion regulation, or behaviour that reflects a dysfunction in the psychological, biological, or developmental processes underlying mental functioning. Mental disorders are usually associated with significant distress in social, occupational, or other important activities.

An expectable or culturally approved response to a common stressor or loss, such as the death of a loved one, is not a mental disorder. Socially deviant behaviour (i.e. political, religious, or sexual) and conflicts that are primarily between the individual and society are not mental disorders unless the deviance or conflict results from a dysfunction in the individual, as described above (DSM-5).

But all this distinction (still imperfect and amenable to improvement) was not such until recent times, in particular up to the discovery of effective treatments on one side and, on the other, to the changes in the society which led to the widening of the personal freedoms and to recognize higher minimal levels of human dignity and rights, at least in western countries.

The border between mental insanity and a number of 'out of standard' behaviours not related to it might still look subtle and it is so since unmemorable times.

Ideas of mentally 'normal' and 'abnormal' are largely shaped by social standards and can have profound social ramifications. What is considered 'normal' changes with changing of societal standards. Society generally sees normality as good and abnormality as bad. Being labelled as 'normal' or 'abnormal' can have profound consequences for an individual, such as exclusion or stigmatization by society. Stigma and discrimination add suffering to the suffering and disability for who is diagnosed with (or perceived to have) a mental disorder and his family.

Often the psychiatrist or its past ‘prototypes’ is attributed a role of arbiter in establishing the boundary between normality and pathology. This role in the past has sometimes been taken very seriously, so much that it has contributed to the discrimination and stigma related to mental illness.

The consequences of this have been dramatic in history, also with religious or political interpretations going from the attribution of behavioural problems and mental illness to the power of the evil, i.e. during the ‘holy inquisition period’ with hundreds burned, to the eugenetic programmes that obscured the first half of the last century with sterilization of the mentally ill in different countries including not only the Nazi regime, but also the USA, with the model from Harry Laughlin drafter in 1914 and putting together any form of ‘unfitness’ like Mentally retarded, Mentally ill, Deaf, Blind, Epileptic, Physically deformed, Orphans, Homeless, Homosexuals. By 1940, 30 of the 48 states of the Union had enacted eugenical sterilization laws. The Soviet Society of Psychiatrists was forced to withdraw from the WPA in 1983, due to the systematic use of psychiatry (and consequent institutionalization) for political reasons, and returned only in 1989 when the USSR was opening up to the outside world.

Besides these extreme consequences (difficult to believe nowadays), the founding of mental asylums all over the world was itself the cause of much of the suffering of the mentally ill, being characterized by seclusion, isolation from the real world, and frequent psychological and physical abuses. Italy, until the passage of 1978, was ruled by a special legislation for mental illness of 1908, the ‘disposal for asylum and insanity’ according to which commitment to a mental hospital was recorded on the court register in a similar way to a criminal conviction.

Looking at the charts in the archives of those hospitals we can find not only the drama of this procedure on many mentally ills, but also the frequently unmotivated or not persisting reasons for receiving and maintaining the commitment.

Persons were somewhat cancelled from the community, and their long-term experience of the mental hospital with chronic isolation became an illness itself, probably the only (iatrogenic) chronic illness for many people who entered the hospital. Deprivation induced depression, with many suicides, affective flattening with unresponsivity to any stimuli and regression of behaviour to few elementary functions, independently from the starting reasons of conviction.

What led to all this was the lack of a medical explanation, of a therapeutical perspective and of a specific nosography, but this need was anticipated by some brilliant minds even in the late nineteenth century, like Emil Kraepelin, one of the (still) more influential psychiatrists of history, who did some experiments in German hospitals testing with scientific methods the effects of known psychotropics on cognition (i.e. reaction times or memory with substances from caffeine to alcohol to morphine) with specific instruments and study designs, calling this research ‘pharmacopsychology’, but that proposed also at the beginning of the twentieth century a nosography based on a definition and separation of different mental disorders.

That is, putting together the two ‘souls’ of Emil Kraepelin, method: adequate and homogeneous target, known effects of an environmental variable interacting with the brain (including drugs) and reliable measure.

These are the main features of current psychiatric research, but also clinical treatment.

But while nosography and psychopathology evolved since the first decades of the twentieth century, effective medical treatments were developed much later.

Yet, until the 1950s there was no such scientific discipline as psychopharmacology and there was no effective drug therapy for mental illness.

In 1950 two psychiatrists, Jean Delay and Pierre Deniker, working at St Anne's Hospital in Paris used chlorpromazine to treat inpatients, including people suffering with mania and schizophrenia. They concluded that chlorpromazine was highly effective and published a series of reports, the first appearing in 1952. They drew particular attention to the ability of chlorpromazine to control agitation and excitement. Over the following years, use of chlorpromazine in psychiatry spreads and further publications appear in the medical press. Psychiatrists were impressed by its benefits and felt that a new era of treatment was starting. The discovery of antidepressants and lithium was practically parallel in the 1950s.

All of them were a tremendous game changer for people health, their lives, and also society, pressing the need of progress in nosography, and changes in political decisions on psychiatry. Population living in mental hospitals dramatically dropped from this time on in the USA and other countries.

As a result of the development of knowledge starting with serendipity with the discovery of chlorpromazine, psychiatry gained the dignity of autonomous medical specialty separated from neurology, and this induced the development of more refined and reliable nosography, the development of new psychometric and technological instruments and consequently of the ability to study mental illness more scientifically. The many progresses reached in the last 60 years, with theories swinging from a prevalent biological (genetic-familiar) explanation to a psychosocial one, led at a certain time, in the last two decades, to consider the origin of the mental illness like any other medical illness nowadays, with an interaction of the individual genetic-structural features at birth with environmental factors occurring during life, in particular early experience, from gestation to adolescence, as mother malnutrition during gestation, childhood and adolescence distressing familiar and social experiences, like migration and urbanicity, or the widespread use of substances.

More and more literature shows that for these reasons prevention and cure is now possible, identifying 'at risk' populations and managing them with preventive programmes, having an effective secondary prevention with early recognition of illnesses and reducing the burden of illness. We have to consider that according to the last analysis of WHO data on the burden of care six psychiatric illnesses including depression, anxiety disorders, schizophrenia, alcohol abuse, substance abuse, and other mental disorders are in the first 20 positions of 'Years Lived with Disability' list including all medical pathologies. These results should influence policies for mental illnesses prevention and care, reducing hospitalizations and increasing community interventions, but also encourage a basic, but solid knowledge of psychiatry also in all the health professionals, as possible observer of an undiagnosed, untreated, or poorly managed mental illness comorbid with another medical condition.

This can happen only if de-stigmatization occurs, also with the help of a clear differentiation between what is psychiatry and what is not, and last, but not least, letting the psychiatrist out of the role in the area 'social control'.

Nowadays the multiple 'phenocopies' of the psychiatric illness (i.e. maladaptive reaction to stress vs stress-induced psychopathology in comparison to mood disorders) coming from the society changes and crises (i.e. migration, drop of support related to the crises of societal values and family structure and continuity, as well as the widespread diffusion of drugs of abuse) make frequently difficult to separate a psychological or a social problem from psychiatric illness, with opposite consequences of under- and over-treatment. Tautologically this effect is inversely correlated to knowledge and specific education.

A poor knowledge of illness with all the previously cited confounding factors leads to a generalization of the attribution of behaviours that can be seen in the 'vulgarization' of the terms with which these behaviours are also reported not only by media, but also from professional to professional.

Again, this is also related to the more and more frequent observation of cases in which the behavioural picture comes from a large interaction of mainly social, psychological, and environmental factors (like substance abuse comorbidity), with a 'spurious' clinical picture in comparison to 'paradigmatic' (i.e. DSM 5 diagnosed) 'mental illnesses'. This facilitates the practice of 'dimensional' clinical approaches through targeting symptoms with interventions chosen by analogy to those indicated in the 'pure' and full-blown illnesses that in this way loose specificity (an example for all: 'depression' may be a symptom, a syndrome, or an illness, with different therapeutical approaches). Knowing not only symptoms, but also epidemiology, sex, and age-related differences, temporal patterns of manifestation and expected rates of response to specific pharmacological interventions, increases the effectiveness of diagnosis and therapeutical indications.

Recognition or even suspicion of a mental illness must lead the professional to evaluate possible paths to diagnosis, through specialistic evaluation of the situation by a psychiatrist who can discriminate the different possibilities making a diagnosis or not, prescribing treatments that might be pharmacological or psychological, and in this last case sending the patient to a psychologist. Obviously this must happen within a good relationship with the patient to avoid to activate fears about psychiatry, showing at the beginning just the sensibility to a little or big distress and then, through the acquired knowledge, activate a decision-making process on when and how to tell the patient he probably needs some support.

Every patient that passes through the nets of clinical assistance without eliciting attention is somewhat lost, not only because the occasion to relief him from suffering is lost, but also because we miss the target to let him know that a communication about these problems is possible, even if he feels shame or he is not able to ask for help. Self-stigma usually relies on these personal dimensions besides cultural conventions and personal experience.

Moreover even nowadays some health professionals consider psychiatry as something not completely medical, far from everyday life and work experience, as a form of more severe psychological distress or disturbed personality, or, on the contrary, as something scaring and needing 'social' or 'pharmacological' control.

Showing then themselves a kind of stigma for psychiatry.

The need to consider these multiple aspects makes a personalization of the therapy especially true for our patients, much more than for other medical diseases.

Patients with psychiatric disorders cannot be largely described by numbers (like, i.e. glycaemia in diabetes or QT time in heart diseases), but curing them means to master the different interacting and coexisting variables and peculiarities as a whole, which makes every patient different, even when the main diagnosis is the same. This builds the necessary relationship on a common field of interest, comprehension, reciprocal acknowledgement and truth, the reason why, for example, even most of patients affected by schizophrenia take the prescribed therapy for an illness they do not believe to have, when they are accepted, understood, and involved in a therapeutical project considering all the different features (the self, the image of the self, the personal history, the psychological attitudes and affects, the psychopathology). These different parts act, from the subjective point of view, within the 'stream of consciousness' experience like for everybody, with psychopathology manifesting together with normal ideas, affects, emotions, and thought processes.

Reaching this mastering and being able to interact with empathy tuned on each patients, without standards, let the patient feel, 'he understands and takes care of me': many patients say that a major problem in having a psychiatric disorders is that others pretend you can behave like them, but you cannot…

For all these reasons, we decided to resume in this book the main basic elements of knowledge of psychiatry and mental illness, including psychopathology, diagnosis, and general indications for pharmacological and non-pharmacological interventions, including notes on the interaction of drug treatments and psychopathology itself with the physical health and with the management of other medical treatments and conditions.

In fact it is not casual that the more the psychiatric condition is untreated, the more it becomes invalidating also for the physical health. A few years ago the *Lancet*, one of the major journals in the panorama of Medicine entitled an important article 'There is no health without mental health' debating the complex interaction between the two, and very recently a large epidemiological study performed in Great Britain showed that the probability of infection, hospitalization, and death during the Covid-19 pandemic is statistically and epidemiologically higher in people with severe mental illness and in particular for patients affected by Schizophrenia, Bipolar Disorder, and Major Depression Disorder, even considering all the known other risk factors.

The aim was to write an opera useful to health professionals in their formation process, but also as a simple reference to professionals for their professional life once they graduate and start their clinical work.

For these purposes this book is based on the Psychiatry lessons held for students of the courses in Medicine, Psychology, Nursing, Physiotherapy, and Odontoiatry of our University, a big occasion for us to have a continuous interaction with different cultural, vocational, and academic backgrounds, rich of feedbacks which helped us to be aware of different pathways to reach, in our opinion, the 'hot spot': that is to increase the possibility to intersecate psychiatric histories, and to better apply the different professionalities to the delivery of efficient treatment programmes considering also psychiatric disorder.

Milan, Italy R. Cavallaro
Milan, Italy C. Colombo

The original version of the book was revised. The Affiliation has been updated on the copyright page. A correction to the book is available at https://doi.org/10.1007/978-3-031-07715-9_15

Contents

1 **Introduction** 1
L. Fregna, F. Martini, F. Pacchioni, J. Sapienza, and C. Colombo

2 **Psychopathology** 15
F. Martini, L. Fregna, I. Vanelli, G. Bigai, L. Tonet, and R. Cavallaro

3 **Mood Disorders** 49
L. Franchini, B. Barbini, R. Zanardi, L. Fregna, F. Martini, E. Manfredi, A. Sarzetto, B. Maiocchi, N. Ragone, and C. Colombo

4 **Psychotic Disorders** 85
M. Bosia, M. Spangaro, F. Cocchi, J. Sapienza, L. Tonet, F. Martini, L. Fregna, C. Passani, and R. Cavallaro

5 **Anxiety-Related Disorders** 121
M. Locatelli, I. Vanelli, L. Fregna, F. Martini, E. Manfredi, F. Pacchioni, G. Bigai, L. Tonet, C. Passani, and C. Colombo

6 **Personality Disorders** 181
A. Somma, L. Fregna, F. Martini, and A. Fossati

7 **Adult Consequences of Neurodevelopmental Disorders** 199
M. Bosia, F. Seghi, G. Bigai, F. Martini, L. Fregna, V. Fazio, and R. Cavallaro

8 **Eating Disorders** 229
S. Erzegovesi, D. Pratesi, F. Martini, L. Fregna, M. Carminati, and M. C. Cavallini

9 **Substance-Related Disorders** 263
F. Martini, L. Fregna, M. Bosia, G. Perrozzi, and R. Cavallaro

10 **Organic Mental Disorders and Psychiatric Issues in the Elderly** 297
M. Bosia, F. Cuoco, G. Bigai, F. Martini, L. Fregna, C. Congedo, and R. Cavallaro

11 Psychopharmacology 333
R. Zanardi, M. Spangaro, F. Attanasio, J. Sapienza, F. Martini, L. Fregna, R. Cavallaro, and C. Colombo

12 Non-Pharmacological Treatments 389
B. Barbini, F. Seghi, M. Bosia, L. Fregna, F. Martini, F. Attanasio, E. Manfredi, G. Vassena, C. Colombo, and R. Cavallaro

13 Psychiatric Emergencies 427
M. Locatelli, E. Manfredi, L. Fregna, F. Martini, D. Pratesi, G. Perrozzi, and C. Colombo

14 Psychiatric Legislation and Forensic Psychiatry 441
G. Travaini, R. Zanardi, L. Fregna, F. Martini, D. Pratesi, A. Sarzetto, G. Perrozzi, and C. Colombo

Correction to: Fundamentals of Psychiatry for Health Care Professionals C1

Introduction

1

L. Fregna, F. Martini, F. Pacchioni, J. Sapienza,
and C. Colombo

1.1 Definitions

1.1.1 Mental Health

The World Health Organization (WHO) defines mental health as "a state of well-being in which every individual realizes his or her own potential, can cope with the normal stresses of life, can work productively and fruitfully, and is able to make a contribution to her or his community."

It should be noted that the concept of mental health goes beyond the simple absence of disease, extending it to all the aspects of life. It is a human condition that takes shape in personal development and social relationships, in the ability to adapt, in the emotional and affective awareness and whose definition is inextricably linked to the cultural context.

L. Fregna (✉) · F. Martini (✉)
IRCCS San Raffaele Scientific Institute, Milan, Italy
e-mail: fregna.lorenzo@hsr.it; francesca.martini1@hsr.it

F. Pacchioni · J. Sapienza
Vita-Salute San Raffaele University, Milan, Italy
e-mail: pacchioni.federico@hsr.it; sapienza.jacopo@hsr.it

C. Colombo
IRCCS San Raffaele Scientific Institute, Milan, Italy

Vita-Salute San Raffaele University, Milan, Italy
e-mail: colombo.cristina@hsr.it

R. Cavallaro, C. Colombo (eds.), *Fundamentals of Psychiatry for Health Care Professionals*, https://doi.org/10.1007/978-3-031-07715-9_1

1.1.2 Psychiatry

According to the American Psychiatric Association (APA), psychiatry is the branch of medicine focused on the diagnosis, treatment and prevention of mental, emotional and behavioral disorders.

Mental disorders are categorized and grouped in manuals, based on diagnostic criteria. The two most important diagnostic manuals in psychiatry are the *Diagnostic and Statistical Manual of mental disorders* (*DSM*), currently in its fifth edition, and the *International Classification of Diseases, tenth revision* (*ICD-10*).

Psychiatry is therefore a medical discipline, which scientifically deals with the prevention, diagnosis, and treatment of mental illness. It is a relatively young discipline and thus extremely dynamic and constantly expanding. As estimated by the WHO, psychic pathology is overgrowing and will reach the top places in the coming years in terms of social and economic impact, making this branch of medicine even more relevant.

1.1.3 Mental Disorder

In the *Diagnostic and Statistical Manual of mental disorders, fifth edition (DSM 5)*, "a mental disorder is a syndrome characterized by clinically significant disturbance in an individual's cognition, emotion regulation, or behavior that reflects a dysfunction in the psychological, biological, or developmental processes underlying mental functioning. Mental disorders are usually associated with significant distress or disability in social, occupational, or other important activities."

The definition of mental illness, as well as of mental health, is complex and not unambiguous. Over the years, there have been continuous attempts to formalize and make this concept as universal as possible. Numerous human conditions have been included or excluded from the psychopathology chapter based on cultural and social changes. Also for this reason, periodically, a revision of the criteria for defining mental pathologies is carried out. This results in a new edition of the reference manual of this branch of medicine, the DSM.

1.2 History

Although psychiatry is considered a fairly young medical discipline, the interest in the psyche and its disease has accompanied the history of humanity. The stages that led to the modern conception of psychiatry will be summarized, starting from the fifth century BC up to present days.

1.2.1 The Origins: Greece and Roman Psychiatry

Although notions of mental illness seem to trace back to the Neolithic, the history of "properly called" psychiatry begins in ancient Greece. While among the most ancient civilizations of Egypt and Mesopotamia, the concept of mental illness oscillated between natural and supernatural explanations, the Greeks expressed themselves in a rather decisive way on the biological origin of pathologies. The main representatives of the Greek medical school may be considered Hippocrates (460–377 BC) and Galen (129–201 AC). There are few direct testimonies of their approach to mental illness, indeed much of what we know is owed to indirect sources, such as Celsus, Aretaeus of Cappadocia, and Soranus of Ephesus.

The classical theory of Hippocrates and Galen is best known as Theory of Humors: the body and its state of health or sickness depends on the prevalence of one of the four humors which are fundamental constituents of human body.

The prevalence of one of the humors on the others leads to an imbalance, therefore to the disease. Thus, the prevalence of black bile is the source, for example, of melancholy.

Although this represents the best known theory in ancient Greece, starting from the fourth century BC New Greek schools were established, and the most important is probably that of the Solidists. Their greater exponent was Soranus of Ephesus (100 BC). The Solidist School took this name because its opposition to the Theory of the Humors and it focused on the solid components of the body. As mentioned before, the Solidist School recognizes as its best-known representative in methodist Sorano of Ephesus. In the methodist conception, mental disease would take origin from excessive tension (*status strictus*) or loosening (*status laxus*) of tissues. In his treatise on general medicine, Sorano reserves at least three chapters to the description of mental illnesses: frenite (from "fren," diaphragm, seat of the soul according to the Greeks), mania, and melancholy. None of the Greco-Roman interpretations correspond to any current clinical picture, in fact the syndromes described included features which belonged to the spectrum of mood disorders and psychosis, as well as symptoms of pathologies that (nowadays) we can define organic. Despite all, the Greek and Roman psychiatry earned the merit of giving the first descriptions, as well as the first attempts of classification and treatment of mental illnesses. Among the methods of treatment, the most applied were rudimentary and totally empirical, based on physical techniques, however, during those times it was already possible to find traces of what will be further known as psychotherapy.

1.2.2 Middle Ages and Renaissance

During the Middle Ages, psychiatry (and the whole medicine in general), underwent an involution. Despite in classical culture it was considered to all effects as a subject related to medical profession, in Middle Ages psychiatry had been brutally dismembered: such as surgery was yet executed by barbers, so psychiatry was related to exorcists and inquisitors. However, this kind of view may be reductive and defeatist, considering how long did Middle Ages lasted: in this time, especially in Middle East and Arabian countries, the birth of the first psychiatric wards of hospitals (Baghdad, Il Cairo, but also London, Paris, Basel, and Munich) can be placed. Doubtless, Middle Ages represents the historical period in which people left space to mystical-religious interpretations of mental illness, rather than to natural-related explanations which were most popular in ancient Greece. According to this kind of theories, fools were people possessed by demons or evil. Actually, the term "fool" comes from "fallitatem," which was coined in that period and literally means "bag fool of wind." Mental illness became a matter of religion, and the most representative textbook about this concept is the "Malleus Maleficarum" (1487 AC), written by Dominicans H. Kramer and J. Sprenger.

Renaissance was an historical age characterized by huge contradictions. On the one side, the so-called witch hunting reached its edge of glory; on the contrary, some brilliant minds were emerging from the community, whose intent was to contrast the inquisitorial tradition. Shortly later, there will be those who will lead the "First Psychiatry Revolution." Among the greatest of them, we can mention Cornelio Agrippa, Paracelso, and Johannes Weyer.

Although inextricably linked to their time, to them is recognized the merit of rediscovering the presence of a natural component among the causes of mental illnesses. This conception had been related exclusively to evil possessions for almost a millennium. Paracelsus (1491–1541), undoubtedly the most famous physician of his time, devoted himself profusely to the study of psychiatry. In his text "Von den Krankheiten so die Vernunfft Berauben" (1567), he describes and classifies mental illnesses in five main categories and then he focused on possible natural causes, taking the distances from both the religious conception proper of his time and the "humoral" Greek one. The most important of Paracelsus' achievements was the introduction of chemical treatments of mental illness. Although the use of substances was still mostly linked to the alchemical tradition, it represented the beginning of new concept of treatment.

1.2.3 Seventeenth and Eighteenth Centuries

The seventeenth century was dominated by the social crisis. Epidemics, wars, and economic crises produced severe consequences on social system, to which the absolutist regimes responded with the internment of the marginalized people. The "hôpital general" in France, "Zuchthaus" in Germany, and "workhouse" in England were filled with beggars, orphans, prostitutes, homosexuals, heretics, criminals, and mentally ill. At that time, mental illness was considered as delinquency and treated as such. When hospital facilities were not enough, prisons became the place dedicated to receiving the alienated, including psychiatric patients. Although it is difficult to classify the descriptions of these (alienated) patients referring to modern categories, in most cases the clinical pictures corresponded to dementia, psychosis, and cerebral consequences of alcoholism.

In concomitance with the internement (or imprisonement) of most severe psychiatric patients, considered socially unacceptable, the psychiatry of the seventeenth century almost completely loss its interest in psychosis, focusing on neuroses. Exactly in this period, Sydenham (1621–1689) coined terms such as "hysteria" and "hypochondria." He described this conditions as multiform, affirming that they lie on the border between physical and mental disorder and that they can mimic a wide spectrum of medical conditions, which goes from migraine to "iliac passion" (chronic inflammation of the small intestine) and which would have benefited from therapies based on ferrous compounds, milky diet, and horse riding.

Worthy of mention are also the works of the neurologist and anatomist Thomas Willis (1621–1675), who was the first to have considered hysteria as a mental pathology and not as related to uterus ("hysteria" derives from Greek "hystera," which means uterus). If the seventeenth century was a century of crisis, the eighteenth was a period of revolutions. The most important was certainly the cultural and philosophical revolution that took the name of Enlightenment. The primacy of reason and scientific positivism characterized this period and inevitably influenced also psychiatry, so that it was finally freed from the medieval concept of demonic possession and became an autonomous discipline. The trust toward mind capacities and in its healing abilities that characterized this historical period reflected at a social level in a philanthropic movement, which aimed to take care of the mentally ill patients (including psychotics).

During this historical period, G. E. Stahl (1660–1734), with his nosological theory, reintroduced the concept of "soul" in psychiatric discipline, which had been substantially based on a somatic orientation, starting from Hippocrates until that moment. According to his theory, mental illness represented a reaction of the soul to harmful stimuli. In parallel to the somatic orientation (which was constantly evolving, thanks to the recent contributions of anatomopathology and neurophysiology), the psychological orientation was about to blossom. However, the major contribution to the psychiatric revolution was given by the French school, in particular by Philippe Pinel (1745–1826) and his main student Jean Etienne Dominique Esquirol.

Pinel's contribution to the development of psychiatry was enormous and eclectic. In particular, as a director of Salpetriere's and Bicetre's Psychiatric Hospital, he decided to release psychiatric patients from the chains they were obliged to wear in the psychiatric wards, allowing the foundations (at least symbolic) for the re-evaluation of the psychiatric patient from the human point of view. Nevertheless, his contribution was not limited to philanthropy: he introduced the concept of heredity into mental illnesses, investigated the role of social institutions and lifestyle, physical factors (such as trauma), and alcoholism. The French psychiatrist, although rooted in the eighteenth century, pushed psychiatry toward modernity.

1.2.4 Nineteenth Century

In the nineteenth century, the heart of the debate and evolution of psychiatry shifted from Enlightenment France to Germany. The first half of the nineteenth century was characterized by the movement of thought that takes the name of Romanticism and that, precisely in contrast with the Enlightenment, marks the primacy of spirituality over rationality. In the psychiatric field, this current of thought is structured in the so-called Psychiker (who considered mental illness as a pathology of the soul, with a solid moral connotation) as opposed to the Somatiker (supporters of a somatic genesis).

Alongside the ideological and theoretical debate on the nature of psychiatric pathology, an important work of institutional reorganization was carried out: modern care institutions were founded, and asylum psychiatry was replaced by university psychiatry. Among the architects of this change, we find Wilhelm Griesinger (1817–1868).

Griesinger represented a turning point for psychiatry, whose full autonomy he advocated as a medical discipline. He was a fervent supporter of the cerebral localization of all psychic pathologies. In "Die Pathologie und Therapie der psychischen Krankheiten" (1845), he wrote: "Psychiatric diseases are diseases of the brain" and again "Madness is only a complex of symptoms to be traced back to different abnormal states of the brain."

Several decades before Freud, he also hypothesized that most psychic activities were unconscious.

According to Griesinger, sensory perceptions constitute, at the cerebral level, abstract concepts, representations. These representations can be abnormal or false (delusions) and lead the subject to consequent abnormal behavior: a kind of "theory of psychic reflexes" that was the basis of the work of the German psychiatrist. According to Griesinger, the set of representations of a subject are structured in his ego. When the subject's ego is in a situation of equilibrium, he can be free and self-determining. From the ego's imbalance arises, instead, the psychic pathology that in the most serious forms can lead to the shattering of the ego itself.

Griesinger's work was revolutionary, he made a synthesis of different disciplines, describing mental pathology from an anatomical, neurophysiological, psychological, and clinical point of view.

Like the first, the second half of the nineteenth century was led by German Psychiatry, the works of K. Westphal (1833–1890), T. Meynert (1833–1893), B. K. Wernicke (1848–1905) were representative of the so-called Psychiatry of the brain, whose basic idea was the close correlation between psychic pathology and cerebral anatomical alteration.

The second half of the nineteenth century, however, sees as protagonists two of the most famous names in the history of psychiatry: Emil Kraepelin (1856–1926) and Eugen Bleuler (1857–1939), the two major representatives of the "clinical school" thanks to which the transition from the simple classification of symptoms to the evaluation of the pathology in a longitudinal sense, along its overall course, was realized.

The German psychiatrist Emil Kraepelin is considered the founder of modern biological psychiatry and psychiatric nosography, but he was also a pioneer in the field of psychopharmacology. One of the most famous Kraepelin's contributions for psychopathology was the distinction of two distinct clinical entities of psychosis, namely "Dementia Praecox" and "Manic Depression." If the first condition was associated with a progressive loss of cognitive functions, the second in contrast was considered as an episodic disorder. Even if in the recent years these conditions have been considered more as a continuum rather than completely separated, this observation is valid still today and represents the fundamental difference between schizophrenic psychosis and affective disorders (major depressive disorders and bipolar disorder).

The Swiss psychiatrist Paul Eugen Bleuler is considered one of the most important clinicians who contributed to define modern psychiatry. Bleuler is mainly known for having coined the term "schizophrenia." He argued that the "Dementia Praecox" described by Kraepelin was not necessarily associated with dementia, pointing out that the splitting ("Spaltung") of psychic functions was the central feature of the disorder.

Although in the nineteenth century the greatest scientific advances in the psychiatric field have German nationality, the work of the French Jean-Martin Charcot (1825–1893) cannot be forgotten. Director of Salpetrière for about 30 years, in addition to being considered the father of modern neurology, he made important contributions in the study of hysteria and in the application of hypnosis techniques. His work will be used as the basis for the work of Freud in the first half of the 1900.

1.2.5 The Twentieth Century and the Contemporary Era

Sigmund Freud (1856–1939) was one of the most famous names in the field of psychiatry: the Austrian psychiatrist and neurologist, starting from the work of Charcot and Breuer on hysteria, devoted himself to the elaboration of a full-fledged scientific philosophical theory based on the role of unconscious processes in the determination of human thought and behavior. In the clinical field, the application of these principles led to the birth of psychoanalysis, of which Freud is considered the father. Freud introduced (and disseminated with extreme effectiveness) a new therapeutic approach based on free mental associations, concepts such as drive, libido, lapsus, missed and unintentional acts, the interpretation of dreams. He placed psychosexual development at the center of his investigation and formalized the three intrapsychic instances of Ego, Es, and Super-Ego. Freud's ideas were enormously popular, although there were critics, first of all Alfred Adler and Carl Gustav Jung. The greatest criticism directed at Freud is probably the excessive dilation of the concept of sexuality, the overestimation of the importance of childhood experiences and his extreme determinism. The work of the Austrian doctor has been partly surpassed over the last century with the advent of more modern "theories of mind"; however, the impact of his thought, even in popular culture, is undeniable.

The first half of the 1900 was sadly marked by the application of psychiatric knowledge for political purposes. In Nazi Germany and later in the Soviet Union, the concept of mental illness became an instrument of eugenic doctrines and political repression. The diagnosis of mental illness woefully became an instrument of the regime, and the psychiatric patient became an enemy to be eliminated.

The end of the Second World War and the second half of the twentieth century marked a radical turning point in the psychiatric field: the advent of psychotropic drugs.

In 1951, chlorpromazine, the first antipsychotic medication, was synthetized in the laboratories of Rhône-Poiulenc by Paul Charpentier, and in 1952, its role in psychiatry as sedative was first recognized by the physiologist Henri Laborit. At the end of the 1940s, the Australian physiologist John Cade published one of the first papers on the use of lithium in the treatment of bipolar disorder. Eventually, in the 1950s, the clinical introduction of the first antidepressants occurred, namely iproniazid, an antitubercular compound and imipramine, the first tricyclic antidepressant.

Advances in the pharmacological field (alongside those in the psychotherapeutic field) provided new tools for the treatment of psychiatric patients. With the increase in the available molecules and with the advent of generations of new and more effective drugs, the conception of the psychiatric patient gradually changed, which from "alienated" returned, not without difficulty, to be considered an integral part of society. This process had in Law 180 of 1978 (passed into history as the "Basaglia Law") its maximum political and social expression. The Italian psychiatrist Franco Basaglia was a pioneer in the process of deinstitutionalization of the psychiatric patient, no longer destined for the asylum all life long but reintegrated into the social fabric and treated primarily on an outpatient basis.

Parallel to the paradigm shift in the treatment of psychiatric patients and thanks to advances in the fields of genetics, pharmacology, instrumental investigation techniques, a new theoretical model for psychiatric pathology was finally developed: the biopsychosocial model.

Conceptualized in 1977 by Dr. George Engel, the biopsychosocial model is today the role model for mental health. It tries to explain mental disorders as the result of the complex interaction between biological correlates, psychological factors, and the socio-cultural background.

To conclude, a consideration on the evolution of the conception of mental illness throughout history. Advances in the field of psychiatry, more than in other medical disciplines, have been accompanied by a progressive change in the way of seeing the psychiatric patient. Over time, the mentally ill person has been considered cursed, possessed by the devil, has been considered and treated like a criminal, a social burden. In recent decades, we are witnessing a progressive reintegration of these patients into the social tissue. However, the prejudice with respect to this condition is still strong in the popular imagination. The notion of mental illness is still too often compared to the concepts of social danger, unproductivity, and incurability that overall fall under the definition of stigma.

It is to be hoped that, like the great effort aimed at research in the field of psychiatry, which has led to a real revolution in the diagnosis and treatment of mental pathology, we will also invest in the information and communication necessary to erase the social stigma that often, like the disease itself, is a source of suffering for the patient.

1.3 Clinical Interview

The psychiatric interview is the central element that guides the diagnosis and therapy of the patient suffering from mental pathology. The interview is an active, dynamic, and multidimensional process of gathering information in which the actors establish a relationship based on communication and trust.

The formal structure of the clinical interview will be described in the following paragraphs.

1.3.1 Stages of the Interview

We can identify three main phases

- *Initial or "exploration" stage*: It is the first contact with the patient and includes the presentation phase (by both parties) and the manifestation, by the interviewee, of the reasons that brought him to the interviewer's attention.
- *Intermediate stage:* It is the phase of the detailed interview, in which the professional will complete the collection of information.
- *Final stage:* It is the moment in which the information previously gathered is summarized, possibly completing it with additional questions. Finally, the treatment plan is explained and arranged.

Box 1.1: Setting the Interview

- Duration of the interview: 30–90 min
- Environment: quiet, comfortable
- Language: accessible, nontechnical, nonjudgmental
- Participants: interviewee, interviewer, maximum 1–2 family members (or close acquaintances)

1.3.1.1 Initial Stage

The initial phase includes the formal identification of the patient with the collection of his personal data and first exposure of the problem. This is the foremost, delicate moment in which the doctor–patient relationship is established.

- Identification
  - Personal data: name, age, occupation, marital status, nationality, place of residence
  - Identification of any accompanying people
  - Acquisition of any documentation related to the patient's clinical history
- Chief complaint (CC): "What brings you here today?" The patient spontaneously describes his experience and what brought him to the interviewer's attention.

This first approach allows information to be obtained on the patient's idea of his condition, on the path that brought him to the interviewer's attention and on the attitude toward the professional.

1.3.1.2 Intermediate Stage

The intermediate part of the interview is the quantitatively most relevant. The professional reconstructs the patient's history, from birth up to the present. The information gathered in this phase falls into two broad categories:

- *The psychiatric history*: it consists of the detailed study, conducted with method, of what the patient has freely described in the exploratory phase of the interview. Furthermore, all the information necessary to precisely reconstruct the interviewee's personal history will be collected.
- *The Mental Status Examination (MSE)*: it is the set of psychic phenomena observed by the interviewer.

> **Box 1.2: The Cone Technique**
> During the interview, it is often useful, in order to let the patient feel at ease, to begin with open questions and move gradually on to closed questions.

Psychiatric History

- Family history
  - Description of the family of origin: information on the family members with particular attention to a possible history of psychiatric illness, reconstruction of the family tree (genogram).

    Many of the most frequent psychiatric diseases have a hereditary component, and in the same way, the response or tolerance to many drugs can have a genetic basis.
  - Description of the patient's current family unit: spouse, children, grandchildren.

    It provides information on the patient's life context and closest social relationships.
- Physiological history
  - Reconstruction of the patient's personal history

    Youth: delivery, full-term or preterm, milestones of somatic and psychic development, childhood diseases, social relationships, temperament, schooling, family relationship. Events or conditions that occurred in childhood can have long-term consequences.

    Factors related to temperament can be identified already at a young age, as well as psychomotor deficits.

    Adulthood: work activities, social and emotional relationships, sexual development, menstrual cycle, pregnancies, menopause, eating habits, hobbies, and peculiar interests.

Social relationships and work activity is an important reference parameter for evaluating the patient's functioning over time. Sexual habits, eating habits, or hobbies complete the evaluation of the patient's lifestyle, whether physiological or pathological.

- Pathological anamnesis
  - Past medical history

    Organic: organic pathologies of which the patient has been or is suffering, taking specific therapies in the past or currently, surgery, previous hospitalizations, trauma or accidents, allergies.

    Any organic pathology can give manifestations comparable to the most common psychiatric conditions. Conversely, many of the most frequent psychiatric diseases can have physical expressions or consequences.

    Psychiatric: reconstruction of the patient's psychiatric history, previous psychiatric symptoms, duration of the illness, times of remission or recovery, periods of remission, duration and dosage of the medications, previous hospital admissions, and psychotherapy.

    A detailed psychiatric history significantly reduces errors in the diagnostic phase and in the choice of the therapeutic path, both in the present and in any future episodes.
- History of current disorder: Information about the condition that brought the patient to the interviewer's attention. Type of symptoms, chronology of manifestations, duration and intensity, impact on work and social functioning, changes in habits, medications prescribed. This is often the most challenging part of the

Table 1.1 Steps of the mental status examination

General description	• General appearance • Level of consciousness and vigilance • Awareness of the environment • Posture and gait • Distinguishing features (scars, tattoos) • General behavior (restless, agitated, aggressive) • Facial expression • Motor activity (hypoactive, hyperactive, presence of abnormal motor activity)
Language and speech	• Rate (slow, normal, rapid) • Quantity (absent, laconic, rapid) • Organization (logical, organized, disorganized)
Thought	• Form (logical, tangential, circumstantial, flight of ideas, perseveration, thought blocking) • Content (preoccupations, overvalued ideas, delusions, phobias, suicidal ideas)
Perception	• Illusions, hallucinations (visual, auditory, tactile, olfactory)
Mood and affect	• Mood (subjective emotional state: depressed, normal, euphoric, dysphoric) • Affect (objective emotional state: normal, blunted)
Cognition	• Attention and concentration • Orientations (space, time, person) • Memory
Insight	• Ability to describe personal, psychological, and physical status

interview in which the professional, with method and empathy, investigates the most central aspects of the patient's suffering or discomfort by carrying out an initial processing work that will lead to a diagnosis based on the criteria.

Mental Status Examination

The mental status examination (MSE) is a structured assessment of the patient's level of general behavior, speech, mood, perception, thought, and cognitive (knowledge-related) function. See Table 1.1 for further details, the single items and terms listed in the table will be thoroughly described in Chap. 2.

1.3.1.3 Final Stage

In the last phase of the interview, all the information collected is processed and used to complete the last three steps of the clinical interview. A systematic review is carried out on the information collected to allow a diagnosis based on criteria. It is not always possible to make a diagnosis after a single interview. In this case, the professional can formulate a provisional diagnosis (or not formulate at all) and, in subsequent interviews, will collect the data necessary to complete the diagnostic process. At that point, he will be able to provide information on the therapeutic path.

1. *Summary*: The interviewer gives a brief summary of the patient's history, describes the main problem, and discusses the biological, psychological, and social factors that may play a role in the interviewee's condition.
2. *Formulation*: The medical interviewer can formulate a diagnosis (albeit provisional).
3. *Therapy*: In this phase, the doctor describes the treatment plan identified on the basis of all the points described above and the possible pharmacological or non-pharmacological strategies. Instrumental or laboratory studies may be recommended (Box 1.3). The patient can also be referred to another professional (doctor, psychologist, social worker) if useful or necessary.

Box 1.3: Additional Investigations

In some cases, it may be necessary to supplement the information gathered during the interview with further investigation. We mention the main ones:

- Physical parameters and vital signs: blood pressure, heart rate, temperature, weight, height, BMI, waist circumference
- General physical examination and neurological physical examination
- Blood tests: full blood count, liver, thyroid, and renal function
- Instrumental investigations: CT scan, MRI, ECG, EEG
- Psychological assessment/psychometric tests

Further Reading

Ackerknecht EH. Breve storia della psichiatria. 3rd ed. Bolsena, VT: Massari; 1999.

American Psychiatric Association. Diagnostic and statistical manual of mental disorders. 5th ed. Arlington: American Psychiatric Association; 2013.

Carlat DJ. The psychiatric interview: a practical guide. 4th ed. Philadelphia: Lippincott Williams & Wilkins; 2016.

Kendler KS. An historical framework for psychiatric nosology. Psychol Med. 2009;39(12):1935–41. https://doi.org/10.1017/S0033291709005753.

McGlashan TH. Eugen Bleuler: centennial anniversary of his 1911 publication of dementia praecox or the group of schizophrenias. Schizophr Bull. 2011;37(6):1101–3. https://doi.org/10.1093/schbul/sbr130.

Messias E. Standing on the shoulders of Pinel, Freud, and Kraepelin: a historiometric inquiry into the histories of psychiatry. J Nerv Ment Dis. 2014;202(11):788–92. https://doi.org/10.1097/NMD.0000000000000208.

Nobile M. The WHO definition of health: a critical reading. Med Law. 2014 Jul;33(2):33–40.

World Health Organization (WHO). Strengthening mental health promotion, fact sheet no 220. Geneva: WHO; 2001.

Psychopathology

2

F. Martini, L. Fregna, I. Vanelli, G. Bigai, L. Tonet, and R. Cavallaro

2.1 Signs and Symptoms

In all branches of medicine, it is essential to use precise terms to describe signs and symptoms in order to guarantee optimal communication between practitioners.

A *symptom* is usually defined as a subjective experience described by the patient, while a *sign* is defined as an objective finding observed by the examiner. For instance, a patient with anxiety may complain of a sense of internal restlessness, an inability to relax and excessive worry. The related signs, in this case, may be visible motor restlessness, tremor, sweating, or a request for reassurance.

Finally, a *syndrome* is a set of signs and symptoms which co-occur and together make up a recognizable disorder.

Unlike other medical branches, in which the distinction between symptoms and signs is clear, in psychopathology the terms are sometimes used as synonyms because mental state alterations are largely elicited by exploring, with the patient, his internal experiences.

Whenever feasible, the examiner should try to corroborate symptoms with signs. Behaviors may be consistent with the symptom. For example, a patient who reports

F. Martini (✉) · L. Fregna (✉) · I. Vanelli
IRCCS San Raffaele Scientific Institute, Milan, Italy
e-mail: francesca.martini1@hsr.it; fregna.lorenzo@hsr.it; vanelli.irenelucia@hsr.it

G. Bigai · L. Tonet
Vita-Salute San Raffaele University, Milan, Italy
e-mail: bigai.giorgia@hsr.it; tonet.leonardo@hsr.it

R. Cavallaro
IRCCS San Raffaele Scientific Institute, Milan, Italy

Vita-Salute San Raffaele University, Milan, Italy
e-mail: cavallaro.roberto@hsr.it

R. Cavallaro, C. Colombo (eds.), *Fundamentals of Psychiatry for Health Care Professionals*, https://doi.org/10.1007/978-3-031-07715-9_2

hearing voices (symptom) may appear worried, in an attitude of listening or even mumbling to himself (signs).

However, it is not always easy to recognize a psychopathological phenomenon, especially concerning a mental activity, such as a delusional thought, that may not have direct behavioral or somatic equivalents and may be not spontaneously described or even denied by the subject.

The collaboration between the patient and the examiner, in this context, is therefore fundamental to examine accurately all psychopathological phenomena.

In the next pages, psychopathologic phenomena will be described, divided according to the main mental functions (consciousness, motor activity, thinking, perception, mood and affect, cognition and insight).

2.2 Consciousness

2.2.1 Introduction

Consciousness can be defined as a "state of awareness of the self and the environment." An individual must be conscious to be able to experience the world.

The three dimensions of consciousness are vigilance, lucidity, and self-consciousness.

2.2.1.1 Vigilance (Wakefulness)–Drowsiness (Sleep)

The term vigilance, though used in different ways, refers to the arousal level on the sleep–wake spectrum (alert–drowsy/asleep). This is not a uniform or unvarying state, but a fluctuating one. Factors that favor vigilance are interest, anxiety, terror, or excitement, whereas boredom promotes drowsiness. Some psychiatric conditions may increase vigilance, while many may diminish it. For example, hypervigilance is one of the symptoms of posttraumatic stress disorder.

2.2.1.2 Lucidity–Clouding

The sensorium that may be defined as the awareness of all stimuli, internal or external, presenting to the organism can be clear or clouded. Evidently, lucidity is related to vigilance: only a fully awake person can be clear in consciousness. In clouding, most cognitive functions are impaired, including attention and concentration, understanding, logical thinking, judgment, verbal communication, and purposeful action.

2.2.1.3 Consciousness of Self

Jaspers and Sharfetter described self-awareness, that is, the ability to distinguish *I* from *not I*, as having the following features

- Awareness of being or existing (ego vitality): I know that I am alive and exist.
- Awareness of activity (ego activity): I know that I am an agent who starts and executes my thoughts and actions.

- Awareness of unity (ego consistency and coherence): I know that I am one person, at any given moment.
- Awareness of identity (ego identity): I have been the same subject all the time.
- Awareness of the boundaries of self (ego demarcation): I am distinct from others and I am aware of the boundary between self and non-self.

2.2.2 Disturbances of Consciousness

The terminology in this area is quite unclear, with the same term sometimes representing different concepts and analogous phenomena being described by different words.

2.2.2.1 Quantitative Changes of Consciousness

Consciousness may be considered as a continuum from full vigilance and awareness to coma. In that sense, it may be considered as quantitative.

Enhanced consciousness is characterized by a subjective sense of richer perception, of increased alertness and a greater capacity for intellectual activity, associated with mood changes, usually exhilaration. Such a state may occur in healthy individuals, especially during intense emotional experiences (e.g., falling in love, sudden religious conversion). Heightened awareness is common with certain drugs (hallucinogens, stimulants) and may occur occasionally in early psychotic phases, in particular mania, or less frequently in schizophrenia.

However, more frequently, altered states of consciousness are represented by a lowering of consciousness. Different levels or stages of diminished consciousness lie on a continuum from full alertness to coma. These levels of lowered consciousness are quite nonspecific and can be determined by various causes: head injury, infection, tumor, cerebrovascular disorder, epilepsy, metabolic disorder, or toxic state.

Clouding of Consciousness

Clouding of consciousness refers to a stage of impairment of consciousness characterized by a minor deterioration in thinking, attention, perception and memory, and, usually, slight drowsiness and diminished awareness of the environment. The individual may be drowsy or agitated.

Drowsiness

It is the subsequent and persistent level of reduced consciousness. The drowsy patient is "awake" but will laps at times into unconsciousness if not stimulated and subjectively feels sleepy. His actions are slowed, his speech is slurred, his intention is sluggish. Basic reflexes, including coughing and swallowing, are diminished, muscle tone is reduced, and the patient tries to avoid painful stimuli. In psychiatric practice, it may appear following overdosage with central nervous system depressant medications or drugs.

Coma

Unlike the drowsy patient, in coma the subject is mainly unconscious, although in early stages, with strong stimuli, he may still be temporarily arousable. The patient does not respond to painful stimuli, he/she has lost the righting response of posture, reflexes, and muscle tone are markedly diminished, and breaths are slow and deep. In later stages, the patient is unconscious and cannot be awakened.

2.2.2.2 Qualitative Changes of Consciousness

These are always associated with some degree of quantitative alterations.

Delirium

According to *DSM-5*, delirium is a transient, rapid-onset, clinical condition characterized by deficits in attention, awareness, and cognition that fluctuate in severity over time. Primary signs and symptoms of delirium include global impairment of cognitive functions, reduced and/or restricted awareness of the environment, attentional abnormalities, increased or decreased psychomotor activity, disordered sleep–wake cycle, and emotional disturbances. Cognitive impairments are highly variable and affect several domains, such as memory, orientation, language, visuospatial ability, and perception. Delirium typically occurs with pronounced diurnal fluctuation: in the late evening, the patient becomes more disorientated, with lower mood and more perceptual disturbances.

Confusion

It refers to subjective symptoms and objective signs indicating reduced capacity for clear and coherent thinking. Both the individuals describe himself as confused, and the external observer considers the subject's thinking as altered and confused. It is a descriptive word and should not be used as a synonym for clouding of consciousness. It may occur with impairment of consciousness or disruption of thought processes in acute and chronic organic states, but it may be observed also in nonorganic disturbances, such as functional psychoses and neurotic disorders.

Twilight State

It refers to a state of altered consciousness, which is characterized by a combination of features: sudden onset and end, variable duration, ranging from hours to weeks, and the occurrence of unexpected irrational and sometimes aggressive acts. Besides altered consciousness, delusions or hallucinations can occur. The individual is temporarily unaware, and he/she has little or no recollection of what happened. It is generally secondary to an organic condition, such as temporal lobe epilepsy, alcoholic intoxication, and brain trauma. However, it may also occur with dissociative states, or as an acute reaction to massive catastrophe.

Dream-Like (Oniroid) State

It is an altered state of consciousness with a prevalence of psychotic symptoms.

It is not easy to differentiate this state from twilight state or delirium. The subject is disoriented and perceives elaborate hallucinations, often visual but also auditory or tactile, that may determine intense emotional change, congruent with hallucinatory experiences. The patient may appear to be living in a dream world.

It is crucial to investigate other symptoms or organic states to distinguish between organic illness and a dissociative nonorganic condition. In physiological condition, it characterizes phases of awakenings or falling asleep, possibly associated with hallucination.

Stupor

This term indicates a syndrome that is characterized by the presence of mutism, akinesia, and unresponsiveness to stimulation. The patient appears awake and even alert but is unable to speak or act. It is different from coma and does not lie on the continuum from vigilance to coma, but it is qualitatively distinct. There is usually some degree of clouding of consciousness, not just in terms of lowering. Subsequent amnesia is common.

It may occur with organic conditions, such as lesions in diencephalon and upper brainstem, frontal lobe and basal ganglia, but also with psychiatric conditions, such as schizophrenia, affective psychoses (both depressive and manic), and dissociative states.

Psychogenic (functional) and neurological (organic) stupor can be particularly difficult to differentiate.

Automatism

Automatism implies actions, simple or complex, taking place in the absence of intentionality. It is generally inappropriate to the context and may be out of character for the subject, although usually purposeful and directed. The individual has an impaired awareness of the environment, seems to be only partially aware of being spoken to, and does not reply pertinently. Afterwards, the patient may not be able to recall the episode or may only partially remember his actions. Automatism may occur in epilepsy, usually arising from temporal lobes, during, or immediately after, a seizure.

2.2.2.3 Disorders of Self

In descriptive psychopathology, self-disorders, also called ipseity disturbances, describe the abnormal inner experiences of *I-ness* and *my-ness*. Each of the five formal characteristics of the self (ego vitality, ego activity, unity of the self over time, self-identity, and boundary of the self) can be impaired by psychiatric disorders.

The sense of vitality can be impaired to produce a feeling of deadness, the extreme example being nihilistic delusions, which may occur in affective psychoses. In disorder of activity, the sense of being an agent executing one's will can be impaired as occurs in passivity phenomena. *Autoscopy* and *dissociative identity* disorders are examples of impairment of the unity and consistency of self. Disorder of self-identity and continuity can be observed in psychotic or affective states. Abnormalities of self-demarcation may occur in LSD intoxication or in schizophrenia as passivity experiences, thought insertion, and thought withdrawal.

Depersonalization is the most frequent symptom of change of self-awareness occurring in psychiatry. The term depersonalization refers a peculiar, usually unpleasant, alteration in the awareness of self, characterized by feelings of unreality, detachment or estrangement from oneself or one's body. Insight is preserved; the *as if* prefix is used by the individual to denote that he/she is trying to describe how he/she feels using metaphors. The individual may report feeling as if he/she is an outside observer of his/her own thoughts or body, sometimes with a loss of control over his/her thoughts or actions. It is frequently associated with the symptom of *derealization*, a term indicating a change in the awareness of the external environment, perceived as foggy, dreamlike/surreal, or distorted. The symptom is described in many different ways by the patient, and it is not always easy to differentiate between depersonalization and derealization. Depersonalization may occur in healthy, normal people, at times of strong emotional stimuli. It may occur with organic conditions including traumatic brain damage, epilepsy, and migraine and with substance intoxication such as cannabis, hallucinogens, and ecstasy. It can be associated with mood disorders (depression) and anxiety disorders including agoraphobia, panic disorder, social anxiety, and other phobic states.

Since the body constitutes the primordial essence of self and carries information about identity, disorders of self may also manifest as disturbances in body awareness.

Disorders of body awareness include neurological conditions such as anosognosia for hemiplegia, in which patients with parietal brain damage deny the presence of their contralesional motor deficits, or somatoparaphrenia, which is characterized by the delusion that the paralyzed limb does not belong to oneself. A peculiar case is that of phantom limb, a condition in which patients experience sensations, usually painful, in a limb that has been amputated. Although pain is spatially located outside the subject, this cannot be considered a hallucination as the subject knows that he has lost his limb and is aware that the feeling of pain is inside himself.

2.3 Motor Disorders

2.3.1 Introduction

Movement disturbances may have crucial diagnostic significance, especially when verbal explanation is compromised.

Movement may be altered quantitatively, e.g., increased or reduced, or it may show different qualitative abnormalities. While some of these motor disorders are involuntary and are appropriately considered neurological, some are voluntary but enacted unconsciously, and some are intentional actions. In some cases, observation of the patient's spontaneous activity will be sufficient do detect a motor disorder. However, if the motility is poor, it will be necessary to ask the patient to perform tasks that exacerbate any alterations in motor activity.

2.3.2 Quantitative Abnormalities of Movement

The increase in motor activity can be differentiated into hyperactivity and agitation. The reduction in motor activity is also defined as motor retardation.

2.3.2.1 Hyperactivity

It describes a state in which motor activity is increased, movement and gestures are numerous and quick, but have a specific purpose. Typical is the increase in gesticulation, readiness, and speed in carrying out activities. This type of hyperactivity is characteristic of hypomanic/manic states and ADHD.

2.3.2.2 Agitation

It describes a subjective mood state (the patient may describe his affect as "feeling agitated") associated with and resulting in physical expression (restlessness and increased arousal). Unlike hyperactivity, the excessive motor activity of the agitated patient may appear purposeless and characterized by movements such as drumming the fingers, rubbing the hands, getting up and sitting down nonstop, scratching the head, and so on. Psychomotor agitation is one of the conditions that most often lead to an emergency psychiatric visit. It may occur in many different psychiatric conditions, including affective psychoses, schizophrenia, organic mental disorders such as senile dementia, or personality disorders, in particular if associated with anxiety. Agitation is often also a typical symptom of organic illnesses, such as hyperthyroidism.

2.3.2.3 Retardation

Motor retardation implies slowness of the initiation, execution, and completion of actions. It is often associated with thought and speech retardation, as in the case of severe depressive illness. The patient may describe him/herself as struggling with thinking and also with initiating and carrying out spontaneous activities. Besides psychiatric disorders, it can also occur in other conditions, such as organic mental disorders or physical illnesses as hypothyroidism. Akinesis is a condition characterized by extreme motor retardation and paucity or absence of voluntary movement. It may be associated with mutism in stupor.

2.3.3 Qualitative Abnormalities of Movement

2.3.3.1 Automatic Obedience

In this condition, the patient executes every command in a literal, concrete way, like an automaton.

2.3.3.2 Catatonia

It is a state of increased muscular tone at rest. It can be distinguished from extrapyramidal rigidity because it is abolished by voluntary activities. Catatonic schizophrenia is a syndrome characterized by the presence of very varied symptoms involving motor activity and posture: hyper- or hypoactivity, mutism, stereotypical posturing and movement, waxy flexibility, mannerism, echopraxia, echolalia, negativism, automatic obedience. Catatonic symptoms may occur in patients with mental disorders, such as psychosis, mood disorders, and autism spectrum disorder. However, they are nonspecific may be observed in other medical conditions.

2.3.3.3 Catalepsy

It is characterized by muscular rigidity and maintenance of a fixed posture regardless of external stimuli.

2.3.3.4 Echopraxia

It is characterized by the involuntary repetition or imitation of the interviewer's actions.

2.3.3.5 Grimace

It is characterized by facial contortions not made in response to a noxious stimulus. For example, Schnauzkrampf (literally "snout spasm") refers to a facial expression in which nose and lips are drawn together in a pout. It may be observed in catatonic patients.

2.3.3.6 Obstruction

It can be considered as the equivalent in the flow of action of thought blocking in the flow of speech. While performing a motor act, the subject stops. After a pause, the patient proceeds with the same action or takes a different action.

2.3.3.7 Mannerisms

They are odd, stilted, voluntary movements and patterns of behavior that seem to have a goal but are excessive for the purpose.

2.3.3.8 Negativism

In this condition, the patient offers resistance to passive movement or repeatedly turns away from the examiner. Sometimes, he does exactly the opposite to what is asked in addition to offering resistance. Opposition may sometimes manifest itself in muteness.

2.3.3.9 Posturing

It is characterized by voluntarily assuming an unnatural or uncomfortable body position or attitude for an extended period of time.

2.3.3.10 Stereotypy

It is a repetitive, abnormally frequent, non-goal-directed movement.

2.3.3.11 Tics

They are sudden, brief, repetitive, hard-to-stop motor movements or vocalizations involving discrete muscle groups. In Gilles de la Tourette's syndrome, multiple tics are associated with involuntary vocalizations that often consist of obscene words or phrases (coprolalia). Tics often occur or worsen with emotional stress.

2.3.3.12 Tremor

It is defined as a rhythmical, involuntary, oscillatory movement of one or more body parts. Tremor is present in acute and chronic manifestations of anxiety and in states of particular emotional tension and alertness. Moreover, it represents a typical sign of alcohol withdrawal. Several psychotropic drugs (lithium, tricyclic antidepressants, antipsychotics) can cause tremors as a side effect.

2.3.3.13 Waxy Flexibility

It is characterized by maintaining a position imposed by the interviewer for a sustained period, even if uncomfortable and even if asked to change it. It can be observed in catatonia.

2.4 Thought Disorders

2.4.1 Introduction

It is customary to divide thought disorders into disorders of *content* and disorders of *form*, despite this division is somewhat arbitrary because belief and thinking process cannot be clearly separated. Moreover, some classifications include among disorders of thinking also disorders of the *control* of thought.

Thought content differs from thought form in that it does not describe how the thoughts are formulated, organized, and expressed but rather what the person is thinking.

In formal thought disorders (FTD), the organization, associative process, and expression of thinking are impaired, whereas in content thought disorders, the disturbance lies on the subject of the patients' thought (e.g., delusions).

A patient can have a normal thought content with significantly impaired thought form. Conversely, there may be generally normal thought form but delusional thought content.

2.4.2 Formal Thought Disorders

In the model of thinking based on Jaspers (model of associations), thoughts can be seen to flow in an uninterrupted sequence in which one or more associations may arise from each thought. There is a vast number of possible associations, but, in normal conditions, the flow of thinking proceeds in a certain direction, toward a goal (determining tendency). This model can be useful to better understand the disorders of thinking and speech that occur in mental illness.

Disorders of the form of thought consist of alterations in the organization and associative process of thinking.

In clinical practice, formal thought disorders are assessed by engaging patients in conversation and observing their speech but also by looking at their writing.

2.4.2.1 Acceleration of Thinking (and Flight of Ideas)

It is an FTD in which associative processes take place more quickly and easily—increased speed of association—at the expense of communicative effectiveness. There is an increase in verbal productivity, and associative links may be marginal or creative, but all of the ideas are logically connected. It is defined as a flight of ideas when connections between ideas are so loose that communication is inefficient or fails. The patient rapidly and continuously moves from one thought to another, because of the effect of labile affect and a very high degree of distractibility, at a pace that is difficult for the listener to keep up with. In this, the goal of thinking is not maintained for long. The flight of ideas is not only a quantitative extreme of acceleration: when the flow of thought is so fast, the determining tendency is weakened or lost and the associations between thoughts are determined by chance relationships, verbal associations (such as assonance, alliteration, and so on), clang associations, or proverbs. Flight of ideas occurs typically in manic states, occasionally in schizophrenia and organic conditions.

2.4.2.2 Retardation of Thinking (and Thought Blocking)

It refers to an FTD in which the associative processes are carried out with marked slowness, low productivity, and reduced effectiveness. Thinking, although goal-directed, proceeds so slowly that the person may fail to achieve those goals. When asked a question, he/she will ponder over it and answers, eventually, after considerable delay. This situation is experienced by the patient as difficulty in making decisions, lack of concentration, and loss of clarity in thinking. It usually occurs in depression or organic mental disorders. It may occur also in schizophrenia, in which, however, it may be difficult to separate this from thought poverty.

Thought blocking occurs when the patient appears to be unable to complete a thought, as his chain of thought unexpectedly and unintentionally breaks off or stops. The patient may stop mid-sentence for seconds or minutes, and after the block, he/she will often start talking about an unrelated subject. Thought blocking is also described as an experience of unanticipated, quick, and total emptying of the mind. It usually occurs in schizophrenia and may be interpreted as a thought withdrawal.

2.4.2.3 Circumstantiality

It is an FTD in which the contents of thought are communicated in an indirect, redundant, repetitive, and stereotyped way. The circumstantial patient includes unnecessary details, parentheses, and digressions that are not directly relevant to the subject or essential to answer a question. However, the determining tendency is preserved, and the patient does eventually answer the question.

Characteristically, it occurs in patients with epilepsy, other organic states, and intellectual disability. An individual with an obsessional personality may be circumstantial due to the anxiety of omitting important information.

2.4.2.4 Perseveration

This FTD is characterized by the repetition of words and ideas, even if they have been enunciated shortly before. The patient retains a constellation of ideas long after they have ceased to be appropriate or relevant, without the ability to move on to other topics. In perseveration, the patient may repeat the same response to different stimuli and may need to be prompted repeatedly before giving an adequate response. It is characteristically an organic symptom of dementia.

2.4.2.5 Tangentiality

The patient replies to a given question in an oblique, digressive, or even irrelevant manner. It refers to responses and not to transitions in spontaneous speech. The patient may move further and further away from the main topic without ever answering the question or he/she may give an answer that is formally correct and marginally related to the question. However, in the context of a conversation, the communication results less effective. This FTD may occur in many conditions, such as schizophrenia, dissociative disorders, and organic conditions, including dementia and delirium.

2.4.2.6 Loose Associations

It is an FTD in which the associative links between the various ideas are altered and poorly decoded, causing a reduction in communication effectiveness.

Loose associations differ from circumstantiality and tangentiality in that it is difficult or impossible to see the connections between the sequential ideas. This is common in patients in psychotic states.

2.4.2.7 Derailment (Incoherence)

It refers to an FTD in which spontaneous speech is characterized by a complete loosening of association and marked impairments in topic maintenance. Derailment describes an unexpected change of direction of a train of thought, which "derails" onto a subsidiary idea. Incoherence (*word salad or schizophasia*) represents the extreme degree of loose associations and refers to a speech characterized by a severe lack of cohesion at the basic level of syntax and/or semantics within sentences. Speech appears unintelligible since each word is dissociated from the one that precedes it. Other abnormalities of thought or speech may occur, such as neologism, fusion, assonance, echolalia.

2.4.2.8 Concrete Thinking

FTD characterized by a reduction or absence of the ability to think abstractly or operate generalizations and a general proneness to accept the literal meanings of figurative expressions such as irony, proverbs, metaphors, and idioms. It is observed characteristically in schizophrenia, but also in mental retardation, autism spectrum disorders, and organic disorders.

2.4.2.9 Illogicality

It implies drawing conclusions from a premise by inference that cannot be seen as logical because the patient uses alternative reference logics (e.g., paleologic thought). Thus, the concepts link with each other according to correct association criteria but the conclusions they lead to are not acceptable.

2.4.3 Other Common Disorders of Speech in Psychiatric Disorders

- *Stuttering* is the interruption of the normal flow of speech by pauses or by the repetition of fragments of words. It often improves with time and occurs when the person is anxious.
- *Mutism* is the complete loss of speech while the ability to speak and understand spoken language is not impaired. Psychiatric disturbances associated with mutism include schizophrenia, affective disorders, conversion reactions, dissociative states, and organic mental disorders. Mutism is an essential element of stupor and is often associated with other signs of catatonia. *Selective mutism* is an anxiety disorder in which a child speaks only in certain situations or with certain people.
- *Echolalia* is the repetition of words or parts of sentences pronounced by another person or by the patient himself (auto echolalia). It is a rare observation more frequent in schizophrenia with high levels of disorganization, in intellectual disabilities, and in organic states such as dementia.
- *Neologism* refers to a new word or condensed combination of several words, with an idiosyncratic personal meaning, or an existing word used in a new sense. Although it is not a true word and is not readily understandable, sometimes the intended meaning may be apparent. It can be observed in schizophrenia or dementias, but also in severe manic states.

2.4.4 Content-Thought Disorders

Disorders of the thought content are abnormalities in beliefs and in the interpretation of experiences. The most common disorders in content of thought are delusions.

2.4.4.1 Overvalued Idea

It refers to a single abnormal belief that is neither delusional nor obsessional in nature, but which is extremely worrying and predominant in the patient's mental life. Overvalued ideas may have an element of truth and are maintained with less conviction than delusions. Nonetheless, they pervade mental activity, take precedence over all other ideas for long periods, and may become so dominant that the patient's whole life comes to revolve around this one idea.

They are strongly toned by affect and differ from obsessions in that they are not experienced subjectively as "senseless."

They are usually associated with an abnormal personality and occur in various psychiatric disorders such as paranoid states, morbid jealousy, hypochondriasis, dysmorphophobia, anorexia nervosa.

2.4.4.2 Delusion

According to *DSM-5*, delusions can be defined as fixed beliefs that are not amenable to change in light of conflicting evidence. Their content may include several themes. Delusions are defined as bizarre if they are evidently implausible, not understandable to peers from the same cultural background, and do not derive from everyday life experiences.

Jaspers (1913) was the first to define the three main criteria for a belief to be considered delusional:

- *Subjective certainty*: Delusions are held with unusual and extraordinary conviction and certainty.
- *Incorrigibility*: They are not amenable to correction by logic; there is imperviousness to other experiences and to compelling counterargument.
- *Impossible or false content*: The absurdity or erroneousness of their content is manifest to other people and cannot be accepted by people of the same culture, education, and period of life as the person who experiences it.

Each of these criteria has been subjected to criticism, and none of them, taken individually, is sufficient to distinguish a normal thought from a delusional one.

Moreover, another important feature in order to recognize delusion is that its content is in most cases centered on the subject.

Delusions may be present in various mental disorders: Schizophrenia and other psychotic disorders, bipolar disorder, major depressive disorder, organic mental disorders as delirium, and dementia.

2.4.5 Semiotics of Delusion

The phenomena through which delusion occurs are:

- *Delusional percept*: The attribution of an abnormal and new meaning, usually in the sense of self-reference, to normal perception. For example, a cloud in the sky may be misinterpreted as meaning that someone has sent that person a message to save the world.
- *Delusional intuition*: A new, sudden, and understandable idea, not derivable from an external percept or internal thought, appears fully formed in the patient's mind. It is sometimes known as "autochthonous delusion."
- *Delusional memory*: It is the delusional interpretation of a normal memory.
- *Delusional system*: The set of primary delusional experiences (perceptions, intuitions, and memories) are commonly integrated into some sort of delusional system and linked together by logic and coherence. This elaboration of delusions has been called "delusional work."
- *Delusional atmosphere (or delusional mood):* It denotes a predelusional state in which the patient experiences something (not yet defined) around him as uncanny, sinister, strange, impending. The environment appears to be changed in a threatening way but the significance of the change cannot be understood by the patient

who appears tense, apprehensive, and perplexed. Delusional atmosphere is part of the underlying process and usually its meaning becomes obvious when a sudden delusional idea or a delusional perception arises and the patient often experiences a sense of relief.

2.4.6 Features of Delusion

In clinical practice, it is also important to describe some qualitative and quantitative features with which delusion can occur in various psychiatric disorders.

2.4.6.1 Content

Delusions are variable in their content that may be determined by social, cultural, and biographical background of the patient, but certain general characteristics commonly occur. The great majority of delusional contents can be traced back to some basic themes: fear, dominance, guilt, sexuality.

Delusions of Persecution

The individual is convinced that someone, an organization, or an institution wants to harm or kill him. This is the most frequent content of delusions, and it occurs in many different conditions: in schizophrenia, in affective psychoses, and in organic states, both acute and chronic.

Delusions of Reference

The patient believes that innocuous events or mere coincidences have strong personal significance that people allude to him in phrases or gestures or that radio, television, or newspapers refer to him.

Delusions of Control

The patient may believe that there are machines or people with particular powers that influence his will, emotions, actions, or thoughts. These delusions are otherwise known as passivity or made phenomena and represent hallmark symptoms of schizophrenia. Sometimes the patient states that the operator of control is installed inside his body, which leads to the narrative of being possessed. At other times, the patient is convinced that his thoughts are inserted in his/her mind by someone else (*thought insertion*), removed by his/her mind by an external source or entity (*thought withdrawal*), or projected and perceived by others (*thought broadcasting*).

Moreover, he may experience feelings, which do not seem to be his own and are imposed upon him by an external source (*made affect*). Otherwise, he may experience the movement of his own limbs like a passive observer rather than the author of the action (*made act*) or that the action he perform as his own is precipitated by a powerful impulse that is imposed upon him by an external source (*made impulse*). Finally, the subject may believe that also his physiological functions are controlled by an external agent. In this case, the patient believe is a recipient of bodily sensations imposed upon him by some external influence (*somatic passivity*). These delusions may be associated with somatic perceptions, such as haptic, thermic, or kinesthetic hallucinations.

Delusions of control are considered also a disorder of self, as the subject lack the normal sense of self agency, ownership, and demarcation.

Grandiose Delusions

The patient may believe himself to have special capabilities or supernatural powers, to be a celebrity, to be involved in some special secret mission, or to be an inventor of extraordinary machines (delusion of invention). This type of delusion typically occurs in schizophrenia and manic states.

Delusions of Guilt, Unworthiness, and Poverty

The patient is convinced that he is an evil person, has ruined his family, or is responsible for serious crimes for which he must be punished. Such delusions are common in depressive illness and may increase the risk of suicidal attempts.

Nihilistic Delusions

They may range from negation of existence of parts of the body to negation of self-existence or belief of being dead. It is typically associated with depressive symptoms. *Cotard syndrome* is a rare condition characterized by nihilistic delusions, which typically affects middle-aged or older people; it can be found in numerous psychiatric conditions, such as depressive disorders, psychotic disorders, but also neurologic conditions.

Delusions of Infidelity (or Jealousy)

The patient unreasonably believes him to be the victim of their partner's unfaithfulness. It may be difficult to differentiate it from nondelusional jealousy. Delusions of jealousy are commonly associated with alcohol abuse.

Erotomanic Delusions

The patient believes that another person, often someone famous or of a higher social class, is in love with him/her. It is more common in women than in men, and it occurs in the context of various psychiatric disorders such as bipolar disorders (manic states), delusional disorder, schizophrenia.

Hypochondriacal Delusions

The patient is convinced that he/she has a serious disease, often incurable, and that doctors hide it to him/her. Hypochondriacal delusions may occur in depression, but can also be found in other conditions, such as schizophrenia and delusional disorder.

Delusions of Infestation

The patient believes that he is infested with insects or other organisms under the skin. It is also known as Ekbom syndrome. It may occur in affective psychosis, in schizophrenia, in delusional disorder, in organic states such as delirium tremens during alcohol withdrawal and other brain diseases.

Religious Delusions

The patient may believe that he/she is a saint, a prophet, or God himself, or (in women) that she is pregnant with God. Religious delusions are common and occur in many different psychiatric disorders. Individuals experiencing religious delusions are preoccupied with religious subjects that are not within the expected beliefs for an individual's background, including culture, education, and known experiences of religion.

Delusional Misidentification

It includes *Capgras' delusion*, *Fregoli' syndrome*, and *the syndrome of subjective doubles*. Capgras' delusion consists of believing that a friend or family member has been replaced with an identical impostor, while Fregoli's syndrome is the delusional misidentification of an unfamiliar person as a familiar one, even in the absence of physical resemblance. The syndrome of subjective doubles is a rare delusional misidentification syndrome characterized by the belief that the patient has a double or Doppelgänger with the same appearance, but generally with different personality, that is leading a life of its own.

Bizarre Delusions

Delusions that in the person's culture would be regarded as totally implausible, that is not consensually shared in a given social or cultural context. For example, the patient may believe he/she is being cloned by aliens or that someone replaced all of one's internal organs with someone else's without leaving a scar. A nonbizarre delusion is, although false, at least plausible. For instance, the subject may believe that he/she is under constant police surveillance.

2.4.6.2 Content Variability

In some psychiatric conditions, such as in delusional disorder, delusional contents do not vary over time. In other cases, there may be extreme variability and fragmentation of delusional contents, even in a short space of time. Extreme variability of delusional contents is more common in acute psychosis, and in conditions where consciousness is altered, like in organic mental disorders.

2.4.6.3 Derivability

There is also a distinction between primary delusions and secondary delusions (delusion-like ideas).

Primary Delusions

The term primary implies that delusion is not occurring in response to another psychopathological phenomenon. It cannot be understood from previous experiences, it does not appear to be related to changes in mood, and it does not appear secondary to other psychopathology. Indeed, primary delusion is something new, irreducible, and not understandable.

Secondary Delusions (Delusion-Like Ideas)

Secondary delusions, differently from primary delusions, arise from some other psychopathology and are ultimately understandable. They can be secondary to abnormal mood, hallucinations, primary abnormal belief, personality, or circumstances of life. Typical examples of secondary delusions are those that occur in mood disorders, such as delusion of guilt in depressive episodes or grandiose delusion in manic episodes (*mood-congruent delusions*). Secondary delusions may occur in the background of personality disorder or abnormal personality traits, for example, abnormally suspicious or sensitive personalities can react to difficulties and specific events with deepening ideas of persecution.

2.4.6.4 Degree of Systematization

It is still common to divide delusions into systematized and nonsystematized. In the completely systematized delusions, there is one basic delusion, and the whole system is logically built on this error. All ideas are coherently and understandably interconnected. There may be various degrees of systematization in different patients, and the level of systematization may change over time, with systematization being usually more common in older subjects or in patients whose delusions are persistent, such as those affected by delusional disorder or paranoid schizophrenia.

A nonsystematized (or fragmentary) delusion is a disorganized false belief or a series of such beliefs that are disconnected, inconsistent, and illogical. Fragmentary delusions are common in disorganized schizophrenia.

2.4.6.5 Behavioral Pervasiveness and Affective Response

The delusional idea can be inferred from the behavior of the subject, even if the individual does not verbalize it. Patients sometimes may act on their delusional beliefs. For example, a patient with depressive delusions of guilt, unworthiness, or hypochondriasis may make a suicide attempt or may try to give him/herself up to the police. Delusions of persecution or reference may lead to aggressive behavior and delusions of jealousy may be even associated with violence.

However, the intensity and pervasiveness with which delusions affect behavior are highly variable, depending on the thematic content and the degree to which the delusional belief involves and influences areas of the patient's life and reality of the world.

The degree to which the patient's emotions are involved with such beliefs (affective response) is variable too. Delusions may generate an intense and proportionate emotional response or a reduced or even discordant emotional response.

Delusions occurring in affective episodes may be mood-congruent or incongruent. For example, delusion is considered congruent with the depressive state if it concerns typical depressive contents (poverty, guilt, illness, unworthiness).

Delusions may vary over time, spontaneously or secondary to treatment, in terms of behavioral pervasiveness and affective response.

2.4.7 Disorders of Control of Thinking

According to Fish's classification of thought disorders into four groups (disorders of the stream of thought, disorders of the control of thought, disorders of the content of thought, disorders of the form of thought), obsessions and thought alienation may be considered as disorders of the control of thinking. In some psychiatric illnesses, the subject experiences a loss of control or sense of possession of his thoughts, which may even appear as foreign to him.

2.4.7.1 Obsession

An obsession is a thought that persists and dominates an individual's thinking despite the individual being aware that the thought is unreasonable, excessive, senseless or, at least, that it is persisting beyond the point of relevance or usefulness.

The essential features of the obsession are its appearance against the patient's will (it is involuntary), its persistence or recurrence, its intrusiveness, its irrepressibility, its interference with mental functioning, the preservation of subject's insight. The patient is compelled to think about things and feels unable to prevent the repetition of thoughts. He tries to resist his thoughts, but unsuccessfully. His psychic functioning is altered, slowed down, less effective, and productive. There is no loss of contact with reality: the subject recognizes that the thought, although unwanted, is his own and that it arises from within himself. Obsessions usually cause marked discomfort that is experienced as a general unease, tension, anxiety, and even a sense of guilt. Obsessions are usually defined as egodystonic, since they are felt to be disgusting, unacceptable, or inconsistent with one's self-concept (e.g., sexual, blasphemous, or aggressive thoughts). Although insight is typically preserved, levels of insight and egodystonicity about one's own symptoms may vary between subjects and in the same subject during the course of illness. A percentage of subjects show poor insight into their obsessions, failing to recognize their excessiveness or irrationality. This is true especially for those obsession which develop early in life and whose content is considered reasonable, somehow functional and egosyntonic by the subject (e.g., symmetry/perfection, contamination). These subjects may accommodate their symptoms for long periods before seeking treatment, which would associate poor insight to longer untreated illness duration and worse prognosis.

Obsessions may occur as images, ideas, impulses, and fears.

Obsessional images are images that persistently occupy the patient's mind. They may be vivid but are always recognized by the patient to be products of his own mind.

Obsessional ideas take the form of ruminations on different topics, often pseudo-philosophical.

Obsessional impulses may be impulses to touch, count, or tidy up objects.

Obsessional fears consist of a groundless fear that the patient realizes is dominating without a cause, such as contamination fears, fears about one's well-being, concerns about disasters happening to him or those close to him, fear to commit aggressive actions toward self or others.

It is customary to distinguish between obsessions and *compulsions*: the word obsession is usually reserved for the thought, while compulsion for the act.

Compulsions are repetitive, intentional, and purposeful acts or behaviors, secondary to obsessions and acted out in order to neutralize them and reduce anxiety. The patient usually recognizes that his behavior is excessive or unreasonable, he does not feel any pleasure from this action, although he achieves a momentary reduction of his discomfort. Compulsions may derive from an obsessional impulse, thought, or mental image that leads directly to the action. For example, an obsessional fear of contamination may lead to compulsive cleaning and washing rituals. Compulsions may be complex and ritualistic and are often time-consuming and exhausting.

The patient with obsessive-compulsive symptoms may adopt a reassurance-seeking behavior and avoid stimuli or situations that may trigger obsessions or compulsions.

Obsessions typically occur in the context of obsessive-compulsive disorder (OCD) as the core symptom. However, obsessional symptoms can occur also in other conditions, such as depression (especially in patients with obsessional-compulsive personality), schizophrenia, and, occasionally, organic psychosyndromes.

2.4.7.2 Thought Alienation

Some psychiatric disorders can be characterized by a loss of control or sense of possession of thinking. These phenomena are called *passivity experiences* of thinking and have also been described as delusions of control. The patient may ascribe his internal thought processes to outside influences. Differently from obsession, the thought is experienced as foreign and out of his control. These symptoms are often associated with delusional explanations of how the patient's thinking could be controlled, for instance, by electronic devices or telepathy.

In *thought withdrawal*, it is believed by the patient that his thoughts suddenly disappear and are being taken away from his mind against his will by a foreign influence. It has been suggested that this may be the subjective experience of thought blocking.

In *thought insertion*, the subject believes that thoughts have been inserted into his mind from outside, without his volition, and recognizes them as foreign.

In *thought broadcasting*, the patient experiences his thoughts as leaving his mind and being diffused widely, so that other people can hear them. Thought broadcasting may also be associated with *thought echo*, a peculiar type of auditory hallucination which consists of hearing one's own thoughts spoken aloud.

Thought alienation is commonly associated with schizophrenia.

2.5 Perceptual Disturbances

2.5.1 Introduction

Perception is defined as the conscious awareness of environmental stimuli by the mental processing of sensory stimuli, thus making perception an active process rather than a passive one. Perceptual disturbances are common features of psychiatric and neurologic illness, and differentiating a primary psychiatric disorder from one associated with organic diseases is the main diagnostic challenge presented by the patient experiencing perceptual aberrations. Healthy subjects also exhibit occasional perceptual disturbances, but the high frequency and the intensity of the aberrations are the elements that alert us of the presence of serious nervous system diseases. Perceptual disturbances involve all sensory modalities (visual, auditory, tactile, olfactory, gustatory, kinesthetic, and proprioceptive pathways) and include misinterpretations and distortions of real elements belonging to the environment, as well as self-generated aberrant perceptions. The pathogenesis of these phenomena is still largely unknown, but aberrations in specific sensory modalities may have specific diagnostic implications.

Sensory system is formed by many elements integrated together, creating a complex functioning network. It requires the integrity of all its components:

- Receptors
- Nervous pathways
- Primary sensory areas
- Associative areas
- Awareness and appropriate level of attention, memory, QI, emotional status

In its complexity, sensory phenomenon consists of

- *Sensation*: Elementary psychic manifestation produced by the activation of sensory organs in response to adequate stimuli (e.g., light for vision).
- *Perception*: Complex psychic function that allows the selection, organization, and interpretation of these sensations. Perceptions of the same senses may vary from one person to another because stimuli are interpreted differently based on that individual's learning, memory, emotions, and expectations.
- *Perceptual Act*: Assigning a significant meaning to any perception.
- *Mental Representation*: Complex psychic activity that mentally reproduces contents of previous perceptual experiences. In this case, a proper physical stimulus is absent, so that sensory pathways are not directly activated.

2.5.2 Sensory Distortions

These include an altered perception of the quantitative and qualitative characteristics of the stimulus, an aberrant perception of its spatial form, or an altered experience of time. Sensory distortions are usually caused by an acute organic condition, intoxication, or dissociative states.

2.5.2.1 Changes in Intensity

- *Hypo/Hyperesthesia*: Lower/increased intensity of sensations, i.e., sound (*hypoacusis/hyperacusis*), taste, and colors' vividness. It is typically associated with the assumption of LSD and other hallucinogen substances, alcohol hangover, migraine or delirium. They may also occur in psychiatric conditions such as depression, mania, anxiety, and dissociative status.

2.5.2.2 Changes in Quality

- *Dyschromatopsia*: Visual perception of changes in objects' colors (*xanthopsia, chloropsia,* and *erythropsia*) that can mutate in, respectively yellow, green, and red coloring. It is usually due to an intoxication bye mescaline or digoxin.

The environment may appear *unreal or unfamiliar* during the process of derealization (characteristic of panic attack disorder).

2.5.2.3 Changes in Spatial Form

- *Micropsia*: Visual perception that the objects are smaller than they actually are.
- *Macropsia*: Visual perception that the objects are larger than they actually are.
- *Paropsia*: The perception of retreat of objects into the distance without any change in size.
- *Telopsia/Pelopsia*: Objects appear more distant/nearer than they really are.
- *Dysmegalopsia*: Distortion of shape and size (height, depth) of objects.
- *Paraprosopia*: Perception that face lineaments are distorting, often reaching grotesque or monstrous semblances.
- *Alloesthesia*: Referring to stimuli perceived on one side to the other (left/right).
- *Akinetopsia*: The perception of movement is compromised.

They are frequently associated with drug assumption or organic diseases (delirium, epileptic seizure, brain tumors, cerebral hypoxia, visual disorders).

2.5.2.4 Altered Time Experience

The perception of time's flowing is an individual and subjective experience that can be altered in some psychiatric conditions. During a severe depressive episode, time appears to flow slower and, sometimes, even to be "*frozen.*" On the other hand, during a manic episode, time flows very quickly. In organic brain syndromes (i.e., delirium), temporal disorientation is common.

2.5.3 False Perceptions

These indicate a group of sensory deceptions that let the subject experience something that doesn't exist. This group includes illusions, hallucinations, hallucinosis, and pseudohallucinations.

2.5.3.1 Illusion

It is an error of meaning ascription to a real sensory stimulus. Illusions are not necessarily a sign of psychiatric disturbance. For example, a man walking along a dark road may misinterpret a harmless shadow as an aggressor. Usually, along the brief time of existence of an illusion, the individual corrects his perception due to maintained consciousness and reality sense. This does not happen when illusions manifest in the context of an organic mental disorder with clouding of consciousness.

Illusions can be divided depending on their main features in
- *Affect illusion*: In this case, the stimulus is perceived in the context of the prevailing mood state. For example, a bereaved person could for a moment perceive the face of the deceased person among the crowd.
- *Completion illusion*: In this case, an incomplete perception is filled in by a process of extrapolation from previous experience and prior expectation to produce significance.
- *Pareidolia*: This type of illusion is created out of sensory percepts mixed with imagination. Pareidolic illusions occur for example when a person sees a vivid picture in clouds.

Completion and affect illusions are due to inattention, and they are banished by attention. On the contrary, attention increases the intensity of pareidolic illusions, as they become more intricate and detailed.

2.5.3.2 Hallucination
It is an interceptive percept which does not correspond to an actual object, but occurs at the same time as real perception.

According to Slade (1976), three criteria are essential for an operational definition
(a) Percept-like experience in the absence of an external stimulus.
(b) Percept-like experience that has the full force and impact of a real perception.
(c) Percept-like experience that is unwilled, occurs spontaneously, and cannot be readily controlled by the subject.

Hallucinations are experienced as originating in the outside world and are recognized as different from vivid mental images. They are experienced as certain and independent from their will. Hallucinations are self-centered as their content refers to the patient himself. However, the subject is not always able to tell if what he experiences is a "public" event or not. Often the patient is convinced that everyone can hear or feel what he does hear or feel, while, other times, he understands that what he perceives is subjective and not shared by everyone. Since the person rarely has an insight about experiencing a hallucination, it usually assumes a delusional connotation.

A hallucination can be described depending on specific features:
- Individual sense involved (vision, hearing, taste, smell, touch, and thermoception).
- Content (sometimes depending on patient mood status).

- Complexity: It can be classified as *simple* hallucination or *complex* hallucination. Simple visual hallucinations include unformed flashes of light and color or undefined shapes, while simple auditory hallucinations comprise whirring noises or rattles. Complex visual hallucinations may show faces, animals, and scenes, while complex auditory hallucinations may manifest as people talking or music.
- Emotional involvement and the consequent impact on behavior and global functioning (pervasiveness).

Peculiar types of hallucination are

- *Reflex Hallucination*: Sometimes the same hallucinatory experience may be percept by different sensory modalities (i.e., smelling colors, touching voices).
- *Functional Hallucination*: The hallucinatory event appears when the sensory system has been already activated by a real sensory stimulus (i.e., hearing voices when water is sinking in the basin).
- *Hypnagogic and Hypnopompic Hallucination*: They represent hallucinations that occur while falling asleep and during awakening, respectively. They are not related to mental disorders, so that also non-ill persons may experience them during vigilance state transition.
- *Extracampine Hallucinations*: The patient has the sensation that what he is experiencing is outside the limits of the sensory field (e.g., the subject may see somebody standing behind them when they are looking straight ahead).
- *Autoscopy (Phantom Mirror-Image):* Autoscopy is an experience of seeing oneself in external space, viewed from within one's own physical body. It is a combination of visual, kinesthetic, and somatic sensation, combined together and resulting in a complex hallucination. It has been reported in seizure disorder, focal disease of the parietal–occipital cortex, and toxic states. In rare cases, the subject may elaborate the hallucination with an explanatory delusional story (Doppelgänger).

Causes of hallucinations can be organic or psychogenic, as detailed in Table 2.1.

Table 2.1 Main organic and psychogenic causes of hallucinations

Organic	Psychogenic
– Hallucinogen drug intoxication – Delirium – Central nervous system or sensory system diseases – Sensory deprivation – Epileptogenic focuses (temporal aura) – Diseases associated with neuronal degeneration (dementia, Parkinson's disease, Huntington's and Sydenham's chorea) – Expansive intracranial lesions (tumors, abscess, aneurysms) – Vascular accidents (thrombosis, hemorrhage)	– Psychiatric disorders (psychotic disorders, delusional depression, or manic episode) – Intense emotional experience – Suggestion

Hallucinations of Individual Senses:

- *Auditory Hallucination*: It is the most frequent and common type of hallucination that occurs in psychiatric disorders. Auditory hallucinations may vary in complexity and intensity from vague whispers or muffled voices, as coming through a wall, to clear voices perceived as originating outside the patient's sense of self. There are many types of auditory hallucinations. Usually, voices commenting and conversing refer to the patient (self-centered), sometimes in third person, especially in schizophrenia. They may give a running commentary on the patient's actions, feelings, thoughts, and experiences or may command the subject to perform actions. The contents expressed by the voices can be independent, linked to the patient's delusional ideation (if any) or to the patient current mood, being in this case defined as *holothymic* or mood-congruent (i.e., harmful, blameful, praising).
- *Visual Hallucination*: It consists in *simple* (unformed images such as flashes of light) or *complex* representations (formed images such as people). Characteristically, visual hallucinations result from organic conditions rather than functional psychosis. A peculiar type of hallucination, called *Microzoopsy* (seeing little creepy animals, such as ants or worms, running through the body), typically occurs in case of delirium tremens secondary to alcoholic abstinence. Again, *Lilliputian hallucinations*, referring to visual hallucinations of objects, animals, or people greatly reduced in size, may occur in conditions such as delirium tremens or brain tumors in the temporal lobe.
- *Olfactory Hallucination:* The sudden, intense, and brief perception of an odor manifests most commonly in medical disorders, in particular in temporal lobe epilepsy and on the onset of migraine headaches and brain tumors. Typically, the smell percept is unpleasant, such as burning rubber or decaying fish. Bad smell hallucinations are reported in severe major depressive episodes and psychotic disorders (e.g., smelling rotting garbage). In other occasions, the patient may perceive a pleasant odor, which may have a special and personal significance.
- *Gustatory Hallucination:* As much as smell, taste hallucinations are usually unpleasant (e.g., metallic taste) and may occur in acute organic states, psychotic disorders, and severe mood disorders.
- *Tactile Hallucination:* This type of hallucination is experienced as originating from outside the body (e.g., being touched) or from inside the body (i.e., feeling something crawling under the skin or moving in the inner organs). This second case, also called formication, may lead to secondary skin lesions resulting from the patient scratching himself.
- *Kinesthetic Hallucination:* It involves muscle or joint sense. The patient may describe the feeling that his/her limbs are being bent or twisted and muscles squeezed. They are often associated with bizarre somatic delusions.

2.5.3.3 Pseudo-hallucination

It is an involuntary perceptual experience that is recognized by the subject experiencing it as being subjective and unreal.

Pseudo-hallucinations are perceived as clear and vivid, they are seen in full consciousness and located into the inner subjective space. They are frequently described as "as if" sensations. Hallucination connects to perception as pseudo-hallucination connects to representation.

2.5.3.4 Hallucinosis

This term indicates a hallucinatory perception that is recognized as pathologic by the patient. It is characterized by having precise physical and spatial connotations (located in the external space), but the patient is able to judge the experience as pathological. Usually, hallucinosis has a simple structure, involves visual or auditory areas, and lacks the self-reference connotation that is typical of hallucinations. It is usually related to organic illness (i.e., noises of broken glass in acoustic nerve neurinomas).

2.6 Disorders of Mood and Affect

2.6.1 Introduction

Mood and affect are often used interchangeably. However, they represent different aspects of emotional life, and it is relevant to understand the distinction between the two concepts:

- *Mood*: Persistent and sustained emotional state that influences, in a conscious way, the patient's perception of the world. Mood is an emotional state which usually lasts for some time, is not related to an object, and colors the experiences of the subject.
- *Affect*: Specific and object-related emotional responsiveness inferred by patient's facial mimic, body attitude, and autonomic responses.

Additionally, two more elements complete the affectivity sphere

- *Emotion:* Individual affective experience which has physiological correlates. It is more similar to a percept rather than a conscious process. The phenomenon usually lasts until the triggering stimulus ends.
- *Feeling:* Conscious elaboration of emotional reactions that may assume a positive or negative connotation.

2.6.2 Disorders of Mood

Mood may be subject to fluctuations in normal conditions, while it is considered to be pathological when it is invariable, it is not related to any external cause, and there is no valid explanation, the related manifestations occur with certain intensity and duration.

Altered mood may be primary in some psychiatric illnesses (major depression, bipolar disorder), but also secondary to other psychiatric conditions (psychotic disorder, anxiety disorder) or organic disorders (infections, drugs or alcoholic intoxications, brain tumors).

Mood should be assessed through both the subjective report of the patient and the objective evaluation of the interviewer. It is described as *euthymic* when it lays in the normal range and can be modulated according to the circumstances.

On the contrary, when altered, mood can be described as

- *Dysphoric*: It is characterized by frequent and rapid fluctuation of emotional states, such as sadness, anxiety, or irritability.
- *Anxious:* It is characterized by a constant unpleasant feeling of apprehension and alert caused by anticipation of a possible internal or external danger.
- *Irritable:* It is qualified by poor control over aggressive impulses, usually directed toward others, most frequently to those nearest and dearest. It may lead to outburst episodes, during which a person is annoyed and provoked to anger.
- *Depressed:* It is characterized by the tendency of the mood toward sorrow. The physiological connotation of depressed mood is *sadness,* which is a normal reaction to unpleasant events or a consequence of any kind of loss. However, depression is qualitatively and quantitatively different from sadness, being characterized by higher intensity, lack of modulation in response to external stimuli, and stability for a defined period of time. Depression may represent a symptom, a syndrome, or a proper illness (depression is the most relevant symptom, as in major depression disorder). The presence of severely depressed mood may lead to a reduction of willpower (*Abulia*), inability to enjoy things and feel pleasure (*Anhedonia*), and lack of desires, interests, and activities (*Apathy*). Although frequently associated with depression, these states may be observed also in schizophrenia or organic mental disorders.
- *Elevated:* It manifests with an exaggerated feeling of well-being and an optimistic view of the experienced. It is usually not recognized as pathological by the subject. The patient may push himself into dangerous situations, due to a lack of insight into his/her proper limits. It characterizes the manic or hypomanic episodes that occur in bipolar disorder. In hypomanic episodes, the elation of mood is less severe and initially may lead to an improvement of individual performances (better socializing ability, more self-confidence, more fatigue resistance, and less sleep need). A much more increased sense of well-being, associated with cheerful thoughts, is defined as *Euphoria*, proper of a manic episode. In this case, mood may reach its higher level, in which euphoria leads to faulty judgment, general overactivity, disinhibited behavior, and grandeur (*Elated* and *Exalted* mood).
- *Ecstatic*: It coincides with an intense sense of rapture or blissfulness, typically seen in stuporous mania.

2.6.3 Disorders of Affect

Affect fluctuations may also occur in the non-ill population. However, in physiological conditions, fluctuations occur consistently with time and context and in ranges that depend on the emotional state. These physiological emotional reactions to stressful events may assume pathological connotations with longer duration and negative impact on global functioning, thus determining the emergence of adjustment disorders, reactive (situational) depression, and complicated grief.

Affect status can be assessed by analyzing the following domains.

2.6.3.1 Intensity

It is the strength with which an emotion is expressed. Usually, it varies according to the situation. The emotional expression could be reduced.

- *Shallow affect*: Incapacity to depth in emotional response.
- *Blunted affect:* Greatly diminished emotional response, notable by absent or reduced facial mimicry and a monotone voice, indifference to distressing topics.
- *Flat affect:* Total absence of affective manifestations.

2.6.3.2 Mobility

It refers to how easy and rapidly a subject can move from one type of emotion to another. In normal conditions, mood variations occur gradually. Pathological conditions are:

- *Constricted affect:* Reduced mobility of affect.
- *Fixed affect:* Affect is so extremely constricted to one emotion that it results as immobile.
- *Labile affect:* Pathological increasing of affect's mobility, characterized by rapid shifts from one type of emotion to another without persistence of any affect.

2.6.3.3 Range

It refers to the variety of emotional expression noted in a session. A patient may possess a *full range* (the subject appropriately expresses many emotions, depending on the context) or a *restricted range* (the person shows only a few shades or a unique fixed emotion).

2.6.3.4 Reactivity

It indicates the ability of affect to change in response to environmental stimuli. It is defined as *nonreactive* when the patient does not respond to the examiner's different provocations (e.g., joke, sadness).

2.6.3.5 Communicability

The capacity of connecting with the interviewer in terms of emotional response to events, interaction, behavior, and situation. A higher need for communication usually occurs in mania, while its absence is common in schizophrenia.

2.6.3.6 Appropriateness

It represents the congruence between the quality/intensity of the expressed emotion and the content of speech and thought, within the overall situation. Following are considered abnormalities of this feature:

- *Paramimia:* Presence of disjunction between the modes of expression (mimic and gestures) and the experienced emotion.
- *Parathymia* (or *Affect Dissociation*): The emotional expression is the exact opposite of what is expected under the circumstances (e.g., laughing at tragic news).
- *Ambivalence*: Coexistence of opposite emotional feelings toward the same experience.

2.7 Disorders of Cognition

2.7.1 Introduction

Cognitive functions represent the tools by which a human subject is able to relate with the aspects of its environment. They are strictly related to each other and altogether with consciousness, circadian rhythm, and sleep. Moreover, they allow the subject to integrate information coming from the external world, elaborate appropriate responses, and convert these elements into memories. Cognitive disorders can be caused by an organic medical condition or may represent symptoms of a psychiatric syndrome. They can be generalized (multiple domains overlapped) or specific (one domain involved).

Some altered behaviors may suggest the presence of cognitive deficits:
- Avolition and apathy
- Deficit in planning abilities
- Poor abstract thinking
- Loss of social interest and lack of social relationships
- Loss of motor skills (e.g., playing a musical instrument)
- Loss of short-term memory or long-term memory storage and recall ability
- Fail in facial recognition (even not recognizing familiar people)

2.7.2 Assessment of Cognitive Alterations

The assessment of the cognitive state includes the evaluation of basic abilities such as consciousness, orientation, attention, and more complex abilities such as executive functioning, language, visuospatial abilities.

2.7.2.1 Orientation

It refers to the subject's awareness of the current setting in terms of time, space, self-consciousness, and environment surrounding. It is not a discrete function but is closely related to memory and the ability to learn continually changing facts. Orientation may fluctuate in some organic conditions (e.g., drug intoxication, delirium) or primary psychiatric diseases (e.g., dissociative disorders). When their orientation is investigated, patients with schizophrenia may give erroneous or bizarre answers, but they are usually secondary to delusional ideation.

2.7.2.2 Attention

It indicates the ability to focus on a particular stimulus or activity. A decrease of attention also occurs in non-ill persons (i.e., during sleep, fatigue, or boredom). Early warning signs that suggest a disturbance in attention are easy distractibility or difficulty in finishing tasks or keeping the concentration on work. Usually, attention results impaired in organic states, usually with lowering of consciousness, but also anxiety, depressive, and psychotic disorders. A severe deficit of attention is a major characteristic of the attention-deficit/hyperactivity disorder (ADHD) in children. Attentional disturbances are usually linked to memory deficits, as the ability to learn and remember something depends in part on our ability to selectively focus on the information we want to retain.

2.7.2.3 Memory

It refers to the mental process that allows the individual to store information. What is experienced or learned is recorded in the CNS (registration) where it persists for varying periods of time (retention) and can be recollected or retrieved from storage at will (recall). It can be divided into:

- *Sensory Memory*: It is the earliest phase of memory, allowing to retain information for a very brief time (milliseconds). Pieces of information selected and recorded need to be then processed as short-term memories or they rapidly decay.
- *Short-Term Memory*: It allows to hold, but not manipulate, a small amount of information for a short period of time. For example, it may be used to remember a phone number. Its duration is believed to be in the order of seconds.
- *Working Memory:* It is a limited capacity system that can hold, manipulate, and retrieve information temporarily. It is considered crucial for reasoning and goal-directed behaviors.
- *Long-Term Memory*: After being processed and elaborated, memories may be transferred to the almost unlimited storage of the remote memory, which allows to recall events that occurred even months or years back. It allows the construction of *autobiographical memory* (collection of memories of life events that represent the individual's personal history).

Memory can be compromised for many reasons, and its deficits can reach various levels of gravity. The main memory disturbances are:

- Amnesia
  It represent a partial or total incapacity to recall events from the past. It may have organic origins (acute brain syndrome, traumatic brain injury) or primary psychiatric origins (dissociative amnesia, depression-related cognitive dysfunction). Depending on the cause, it can last from few seconds to much more time. The loss of memory may concern events that occur immediately (or more) before the trauma (*retrograde amnesia*). On the contrary, *anterograde amnesia* compromises the storage of events that occur after the trauma. Alcoholics may suffer from *blackouts*, brief episodes of anterograde amnesia that manifest while or right after drinking.

- Paramnesia
  It is a false memory acquisition, due to a distortion of the information acquired.
  - *Recall distortions* represent a type of paramnesia. Examples of this type of memory failure are *retrospective falsification* (giving to memories peculiar emotional connotations, based on individual experience), *false memory* (remembering something that never occurred), *screen memory* (elaboration of a traumatic memory in order to make it less harmful), *confabulation* (memory falsification secondary to an impairment of retrieval, of which the subject is unaware, typical of organic brain disorders). It should be distinguished by *fantastical pseudology* (fluent plausible lying typical of antisocial and other personality disorders). *Recognition distortions* represent another type of paramnesia. Some examples are *déjà vu* (sensation of having already experienced what is currently happening), *false recognition* (error in identifying environment elements).
- Ipermnesia
  It is defined as an exaggerated recording and storage of information. They could be associated with an intense emotional status (*flashbulb memory*). *Flashbacks* are brief and vivid intrusive memories that are linked to past traumatic events or occur after the assumption of psychedelic drugs (even after months or years later).

2.7.2.4 Executive Functioning

It refers to the person's ability to plan and initiate activities to achieve a specific goal. It allows to identify a problem and its possible solutions, select and implement a strategy, and monitor its effectiveness. It requires the integrity of basic cognitive functions as attention, cognitive inhibition, working memory, cognitive flexibility. It can be impaired in ADHD, schizophrenia, and dementia.

2.7.2.5 Visuospatial Functioning

It reflects people's ability to identify, integrate, and interpret visual form, structure, orientation, and location of objects and spatial relationships among them. Deficits can be frequently observed in dementia.

2.7.2.6 Global Intelligence

It is the combination of the ability to think logically, act rationally, and deal efficiently with the environment. Intelligence testing is used to quantify the patient's general intellectual functioning, which is expressed through the intelligence quotient (IQ). High intelligence is usually associated with better prognosis in several psychiatric conditions.

2.7.3 Cognitive Examination

A cognitive assessment should be considered in all patients aged > 65 years or when cognitive impairment is suspected. It is recommended to use a standardized test, in order to avoid vague descriptions or inaccurate evaluations. To ensure that the cognitive assessment effectively reflects the tested abilities, it is necessary to verify the integrity of sensory pathways, the consciousness accessibility, and the mood status that may influence the execution of tests.

Some tests commonly used to evaluate cognitive impairment in clinical practice are

- *Abbreviated Mental test (AMT):* It takes few minutes to perform, but it is not sensitive to mild and moderate impairment.
- *Mini-Mental State Exam (MMSE):* It covers most cognitive domains, but it results not sensitive to mild impairment and does not consider executive functions. It is deeply influenced by premorbid IQ and scholar level.
- *Clock Drawing Test (CLOX1 and its variants):* It evaluates praxis and executive functions. It can be much influenced by poor motor control or visual impairment, but it does not depend on premorbid IQ.
- *Addenbroke's Cognitive Examination-Revised (ACE-R):* It is a longer and more detailed test that analyzes all cognitive domains, resulting sensitive to mild impairment too. Premorbid IQ impacts on test's outcome.

2.7.4 Intellectual Evaluations

Intellectual premorbid level may represent a prognostic element of better outcomes as higher as it was. Different types of tests are used to evaluate the IQ level, and it could be useful to combine them in order to obtain the most accurate result. The Wechsler Adult Intelligence Scale (WAIS) is one of the most used tools to measure intelligence. Its last version (WAIS-IV) includes four index scores representing the major domains of intelligence:

- Verbal Comprehension Index (VCI)
- Perceptual Reasoning Index (PRI)
- Working Memory Index (WMI)
- Processing Speed Index (PSI)

Two additional scores can be derived to determine general intellectual ability

- *Full-Scale IQ (FSIQ):* It results from the combined performance of the VCI, PRI, WMI, and PSI.
- *General Ability Index (GAI):* It considers only the six subtests that the VCI and PRI include.

2.8 Insight

2.8.1 Introduction

The patient's degree of awareness that his psychic symptoms are due to a mental illness and require proper treatment. Lack of insight is connected with deficits in introspection, self-awareness, self-monitoring, empathy, and communication. It requires the capacity of both inner and outer orientation. In the most severe cases, the patient exhibits complete denial of illness. In other cases, he may show some awareness of the illness, while blaming external factors, other persons, organic factors, or even mystical causes.

Insight can be rated on a 6-level severity scale:
- *Grade-1:* Unawareness and complete denial of illness.
- *Grade-2:* Poor awareness of being ill and needing proper treatments, but denying at the same time.
- *Grade-3:* Awareness of being ill, but the blame is placed on external or physical factors.
- *Grade-4:* Awareness of being ill, but the cause is unknown.
- *Grade-5:* Awareness of being ill and that the symptoms in social adjustment are due to own peculiar irrational feelings or thoughts (*intellectual insight*), but this knowledge is not applied to current or future experiences.
- *Grade-6:* In this case, the awareness of illness leads to significant basic changes in future behavior (*true emotional insight*).

An insight impairment may occur in schizophrenia, delusional disorder, mania, and severe major depressive disorder. Poor insight may also be observed in a subset of subjects with obsessive-compulsive disorder. Moreover, insight may be altered in cognitive impairment.

2.8.2 Assessment of Insight

In order to reach an appropriate measure of insight in clinical practice, it is necessary to consider that
- Insight is a complex and multidimensional concept.
- Cultural factors may have an influence on it.
- It reaches different levels of impairment across the various manifestations of mental illnesses.
- It will be probably necessary to reach information from clinical evidence rather than the interview.

Numerous types of assessment scales are available and represent valid instruments that help the clinician evaluating insight impairment, such as Insight and Treatment Attitudes Questionnaire (ITAQ), Birchwood Insight Scale (BIS), Schedule for the Assessment of Insight-Expanded version (SAI-E), and Schedule for the Assessment of Unawareness of Mental Disorder (SUMD).

Further Reading

Akiskal HS. The mental status examination. In: Fatemi SH, Clayton PJ, editors. The medical basis of psychiatry. 4th ed. New York: Springer; 2016. p. 3–16. https://doi.org/10.1007/978-1-4939-2528-5.

Association AP. Diagnostic and statistical manual of mental disorders (DSM-5®). New York: American Psychiatric Association; 2013.

Bebbington P, Freeman D. Transdiagnostic extension of delusions: schizophrenia and beyond. Schizophr Bull. 2017;43(2):273–82. https://doi.org/10.1093/schbul/sbw191.

David AS. Assessment scales for insight. In: Keefe R, editor. Guide to assessment scales in schizophrenia. Heidelberg: Springer Healthcare; 2012. p. 55–64.

Geddes JR, Andreasen NC, Goodwin GM. New Oxford textbook of psychiatry. Oxford: Oxford University Press; 2020.

Gilleen J, David AS. The cognitive neuropsychiatry of delusions: from psychopathology to neuropsychology and back again. Psychol Med. 2005;35:5–12.

Kiran C, Chaudhury S. Understanding delusions. Ind Psychiatry J. 2009;18(1):3–18. https://doi.org/10.4103/0972-6748.57851.

Marwick KFM, Birrell S, Horton-Szar D, Hall J. Crash course psychiatry. London: Elsevier Health Sciences; 2015.

Ouwersloot G, Derksen J, Glas G. Reintroducing consciousness in psychopathology: review of the literature and conceptual framework. Front Psychol. 2020;11:586284. https://doi.org/10.3389/fpsyg.2020.586284.

Oyebode F. Sims' symptoms in the mind: textbook of descriptive psychopathology. 5th ed. Amsterdam: Elsevier Health Sciences; 2015.

Renn BN, John SE. Mental status examination. In: Segal DL, editor. Diagnostic interviewing. 5th ed. New York: Springer; 2019. p. 77–102. https://doi.org/10.1007/978-1-4939-9127-3.

Sadock BJ, Sadock VA, Ruiz P. Kaplan & Sadock's synopsis of psychiatry: behavioral sciences/clinical psychiatry. Baltimore: Wolters Kluwer; 2015.

Sobin C, Sackeim HA. Psychomotor symptoms of depression. Am J Psychiatry. 1997;154(1):4–17.

Stephen BS. Fish's clinical psychopathology: signs and symptoms in psychiatry. Med J Aust. 1975;221:106317. https://doi.org/10.5694/j.1326-5377.1975.tb106317.x.

Walther S, Strik W. Motor symptoms and schizophrenia. Neuropsychobiology. 2012;66(2):77–92.

Mood Disorders

3

L. Franchini, B. Barbini, R. Zanardi, L. Fregna, F. Martini,
E. Manfredi, A. Sarzetto, B. Maiocchi, N. Ragone,
and C. Colombo

3.1 Unipolar Depression

3.1.1 Definition

Unipolar depression is classified as major depressive disorder in the *Diagnostic and Statistical Manual of Mental Disorders 5* (APA, *DSM-5*), and it is characterized by recurrent disturbances of mood and affect, diminished ability to concentrate, anxiety, fatigue, recurrent thoughts of death, decreased appetite, weight loss and insomnia.

3.1.2 Epidemiology

Among depressive disorders, major depression represents the most prevalent psychiatric disorder in the general population, and it is, according to WHO (2016), the

L. Franchini · B. Barbini · R. Zanardi · L. Fregna (✉) · F. Martini (✉)
IRCCS San Raffaele Scientific Institute, Milan, Italy
e-mail: franchini.lindaannamarina@hsr.it; barbini.barbara@hsr.it;
zanardi.raffaella@hsr.it; fregna.lorenzo@hsr.it; francesca.martini1@hsr.it

E. Manfredi · A. Sarzetto · B. Maiocchi · N. Ragone
Vita-Salute San Raffaele University, Milan, Italy
e-mail: manfredi.elena@hsr.it; sarzetto.alessandro@hsr.it;
maiocchi.beatrice@hsr.it; ragone.nicola@hsr.it

C. Colombo
IRCCS San Raffaele Scientific Institute, Milan, Italy

Vita-Salute San Raffaele University, Milan, Italy
e-mail: colombo.cristina@hsr.it

R. Cavallaro, C. Colombo (eds.), *Fundamentals of Psychiatry for Health Care Professionals*, https://doi.org/10.1007/978-3-031-07715-9_3

leading cause of disability in the world, understanding disability as the restriction of an operational capacity resulting from a physical or mental impairment.

Major depression has a prevalence of 12% in Europe and 7% in the United States. Females have prevalence rates 1.5–3 times higher than males. The age of onset of major depression sees a prevalence in individuals between the ages of 18 and 30 years three times greater than the prevalence in individuals older than 60 years. The onset in juvenile age is associated with greater familiarity for the disorder and an increased risk of personality disorder development and worse prognosis.

3.1.3 Etiopathogenesis

The etiopathogenesis of major depression is multifactorial, and it is the result of a gene–environment interaction.

According to twin studies, about 37% of the variance in depression is explained by genes; heritability becomes higher considering hospitalized patients than outpatients; furthermore, one small adoption study confirms the modest heritability in major depressive disorder.

The monoaminergic role in the etiopathogenesis of depression has represented the main hypothesis of this model. According to this theory, the neurophysiological basis of depression is a reduction in the levels of norepinephrine, serotonin, and dopamine in the central nervous system. In support of this mechanism is the pharmacodynamics of the antidepressants: medications that enhance these neurotransmitters' levels in the brain have all been shown to be effective in ameliorating depressive symptoms. However, further research has failed to confirm the hypothesis of a primary dysfunction of the monoamine system in patients with major depressive disorders. At the present time, a reduced sensitivity of postsynaptic receptors was associated with the development of clinical symptoms, especially in predisposed individuals and/or exposed to early adverse events.

More recently, the involvement of other neurotransmitter systems including glutamate, substance P, TRH (thyrotropin-releasing hormone), CRF (corticotrophin-releasing factor), somatostatin (leptin), acetylcholine and BDNF (brain-derived neurotrophic factor) is reported in the pathogenesis of depression.

Functional MRI studies have shown functional abnormalities of specific functional areas (limbic system, basal ganglia and hypothalamus) responsible for emotion (processing, regulation) and gratification. These abnormalities could be at the basis of the excessive responses to negative emotional stimuli in depressed patients. Similarly, a defect in BDNF expression has been related to an altered neuronal plasticity and cellular atrophy in some brain areas.

In subjects with major depression, some studies have indicated a reduction in connectivity between the medial part of the orbitofrontal cortex, involved in the processes of gratification and other areas involved in the mechanisms of formation and evocation of memory, such as the medial temporal lobe and the para-hippocampal gyrus. This would provide an anatomical basis for some typical behaviours of the patient during a depressive episode such as anhedonia and the tendency to brood over negative thoughts and memories.

The hypothalamic–pituitary–adrenal (HPA) axis is responsible for stress management. In depressed subjects, a general hyper-activation of the HPA axis occurs. It implies a higher CRF, ACTH, cortisol levels in blood and a failure in the dexamethasone suppression test is observed. This determines a dysregulation of the axis with a lack of negative feedback, persistent high levels of blood cortisol and a consequent increase of peripheral pro-inflammatory factors.

Also the exposure to early and chronic stressors (i.e., abuse or neglect in childhood, interpersonal conflicts and social isolation) plays a role in the dysregulation of the HPA axis.

If the hyper-activation of the HPA axis is an adaptive response in front of an acute stress, on the contrary, when the hyper-activation becomes chronic, it represents a vulnerability factor, resulting in many alterations in immune and inflammatory systems. According to these considerations, numerous evidences underline the role of inflammation as one of the pathogenic processes underlying depressive illness. During a depressive episode, high levels of interleukin-6,-1, TNF-alpha, interferon-gamma, PCR have been described, with a directly proportional correlation to the severity of the episode. Moreover, the role of inflammatory processes could be at the basis of the significant comorbidity between medical conditions such as diabetes and cardiovascular disease and major depression.

The subjective response to adverse events depends on both temperamental aspects and different polymorphisms regarding genes involved in the main neurotransmission pathways. Among these, one of the most studied gene is the SERT gene for which it was identified a positive association between stressful events and development of depressive disorders, particularly expressed in individuals with s/s allele.

Among the etiopathogenetic hypotheses of mood disorders, modifications of the circadian rhythms control systems are described. Clock, GSK3, and melatonin genes are involved in the delay, advancement, and desynchronization of this system. In patients with major depression, molecular polymorphisms of CLOCK genes involved in the functioning of the biological clock have been identified, and they could be responsible for some psychopathological features such as the seasonal pattern of the disease, the early age onset, the response to antidepressant therapies, the quality and quantity of sleep, the presence of self-injurious behaviour and suicidal ideation.

Sleep disorders are commonly observed in patients with major depression. In depressed patients, characteristic polysomnographic correlates are described, such as sleep fragmentation, reduction of the interval between sleep onset and first REM phase (the so-called REM latency), reduction of slow wave sleep (N3), increase of REM sleep and increase of rapid eye movements (the so-called REM density).

In general, there are many alterations associated with the depressive condition, neuro-endocrinological, neuroanatomical and neurophysiological and for an in-depth study of the individual topics, please refer to Saveanu and Nemeroff (2012). However, none of them is currently enough for the diagnosis, which remains exclusively clinical.

3.1.3.1 Assessment

The term "depression" is often used to denote clinical conditions of very different meaning and management.

Therefore, it is important to determine whether this term is used to define a symptom (1), a syndrome (2), or an actual disorder coded as a major depressive episode (3).

1. Depressive symptoms can occur in absolutely physiological contexts. In fact, sadness is one of the primary emotions and represents the normal human response to adversity and has an important adaptive function. A depressive symptom is such if isolated, not clinically significant, associated with life experiences or stressful events and of short duration.
2. A depressive syndrome occurs during the course of other medical or psychiatric illnesses or may be iatrogenic (see Table 3.1); it is configured as a recognizable set of symptoms and signs, clinically significant, it is associated with a variable degree of suffering and functional impairment, and its resolution depends on the management of the underlying primary problem.
3. Obviously it is possible that in a patient with a medical pathology, major depression coexists. Faced with this condition, it is important to carefully evaluate the weight that one disease has on the other because the presence of depression can sometimes affect the expression of medical pathology and influence its course. By example, there are numerous clinical evidences of the synergistic link between depression, both major and minor, with cardiovascular diseases, worsening the condition of one and vice versa. Moreover, the diagnosis of depression is made with an average frequency of 21.5% of patients with cardiovascular disease.
4. Major depressive episode is characterized by the primary alteration of mood. It is accompanied by a constellation of signs and symptoms that meet the criteria codified in the psychiatric nosography and are present simultaneously and for at least 2 weeks.

3.1.4 Clinical Presentation

Depressed mood and the loss of interest or pleasure (anhedonia) are the principal clinical phenomena. Patients may say they feel in a "bad mood", "powerless", "down" or "useless"; these symptoms take on pervasive and constant characteristics throughout the day and can hardly be modified by external stimuli.

The facial mimic is marked by sadness: the forehead is frowned, the gaze may be simultaneously off and full of anxiety, the corners of the mouth trend downward, face muscles appear contracted and the tendency to cry is frequent. Apathy may be pervasive, and in some cases, patients present difficulties in externalizing sadness with the feeling of no longer being able to feel emotions.

Depressed patients are characterized by the weakening of the will which also may appear as a psychomotor inhibition/slowing down. Patients may present a

Table 3.1 Most frequent conditions associated with the secondary development of depressive syndromes

Neurological disease	– Parkinson disease – Huntington disease – Progressive supranuclear paralysis – Alzheimer's disease and other dementias – Cerebrovascular alterations – Brain trauma – Brain neoplasms – Brain infections – Multiple sclerosis – Epilepsy – Headaches – Narcolepsy – Hydrocephalus – Night apnoea syndrome – Wilson's disease
Infectious disease	– Bacterial – Virus (Sars Covid-19)
Endocrine disease	– Cushing's disease – Huntington's disease – Hyperaldosteronism – Hypo/hyperthyroidism – Hypo/hyperparathyroidism
Inflammatory disease	– SEL – Rheumatoid arthritis – Temporal arthritis – Sjögren's syndrome
Vitamin deficiencies	– Folate – Vitamin B_{12} – Niacin – Vitamin C – Thiamine
Other diseases	– Cancer (particularly pancreatic and lung cancer) – Systemic neoplasms – Cardiopulmonary diseases – Renal diseases and uraemia – Porphyria – AIDS – Klinefelter's syndrome
Iatrogenic depression	– Corticosteroids – Extrapyramidal disorders secondary to chronic intake of antipsychotics

slowdown in speech and in the execution of movements; in severe clinical condition, it is possible to observe a psychomotor block known as "*stupor melancolicus*". The posture is generally characterized by a folding on himself, the movements are slowed, the gait is uncertain. Manifestations of psychomotor restlessness may also be present: the patient walks back and forth, stands and sits continuously, twists his hands and repeatedly brings them to his face and head. The patient manifests asthenia, the reduction until the disappearance of any vital energy with a tendency to clinophilia, manifests loss of interest, lack of responsiveness to stimuli, reduction/loss of initiative even with regard to self-care and short- and long-term planning.

The depressive condition is also characterized by psychic inhibition with slow and stagnant thinking, difficulty in concentration and memorization, mental monotony with reduced associations, inability and/or difficulty in synthesis. The perception and awareness of inner emptiness and inability to feel pleasure in the presence of positive stimuli, typical of the depressive state, could be defined as "moral pain". Another crucial psychopathological element of the depressed patients is a cognitive distortion in evaluating themselves and their lives. Depressive pessimism is an essential psychopathological phenomenon; it manifests itself with the painful recollection of one's past, with themes of unworthiness, incapacity, self-evaluation, self-accusation, absence of hope and future prospects, also concerning the possibility of recovery. During a depressive episode, in 10–15% of cases, it is possible that the cognitive distortion assumes a delusional quality. Delusions will be either congruent (guilt, impoverishment and hypochondria) or incongruent (control or thought transmission, persecution) with depressed mood.

A minority of patients experiences hallucinations, more frequently auditory, in the form of disparaging voices. The presence of psychotic manifestations during a depressive episode is an index of severity, and it is a poor prognostic indicator. As a matter of fact, the presence of psychotic manifestations represents a decision-making element for the hospitalization of the depressed patient.

Pessimism, re-actualization of negative experiences of the past (faults, failures), intolerability of the present (recognition of its non-functioning and its inability and lack of responsiveness) and absence of prospects for the future generate the idea of death as a positive fact, which can result in the implementation of self-harming behaviours or death by suicide. The presence of suicidal ideation with suicidal planning represents a crucial element for the hospitalization of the depressed patient. Most depressed individuals report the presence of suicidal ideation, 10–15% actually go so far as to commit suicide. Among the various risk factors proposed (male sex, social isolation, feelings of hopelessness) for the identification of subjects most at risk for the implementation of suicidal behaviour, the most established seems to be the positive history of previous suicide attempts.

Insomnia is a phenomenon that characterizes the depressive state, and a large number of abnormalities are reported, often coexisting with each other: the difficulty in falling asleep, to maintain sleep, the presence of early awakenings and, less frequently, hypersomnia. Not infrequent is the tendency to self-medication with sedative drugs or through the use of alcohol.

Alterations in vegetative functions are also central and characteristic: decreased appetite and weight, more rarely increased appetite, decreased libido, constipation and the presence of changes in circadian rhythms with worsening of clinical conditions in the morning.

The depressive symptomatology in some patients can organize itself in peculiar clinical forms: catatonic, melancholic and atypical.

Catatonic depression presents at least two of the following symptoms: immobility and lack of responsiveness to the interlocutor (catalepsy) or waxy flexibility (maintenance of a postural attitude imposed from outside); afinalistic motor hyperactivity; extreme negativism (resistance to any instruction or external input); postural stereotypy or mannerisms; echolalia or ecopraxia (unmotivated repetition of words or gestures).

Melancholic depression presents from at least one of the following symptoms, occurring during the period of greatest severity of the depressive episode: loss of pleasure in all or most activities; loss of responsiveness to habitually pleasant stimuli (not feeling better, even temporarily, when something good happens). In addition, at least three of the following symptoms are required:

- A particular quality of depressed mood (i.e. depressed mood is experienced as distinctly different from the kind of feeling experienced after the death of a loved one).
- Regularly worse depression in the morning; early morning awakening (at least 2 h earlier than the usual wake-up time).
- Marked motor slowing or agitation.
- Significant anorexia or weight loss.
- Excessive or inappropriate feelings of guilt.

Atypical depression shows mood reactivity (i.e. mood rises in response to current or potential positive events) and two (or more) of the following characteristics are present: significant weight gain or increased appetite, hypersomnia, "leaden paralysis" (feeling heavy or having leaden arms and legs), or an enduring pattern of hypersensitivity to interpersonal rejection (not limited to mood-altering episodes) that results in significant social or work impairment.

Box 3.1: Semeiology of Depression

To investigate the presence of depressive symptomatology, the examination of the mental state passes through some simple questions:

"How do you feel?" "How would you describe your mood?" (*They might tell you that they feel tired, uninvolved by what happens around them and worried about such indifference*); "How do you spend your days?" (*Patients tend to spend most of the day in bed or on the couch*); "How long in the day do you feel this way?" "Are there any differences between how you feel in the morning and in the evening?" (*Patients in the most acute phase have no changes in their state, and this is a source of anxiety; on the contrary, the appearance of fluctuations throughout the day may indicate an initial effect of antidepressant treatment*); "Are there any activities that you do with pleasure?" (*They often report that they do not feel like doing anything, are unable to take care of themselves, their home, their work and their interests*); "Is the way you feel different from your usual way?" "How do you sleep?" "Have you noticed alterations in appetite?" "Are there any recurring thoughts during this period?" "What are they about?" "How often do you have these thoughts?" "How do you see your future?" "How would you describe your energy?" (*Recovery of physical energy alone in the absence of cognitive improvement, may increase the risk of enacting self-injurious conduct*); "How long have you been feeling this way?" "Have you ever felt that life no longer makes sense or is not worth living?" (*Rather than encouraging suicidal behaviours, talking openly can give an individual other option or the time to rethinking his/her decision*)

"Has it ever happened to you to feel this way other times?"

3.1.4.1 Major Depressive Episode: Diagnostic Criteria

In order to make a definitive diagnosis of a major depressive episode, in the psychopathological investigation, it is necessary to define the minimum temporal duration of symptoms, to exclude secondary depressive manifestations (medical or psychiatric conditions) and to evaluate the interference of symptoms on the functioning of the subject.

It must also be investigated the presence of significant losses (death of a relative, economic collapse, having survived a disaster, serious illness or physical disability) concomitant with the episode and should never be hastily put the possible depressive condition as a reaction in some way justified, understandable and without indication for treatment, but rather should be included in the history of the patient, investigating the previous mode of reaction to other events of loss.

All these aspects are better summarized in *DSM-5*, where nine typical symptoms of depression are listed, all of them present for most of the day, nearly every day

1. Depressed mood either subjectively reported by the patient or observed by other people
2. Significant reduction in the interest or pleasure in most of the activities
3. Important weight loss not explained by a diet or weight gain or changing in appetite
4. Sleep disturbances (both insomnia and hypersomnia)
5. Psychomotor agitation or retardation observed by other people
6. Tiredness, fatigue
7. Excessive guilt (sometimes delusional) apart from the guilt about the illness
8. Difficulties in concentration and decision-making
9. Recurrent thoughts about death, suicidal ideation both with and without a precise plan, suicide attempt

In order to define a major depressive episode, at least five symptoms are necessary, present for at least 2 weeks and determining a modification in functioning; among the symptoms, depressed mood and/or loss of interest has always to be present. Social, occupational or general functioning has to be significantly affected by the symptomatology, which has not to be possibly caused by a substance or a different medical condition. It is obviously important to exclude the presence of other psychiatric disorders that could explain such a clinical presentation, in particular psychotic disorders, as well as a history of manic episodes.

3.1.4.2 Clinical Course

In the course of mood disorder, we must distinguish the evolution of the acute depressive episode from the longitudinal evolution of the mood disorder.

The acute depressive episode itself can self-resolve with a natural time described between 6 months and 2 years. The goal of acute therapy through the use of antidepressant treatment is to reduce the duration of the acute episode and the achievement of symptomatic remission. In clinical practice, this theoretical goal is achieved in 40–50% of patients; in the remaining cases and especially in the presence of severe depressive episodes with psychotic manifestations or with prevailing manifestations of anxiety, complete remission is hardly achieved. In fact, it is possible that some patients show residual symptoms such as persistence of feelings of insecurity, excessive worry, difficulties in interpersonal relationships and rigid behavioural patterns that preclude them from changing any lifestyle useful for a proper management of the depressive illness. In this sense, depressive condition involves a tendency to poor medical prescriptions adherence and healthy lifestyles, which potentially worsens the course of any concomitant disease.

The longitudinal course of mood disorders needs to define the concept of euthymia. Euthymia is the phase of the mood disorder course during which the patient's mood fluctuations stay within physiological limits, without other relevant symptoms especially of neuro-vegetative type and the subject comes back to a level of premorbid functioning.

It is important to know that mood disorders, after the onset of illness, have a periodic course, characterized by a succession of cycles, consisting of relapses (full return of depressive symptoms once remission has occurred—but before recovery has taken hold) or recurrences (another depressive episode after recovery has been attained) and subsequent return to euthymia, until the onset of the next episode. *With regard to depressive disorders, if there are two or more major depressive episodes, we speak of major depression, recurrent.*

It is important to remember that diagnosis stability is a crucial clinical issue. This is particularly relevant when we are dealing with an ongoing depressive episode. In fact, 5–10% of individuals who present a single episode of depression later develop a manic episode, changing their diagnosis in bipolar disorder. Consequently, also in a non-specialist setting, it is mandatory to perform an accurate anamnesis. In fact various risk factors heighten the risk of the development of a subsequent bipolar form: early and acute onset of major depression, presence of psychotic features, psychomotor retardation and a positive family history of bipolar disorder.

On the contrary, the presence of a past manic episode will easily guide the clinician to a bipolar depression with considerable implication for the acute and long-term treatment. In fact depressed bipolar patients are vulnerable to polarity switch during any antidepressant treatment.

Lifetime periodicity of mood disorders follows non-generalizable rules and individual rhythms, so that for each patient it is possible to identify a particular periodicity. It is frequently observed in clinical practice that some patients present new episodes of disease in specific periods of the year, as well as clinical forms with a real seasonal trend with the beginning in the fall and the end in the spring.

Despite the higher incidence of the disorder in females, this does not imply a worse evolution nor does it lower treatment response rates. In females however, there are a few particular moments when depression recurrence is particularly probable. These periods of increased vulnerability are related to female reproductive cycle and will be described later in further detail.

Mood disorder cyclicity does not remain stable over time, and it shows a natural tendency to increase progressively with a reduction in the duration of euthymia and an increase in new episodes of illness. According to this point, the failure to achieve complete remission from a depressive episode and the persistence of residual symptoms is associated with an increased risk for new episodes of disease or for the chronicity of the disorder. Chronicity of the disorder is also frequently associated with underlying personality, anxiety and substance use disorders.

Regarding the ageing process in individuals with a diagnosis of major depressive disorder, it has been hypothesized that depression can lead to a decreased brain and cognitive reserve during life span. Furthermore, it seems that a positive correlation exists between the burden of depressive illness and cognition impairment in older mood disorder patients. Thus, individuals with early onset or a chronic course of the disease might experience cognitive impairment earlier, reflecting a decreased cognitive reserve. Instead, the association between late-onset depression and increased risk of dementia more controversial. This last topic of large complexity will be treated later in a dedicated section.

Moreover, as previously reported, lifetime depressive disorders are associated with worsening quality of life of patients, increased morbidity and functional disability, as well as increased use of healthcare resources in general.

The periodic course of mood disorder, both in the recurrent major depressive form and in the bipolar form, is the main objective of the stabilizing therapy which aims to lengthen the duration of euthymia. As a general rule, a longer duration of euthymia decreases the risk of new episode of illness.

Consequently for a good clinical management of a patient affected by a major depressive episode, once the clinician has identified an ongoing depressive episode, there are further important elements to be considered for a correct definition of the disorder. In order to do this, clinicians should:

- Consider the severity of the current episode (mild, moderate or severe), verifying the presence of psychotic features
- Understand if the episode is in acute phase or in course of remission

- Clarify if this is the first episode or if the patient has already suffered from depression; in the second case, it is important to investigate how many depressive episodes has the patient experienced and how much time elapsed between them
- Exclude the presence of past manic episodes
- Accurately describe euthymia with attention to its length and the presence of eventual residual symptoms.

Box 3.2: Depression of the Elderly

In the elderly, the onset of the first episode of depression can be particularly subtle because it often shows mild, poorly defined, mainly somatic symptoms. Subjective slowness and pain in movements, difficulties in memory and concentration, reduced interest, digestive difficulties, general somatic complaints and insomnia are easily interpreted as a simple reflection of advancing age. This can result in a diagnostic and treatment delay, with significant consequences. Late-onset depression in the elderly is associated with a high risk of chronicity, relapse (about 40% of cases), drug resistance, cognitive impairment, psychotic manifestations, deaths by suicide, worsening of comorbid illnesses, increased healthcare costs and mortality secondary to concomitant physical illnesses.

Cognitive impairment is always present in depression, to the point that it represent a *DSM-5* diagnostic criterion. In elderly patients, this phenomenon can create problems of differential diagnosis. Especially when cognitive symptoms are particularly severe, with confusion and functional impairment, depression can cause the so-called pseudo-dementia. This term indicates a severe cognitive impairment, secondary to the depressive condition and that resolves only when depression ameliorates. Sometimes it may be harsh to distinguish between dementia and pseudo-dementia, making diagnosis possible only after the treatment of depression. However, these conditions show some clinical differences. From a semeiological point of view, in pseudo-dementia, the patient describes the deficit in detail, emphasizes the difficulties and does not have marked attentional and concentration problems. On the contrary in dementia, generally, the subject does not appear to be concerned about the deficit, tries to hide it and rather emphasizes the positive results; moreover in dementia, especially short-term memory, appears deteriorated.

It remains an open question whether depression in the elderly is associated with an increased risk of developing mild cognitive impairment (MCI) or dementia. Recently, literature showed that subjects who had developed a depressive syndrome after the age of 50 years were about three times more likely to develop dementia than the general population. It has been questioned whether late-onset depression is a real risk factor for dementia, or an early marker of the disease, or even a triggering factor. With the current knowledge, it is impossible to answer the question, but the most credited theories hypothesize that the vascular component is the main pathogenic mechanism of late-onset depression.

Box 3.3: Depression and Female Life Cycle

In the following paragraphs, we will focus on the relationship between depression and the female life cycle: women develop depression more than men with a 2:1 ratio; this difference begins in adolescence and continues throughout life. The hypothesized reasons for this phenomenon are many: first of all, there is a greater propensity of women to seek help than men, resulting in a greater use of health services, a higher number of prescriptions of psychotropic drugs and a higher rate of hospitalization. We should also consider the role of neuroendocrine factors that determine differences in brain structure and in the impact of sex hormones on the brain. Psychosocial factors may also play a role including: coping, personal vulnerability, frequency of exposure and quality of stressful events such as traumatic separation or loss, abuse or violence, dysfunctional attachment relationships, and anxious and inhibited temperamental traits.

Pregnancy is for a woman a period of profound changes, not only physical but also psychological. Changes can be so intense that some women find it difficult to accept the state of pregnancy, feeling conflicting feelings of happiness and fear for what lies ahead. Very important in this period is for a pregnant woman to listen to her feelings and their changes, because sadness, discouragement and anxiety can easily turn into real symptoms of depression.

The relationship between depressive manifestations during pregnancy and in the following period is a debated topic. In particular, it is not clear whether peri-partum depressive symptoms are predictive of postpartum depression. However, it is certainly important to know how to recognize and not to neglect them, as they represent a risk for a "high-risk" pregnancy. In other words, depression during pregnancy is accompanied by risky lifestyles both for the woman and for the unborn child. Treating depressive symptoms in pregnancy is essential; antidepressants are the first choice for most forms of depression and most of them are safe.

With regard to the postpartum period, clinically, there are forms of rapid onset and spontaneous resolution such as the baby blues and more complex forms that require therapeutic intervention. Postpartum depression is the best-known form of female depression. The risk factors are similar to those occurring during peri-partum, with the addition of impairment of sleep quality. In addition, if the woman has already experienced depressive episodes in the postpartum period of previous pregnancies, the risk is further increased.

Postpartum depression can assume a high degree of severity also due to the presence of psychotic manifestations. This clinical situation may require hospitalization of the mother. Hospitalization in a psychiatric setting allows a rapid treatment of the mother's depression and reduces the risk of serious events such as suicide.

3.1.5 Treatment

Mood disorders have a periodic course, characterized by a succession of cycles, consisting of episodes of acuity and subsequent return to euthymia, until the onset of a next episode. This is the fundamental paradigm of mood disorder treatment which includes an acute phase of treatment in order to achieve depressive remission, a continuation phase of treatment in order to prevent relapses and a prophylaxis phase treatment in order to prevent recurrences. Each phase has different duration over time, and it depends on the individual characteristics.

The acute depressive episode, as mentioned at the beginning of this chapter, tends to self-resolve. However, the discomfort and the impairment of functioning caused by depression require an antidepressant treatment that relieves the suffering of the patient, reduces the duration of symptoms and the associated risks. In fact, a depressive episode of prolonged duration or untreated may bring the patient to a persistent cognitive distortion. In other words, the patient may consider that his current (depressive) condition is normal, forget his previous functioning, thinking he can never go back. This cognitive distortion may be accompanied by complications, such as secondary substance abuse (i.e., alcohol and anxiolytics) to reduce the symptoms and a progressive and more structured suicidal intent.

An antidepressant treatment can never be started without a clinical intervention aimed at sharing with the patient essential information about the disorder. It is firstly important to clarify the medical nature of the disorder, then the timing and the way the treatment will lead to psychopathological improvement and the behaviours to be implemented by the patient and family members.

Pharmacological therapy represents the essential component of the treatment of the depressive episode, and its main objectives are:

1. Rapid resolution of acute phase symptoms, that is, resolution of the episode and not only a reduction in symptom's intensity.
2. Research of a treatment as tolerable as possible.
3. Return to a premorbid level of social and work functioning.

The treatment of a depressive episode should always be initiated with antidepressants, and it is usually possible for the patient to take them as an outpatient, with periodic interviews for clinical monitoring.

It is wrong to treat mild forms with benzodiazepines only, as they are not a curative treatment of the depressive core but only symptomatic. Benzodiazepines may be useful in the initial phase in order to reduce the presence of anxiety symptoms or to improve sleep quality. Similarly, it is wrong to treat only with antipsychotics severe forms with psychotic manifestations such as delusions or hallucinations, as the presence of these symptoms only indicates a greater severity of the depressive state, without changing the diagnosis. Low-dose neuroleptics may be prescribed, only in association with antidepressants.

The available antidepressant molecules are numerous, and each pharmacological class has defined peculiarities that may make them more suitable for certain patients.

In general, it is appropriate, when possible, to use only one medication, and, as in all areas of medicine, in the choice of an antidepressant drug, it is necessary to know the presence of diseases and concomitant medications to avoid interactions or side effects. Classically, tricyclic antidepressants are contraindicated in patients with heart disease, prostatic hypertrophy, closed angle glaucoma, cognitive impairment, while antidepressants belonging to the class of serotonin reuptake inhibitors are contraindicated in patients with gastritis and peptic ulcer.

In the choice of antidepressant drug, we can find useful information in patient's medical and family history: in fact, if an antidepressant molecule was effective in the treatment of a previous episode, it is to be considered the first choice and should be prescribed in subsequent episodes. Studies showed that the patient who has responded to a drug has 90–96% of probability to respond to the same drug in the following episode. Similarly, if a relative suffered from depression and achieved remission with a particular antidepressant, the same drug will be more probably effective also in our patient.

Once the treatment is chosen, it is appropriate to provide information to the patient about the latency time of therapy: physicians should explain that antidepressants do not have immediate effect but delayed in time (not before 10–15 days). Also probable size effects should be explained together with their management. It is appropriate then to gradually increase in dose to minimize side effects and identify the minimum effective dosage; afterwards treatment should be continued for an adequate time, never less than 4–8 weeks.

Before an antidepressant medication can be said to be ineffective, it must be administered at an appropriate dosage for an appropriate length of time.

Information must be provided to family members not only about treatment but also about behaviours to avoid. Given the medical nature of depressive disorder, the patient, especially in the most acute phase of the episode, is not always able to respond positively to stimuli, and family members must be instructed not to stimulate the patient too much in order not to increase his feelings of inadequacy or guilt. The evaluation of the context, where the patient lives, is also fundamental: if the family environment is not suitable, it may be indicated to hospitalize the patient and treat him in a psychiatric ward.

Box 3.4: Clinical Intervention and Major Depressive Episode

- Patients must be told that depression is not a sign of weakness or madness but a real disease.
- Do not talk to patients with a reproachful tone, but listen to them and be interested in what they tell you. Be comprehensive and sometimes silence is important.
- Patients must be reassured that therapies are often effective and that the situation deserves time to improve.
- It is important to investigate if relatives are aware of the patient's disease and if they agree or not with his choice of taking care of himself. Consider the possibility that family may represent a source of stress for the patient. Evaluate if there are other stressful agents potentially represented by people or situations.

3.1.5.1 How to Manage an Antidepressant Therapy

- If there is a clear improvement of depressive symptoms, therapy should be continued at the same dosage for 6 weeks. After complete remission, therapy should continue for 4–6 months and then consider prophylaxis therapy.
- In case of incomplete improvement of depressive symptoms, the compliance should be checked. If the maximum dosage has not been reached, it is possible to increase the dosage and continue with the same treatment, to be monitored every 2 weeks and evaluating the response after 2–4 weeks.
- In case of non-response to treatment, that is, persistence or minimal reduction of symptoms after 6–8 weeks of treatment at adequate dosage, it is appropriate to change the therapy and possibly send the patient to a specialist.
- It is possible for every clinician to perform a pharmacological treatment. However, more complicated cases require referral to a psychiatrist. Example of cases requiring expert attention are incomplete response or non-response to therapy, diagnostic doubts, high recurrence of episodes, associated personality disorder, particularly severe pictures (with psychotic manifestations, psychomotor block) and presence of suicidal ideation and/or previous suicide attempts. Once depression recovered, it is necessary to continue the therapy for about 6–9 months (continuation phase) to consolidate the symptoms remission and avoid relapses related to premature discontinuation.

Box 3.5: General Indications for Treatment

Regarding antidepressant treatment, a non-specialized clinicians must:

- Be familiar with at least three to four molecules with antidepressant activity: know the therapeutic dosages, the most frequent side effects, the mode of prescription, titration, suspension, major pharmacological interferences.
- Know the nature of the disease and know how to give correct information to the patient and family members with particular regard to response times, side effects, symptomatic aspects of the disorder, the most useful behaviours and those to avoid.
- Be able to support the patient, listen to him/her in an empathic manner, accept his/her experiences and respond to his/her fears.
- Be able to identify initial responses to therapy and give correct feedback to the patient and family.
- Know how to evaluate the possible inefficacy of the treatment, according to a correct evaluation of timing and dosage and to know how to choose a different therapy.
- Set up a correct evaluation for a preventive therapy and to give exhaustive indications to the patient.

In patients with a history of two or more episodes of depression, the prophylactic long-term therapy (maintenance phase) is indicated, because the greater the number of episodes, the greater the risk of recurrence. Generally, it is continued with the same drug and the same dosage used in the acute phase, although it is possible to reduce the dosage of 10–40% in the maintenance phase. The duration of this phase of treatment is variable, but not shorter than 2 years, and depends on the individual periodicity.

Box 3.6: Long-Term Therapy

In some cases, it becomes necessary to treat patients with long-term therapy. These are some cases:

- High frequency of the episodes
- Severity of the episode
- Occurrence of the disease during youth
- Long duration of the episode
- Alcohol or substance abuse
- Bipolarity
- Familiarity of affective disorder
- Partial remission

Box 3.7: Criteria for Hospitalization During a Depressive Episode

(a) Particular severity of the psychopathological picture, with the presence of agitation or psychomotor arrest, persistent insomnia, self-harming behaviours
(b) Negativism and lack of cooperation
(c) Inadequate family support
(d) Medical complication such as dehydration, malnutrition, hydro electrolytic imbalances, serious physical diseases or abuse of alcohol or psychoactive substances
(e) Lack of awareness and refusal treatment

3.2 Bipolar Disorder

3.2.1 Definition

Bipolar disorder (BD) is the modern name of the manic depression or affective psychosis described in the nineteenth century by Emil Kraepelin. It is characterized by alternating phases of energy profile that goes parallel to phasic changes in mood, neurovegetative and cognitive functions, and sleep, ranging from full excitement (manic phases) to inhibition of psychomotor activity (depressive episodes, generally similar as in MDD).

The distinguishing clinical presentation of BD is mania, which is absent in MDD and other depressive disorders. Historically, an accurate definition of mania had been given by Karl Jaspers in his *General Psychopathology*: "Pure mania is characterized by an unmotivated and overflowing primary gaiety and euphoria, by a modification of the psychic course in the sense of the flight of ideas and by an increase in the associative faculties. The 'joie de vivre' stimulates all the instinctive drives: sexuality is increased, the impulse to move is increased; the impulse to talk is increased, states of excitement are raised. The fugue-like course of psychic life causes every activity to begin briskly and then stop and change. Each new stimulus distracts the sufferer". In BD, patients face alternating phases of mania, depression and in-between well-being (that is euthymia). Indeed, there is a restoring of one person's global functioning including cognitive performances, which differentiated BD from dementia praecox in Kraepelin's classification.

3.2.1.1 *DSM 5* Classification

In *DSM-5*, Bipolar Disorders chapter includes the following disorders:

- Bipolar disorder type I
- Bipolar disorder type II
- Cyclothymic disorder

According to *DSM-5*, BD type I is necessarily defined by the presence of a manic episode lasting at least a week, which may or may not be preceded or followed by hypomanic or major depressive episodes. BD type I patients nearly always experience at least one depressive episode, although there are also some forms of BD which exclusively show manic episodes (the so-called unipolar mania, 4% of subjects in a 20-year prospective study of BD type I patients).

BD type II is defined by the presence of a current or previous hypomanic episode and a current or previous MDE. The distinction between hypomanic and manic episodes is based upon the intensity and duration of symptoms (hypomania is less severe and lasts at least 4 days, while mania strongly compromises global functioning and lasts 7 days at minimum). Moreover, the presence of psychotic features determines a diagnosis of full manic episode by convention.

Cyclothymic disorder is defined by the alternation for at least 2 years of numerous phases of sub-syndromic hypomania and depressive symptoms that do not meet the criteria for a major depressive episode. Even if less intense, these symptoms can cause significant distress or psychosocial impairment at some point.

3.2.2 Epidemiology

The prevalence of BD in the general population ranges from 0.4 to 1.2%, peaking up to 4% depending on the study.

As shown in a retrospective study of 2308 patients, the first presentation of BD was:

- 54% Depressive episode
  - 22% Manic episode
  - 24% Mixed features episode

If the very first episode of illness is an MDE, despite some clinical features might indicate a BD, many patients are initially diagnosed with MDD and receive a proper BD only after the first episode of mania. This implies on the one hand a delay in correct diagnosis (about 5 years, but it can reach 10 years) and therefore in initiating mood-stabilizing treatment, standing that time of untreated disorder is associated with higher risk of suicidal behaviour and a longer duration of the disorder, on the other an important epidemiological distortion: rates of conversion from MDDR to BD are highest in the first years and then gradually decrease (from 3.9% in the first year to 0.8% after 5–10 years), so that institutional registries and researchers might have troubles pacing with BD incidence and prevalence over time. BD carries a burden of difficult differential diagnosis in *primary care settings* (see Assessment), and it has been demonstrated a tendency to over-diagnose BD, in particular some impulsivity and personality disorders that share a common emotion dysregulation (e.g. anxiety disorders, ADHD, alcohol and/or substance abuse disorder, personality disorders) are frequently misinterpreted or co-diagnosed with BD. This has a practical consequence for patients, that is delay in offering proper treatment (pharmacological and psychological) and therefore lower rates of success, so that these patients are more likely to be labelled as "resistant".

In BD the sex distribution shows an equal M:F ratio (1:1), the average age of onset is earlier than MDD (mean age: 25 years). Bipolar disorder patients must face higher risk of psychiatric and organic comorbidities and lower life expectancy (about 7–10 years compared to controls); moreover, earlier onset correlates with greater periodicity and higher frequency of manic episodes.

The WHO ranked BD as the 46th greatest cause of disability and mortality worldwide, ahead of breast cancer and Alzheimer's disease.

BD carries a considerable risk for suicidal behaviour, the highest amid psychiatric disorders, which is 20-fold than general population and about 15–20% of attempted are lethal. Risk factors are early onset, female gender for attempted suicide and male gender for completed suicide, predominant depressive polarity, comorbid anxiety, substance/alcohol abuse or personality disorders, positive family history for suicide (see Suicide).

3.2.2.1 Comorbidity

In general, bipolar patients more frequently show comorbidity of substance use (about 30–50% of BD patients) and anxiety disorders (about 75% of BD patients, e.g. panic disorder, social anxiety disorder, specific phobia) than do patients with unipolar depression; also eating disorders, especially binge eating disorder (10–20% of BD patients) is more frequent in BD subjects. Having these comorbid disorders worsens indeed the prognosis and markedly increases the risk of suicide.

Adults with bipolar I disorder have high rates of co-occurring medical conditions: metabolic syndrome and migraine are more common among individuals with bipolar disorder than in the general population.

3.2.2.2 Differential Diagnosis

- MDD with hypomanic/manic symptoms, which are fewer and last for a shorter time than in full hypomania/mania; this is particularly applicable for irritable mood.
- Schizophrenia and other psychoses: schizophrenia, schizoaffective disorder and delusional disorder are characterized by periods of prominent psychotic symptoms in the absence of strong mood symptoms (which are usually more likely to resemble hypomanic episodes than full mania); diagnostic cues derive from accompanying symptoms, previous course and family history.
- Generalized anxiety disorder, panic disorder, PTSD: these disorders could be a primary disorder or a comorbid disorder. In anxious individuals, ruminations might be mistaken for racing thoughts, and anxious feelings might lead to secondary irritability or impulsive behaviour. Overall, they are frequently present as comorbid disorders with BP type II. In PTSD, it is important to determine the episodic versus continuous nature of the symptoms, as well as the presence of neat stressors or triggers.
- Substance/medication-induced BD (see Table 3.2): there may be an overlap given the high comorbidity rates of substance abuse in BD patients. Substance use might manifest with secondary manic symptoms; response to mood stabilizers may not necessarily be diagnostic for a pure BD. A primary diagnosis of BD must be established based on symptoms that remain once substances are no longer being used.
- BD due to other medical condition (see Table 3.2): some physical condition can present with mood or behavioural changes that resemble a manic episode. A primary diagnosis of BD must be established based on symptoms that remain once the medical comorbidity is treated.
- ADHD: especially in adolescents and children, symptoms such as fast speech, racing thoughts, distractibility, less need for sleep might mislead to a BD diagnosis, but in ADHD there is not a distinct phase of disease rather a persistent presence of symptoms.
- Personality disorders: mainly borderline personality disorder might have substantial symptomatic overlap with bipolar disorders, since mood lability and impulsivity are common in both the disorders, but in BD symptoms are confined in distinct episode, and there is a noticeable change in functioning over baseline. In order not to misdiagnose a patient, a personality disorder should not be diagnosed during an active and untreated mood episode.

Table 3.2 Causes of secondary development of manic symptoms

Neurological diseases	– Extrapyramidal syndrome – Stroke – Encephalitis (syphilis, St. Louis, herpes) – Dementia – Demyelinating disease (multiple sclerosis) – Epilepsy – Temporal lobe syndrome – Traumatic lesions of the right hemisphere – Neoplasm
Endocrine diseases	– Hyperthyroidism – Adrenal diseases (Cushing's, Addison's) – Carcinoid syndrome
Psychotropic substance abuse	– Cocaine – Amphetamines and methamphetamines – Opioids – Phencyclidine
Drugs	– Corticosteroids and anabolic-androgenic steroids – Hormonal replacement therapy – Levodopa and other antiparkinsonian drugs – Isoniazid – Sympathomimetic drugs – Thyroxine – Chloroquine – Baclofen – Alprazolam – Captopril
Other	– Post-infectious mania – Vitamin deficiency (B_{12}) – Uraemia

3.2.3 Aetiopathogenesis

The aetiology of bipolar disorder is multifactorial and yet not fully understood in its complexity: it includes biological alterations in the CNS and, peripherally, environmental influences and genetic predisposition.

3.2.3.1 Biological Alterations

Several observations about the drugs used to treat mood disorders support the theory of the biological aetiology of bipolar disorder, with special regard to monoamine system dysregulation. In particular, it is thought to be an increase in monoaminergic neurotransmitters (noradrenaline, dopamine and serotonin) during manic phases, and a decrease during depressive phases. More recent studies suggest the presence of an altered sensitivity of monoamine receptors rather than their hypo/hyperactivation in depressive/manic phases. The use of drugs that increase monoaminergic transmissions through action on receptors and transporters (e.g. SSRIs, TCAs, and MAO inhibitors) in depression supports this theory. Lithium and other mood stabilizers seem to act in both the directions: they can either increase or reduce the transmission. In particular, reduction in the dopaminergic tone is fundamental in treating acute mania (i.e. the rationale for antipsychotics use in BD).

3.2.3.2 Circadian Rhythm

The hypothesis of circadian rhythm, in particular the sleep–wake cycle, involvement in BD etiopathogenesis arises from the observation of core symptoms in BD patients. In mania, there is a markedly reduced need for sleep (patients can sleep for less than 2 h per night): this reduction is one of the earliest and most predictive symptoms of mania. Nights of spontaneous poor sleep precede manic episodes, and the manic episode is often preceded by "triggering" factors that reduce the time spent sleeping. Therefore, sleep deprivation can be seen as the "common final pathway" in the pathogenesis of mania. On the opposite pole, about 90% of patients during the depressive episode experience insomnia: they usually report troubles in falling asleep, frequent awakenings during the night, early morning awakening at dawn (terminal insomnia) and sleepiness during the day. In most depressed patients, polysomnography revealed a reduction in slow-wave sleep (stages 3, 4) and anticipation of REM sleep cycles.

3.2.3.3 Genetics

The relevance of the genetic contribution is probably what mostly differentiates MDD from BD, in which the genetic predisposition is extremely higher (reaching 85% of heritability). Studies on twins support the hypothesis of a marked genetic component to the disease: in homozygotes, the concordance is highest (80%) for BD type I, while it is about 78% in BD type II, thus it seems that a genetic factor contributes also to disease severity too. In practice, candidate gene analyses, GWAS, copy-number variation and next-generation sequencing highlighted several genes that recur in BD patients; apart from rare variant with high penetrance, each genetic variation alone gives lower risk to develop BD, but the addition of multiple genetic risk factors raises the risk of BD. Genes usually encode for proteins involved in signal transduction (e.g. DGKH), ion channels (e.g. CACNA1C, SCN2A, CACNB2 and KCNB1), neurotransmitters transporters (e.g. GRIN2A, 5-HTTLPR), synaptic dynamics (e.g. ANK3), biological clock (e.g. CLOCK, CRY1, PER3, NR1D1, GSK3-beta), immune system (e.g. MHC complex).

Besides genetic load, epigenetics can play a role in augmenting the risk for BD development.

Nonetheless, as per the state of the art, genetic information cannot predict the individual course of BD development. New insights are coming from integrating genetic information with biological pathways involved in BD pathogenesis, in order to guide development of new therapies.

Moreover, there is a family co-aggregation between BD and schizophrenia, suggesting a common genetic basis. The latter knowledge poses grounds for a theory that put major depression at a pole of a continuous spectrum and schizophrenia at another pole, with BD and other disorders (e.g. schizoaffective disorder) in between, based on whether they manifest with preeminent affective or psychotic symptoms.

3.2.3.4 Environment

Environmental risk factors include higher social provenience, early life (in particular, childhood physical abuse) and later stressful life events (bereavement, divorce, or vocational loss, especially as depressive episode triggers, but also for mania), substance abuse, while no significant differences have been found between ethnic groups. Stressful events seem to play a role especially concerning the first manic or depressive episode, while they lose importance with subsequent episodes.

3.2.3.5 Endocrine Alterations

In addition to the sleep–wake rhythm, circadian mechanisms of hormone release may be altered in bipolar disorder. Manic patients have increased nocturnal levels of melatonin, with advanced timing of the nocturnal peak; depressed patients have reduced nocturnal levels of melatonin, with normal or advanced peak. It is hypothesized that early life stress, through sustained levels of corticosteroids, alters the response of hypothalamus–pituitary–adrenal axis and that persists throughout a patient's life. Cortisol level normally decreases during the night with a nadir in the first few hours of sleep and then increases to a normal peak at morning: a phase delay in the rhythm of cortisol, as well as altered response to dexamethasone suppression test, has been observed in bipolar patients. Also the hypothalamus–pituitary–thyroid axis is affected; in fact, a considerable part of BD patients suffer from thyroid gland disorder, with hormonal imbalances occurring together with acute phases of the illness. Lastly, metabolic changes in glucose and lipid profile are frequent in BD patients, both in naïve and treated patients, and higher risk of metabolic syndrome goes along with higher levels of systemic inflammation and poorer course of the disease.

3.2.3.6 Inflammation

Immune system dysregulation has become one of the main targets in psychiatric research. Studies have found that central and peripheral immune system proteins (e.g. CRP, IL-1β, IL-4, IL-6, COX-2, TNF-alpha and many others), immune cell subpopulations (e.g. T-helper cells, cytotoxic T cells, neutrophils, microglia…) and various membrane and soluble cytokine receptors are altered in patients with BD. Supporting the inflammation theory, anti-inflammatory medications exhibited antidepressant properties and vice versa, and anti-inflammatory capacity has been demonstrated for antidepressants and lithium. Of note, some of such immune system elements are peculiarly altered during active phase of illness, so that a typical immune signature has been proposed and tested as a biomarker for BD diagnosis.

3.2.3.7 Neuroanatomy and Networks

Differences in the activation of brain areas involved in emotion processing became evident between healthy subjects and BD patients since functional neuroimaging techniques availability. Amygdala, which is involved in frightening or sad stimuli processing, and cingulate gyrus are usually hyperactivated. In manic phases, the striatum, which takes part in the reward circuit, is hyperactivated. A model of the functional neuroanatomy in BD hypothesizes a disruption of early developmental processes (e.g. white matter wiring and dendrite pruning) within brain networks that modulate emotional behaviour; this leads to decreased connections among prefrontal and limbic structures, which gives susceptibility to emotional intense stressors that overcome homeostatic capacity and disturbances in mood, thinking and behaviour emerge. The grey matter volume is often decreased in BD compared to healthy controls (e.g. hippocampus, insula, anterior cingulate cortex, fronto-temporal cortex), as well as other mental disorders. Post-mortem studies on BD patients showed a decrease in density and somal size of pyramidal neurons and changes in glial subpopulations in dorsolateral prefrontal cortex and anterior cingulate cortex; a reduction in the number and size of hippocampal neurons; in the amygdala, a variable decrease in the total number and density of neuron and a reduction in the total volume of the lateral nucleus, as well as decrease in neuronal density have been demonstrated; in the thalamus, there is a reduction in the number of oligodendrocytes, and global thalamus volume is significantly smaller in BD patients than in healthy controls. These neuropathological evidence is concordant with in-vivo neuroimaging studies and give support to the neuroprogression theory of BD, where multiple mood episodes load the CNS with glutamate excitotoxicity and neuroinflammation that can lead to accelerated neuronal damage. Such anomalies might explain the cognitive impairment in specific tasks (working memory, visuo-spatial memory) that affect BD patients, not only during active phases, but also in euthymia and can be encompassed in the neuroprogressive course of BD.

3.2.3.8 Assessment

As seen in the Epidemiology section, a crucial issue in BD nosology is making a prompt and correct diagnosis. The clinical evaluation includes collection of psychiatric and general medical history, a complete physical examination, latest administered therapies (including dosages taken and timing of treatment) and a routine set of laboratory tests (e.g. thyroid-stimulating hormone, complete blood count, serum electrolytes, urine toxicology to screen for substances of abuse, etc.). Any co-occurring medical condition or medication or substance use that might give rise to a secondary manic or depressive syndrome must be ruled out as soon as possible.

It is important to ascertain the number, frequency, polarity, intensity and duration of previous mood episodes.

3.2.4 Clinical Presentation

Bipolar disorder, as well as depressive disorders, is characterized by three fundamental dimensions, whose recognition permits a correct diagnosis:

- Cyclicity is the alternation of periods of well-being and phases of illness; phases which by their nature tend to resolve, even if untreated, and then reoccur
- Periodicity is the variable period of time between one mood episode and the next
- Polarity defines the type of illness episode in progress (depressed versus inflated mood)

3.2.4.1 Mania

Manic episodes involve clinically significant changes in mood, energy, activity, behaviour, sleep and cognition. Episodes are often similar within a given patient, but substance abuse is one of the factors that can change the clinical presentation of a manic episode within a patient. The intensity of manic episodes varies widely across patients. The onset of a manic episode may be abrupt or gradual; an abrupt onset is usually followed by more rapid and severe development of the clinical picture and is more frequent if stressors or substance abuse (e.g. amphetamines and cocaine) triggers the episode. When the onset is gradual, there is usually a period of mild prodromal symptoms (e.g. mild hyperactivity, reduced need for sleep, excessive feelings of energy and well-being, talkativeness, involvement in various interests, euphoric mood), with initial maintenance of general functioning. The duration of the episode varies from a few days to 3–4 months, if untreated. The resolution phase varies in duration and may also be abrupt or gradual. Symptomatic resolution is usually complete, although it is possible for some of the symptoms to become chronic, mainly in patients with poor compliance and poor disease awareness.

Core symptoms in manic episodes are:

- Elevation of mood. The patient shows great expansiveness, great energy, and inexhaustible vital energy, a drive to want to live and do things.
- Psychomotor excitement and increased targeted activities: the patient is in constant movement, unable to sit still. Increased planning and activity is typically marked by impulsivity, poor judgement, and disregard for risks. Examples include taking on new business ventures, excessive prodigality, dangerous sex behaviour and driving recklessly.
- Alteration of form and content of thought: thought is accelerated, abundant and connected to great productivity, but there is a discordance between apparent richness and real content, which is poor and fragmented. This feature is at the basis of the "deconstruction of manic consciousness" which gives the patient the illusion of great productivity and positivity. The strong acceleration of thought leads to what is called "flight of ideas": a rapid and disordered flow of ideas that leads to an alteration of associative links and manifests itself in a sequence of disconnected thoughts and words. Manic speech is generally loud, accelerated and difficult to interrupt and may be accompanied by jokes, singing, clanging and dramatic hand gestures. During dysphoric mania, patients often make hostile comments and become easily irritable.
- Exaggerated self-confidence and hypertrophic self-esteem, which can be the ground of delusional ideas. Typical topics in manic episodes are grandiosity, mystical and genealogic delusions. As an example, some patients believe they have a special relationship with God or celebrities or possess skills that make them special or linked with the inner nature of their environment.
- The state of consciousness is intact, but driven by instincts and impulsiveness that make the patient believe that everything is possible.

– Deficits in attention and concentration are striking, as the patients are easily distractible, moves from one topic to another, without ever stopping to reflect or concentrate, with consequent deficits in attentive and memory tasks.

3.2.4.2 Staging

Clinical severity of a manic episode can be assessed through evaluation scales, of which the most widely used is the Young Mania Rating Scale (YMRS).

Historically, the clinical picture of mania was divided into three stages by Carlson and Goodwin in their manual "Manic Depressive Illness", despite that these stages do not reflect the natural progression of the illness, since a patient can pass from stage I to stage III directly.

I. Hypomania. It is characterized by a slight increase in psychomotor activity and talkativeness, with little acceleration of the course of thought and contents of inflated self-perception; there is occasionally an increase in irritability, but ideation is generally coherent and general functioning is still preserved.
II. Mania, intermediate stage. It is characterized by euphoria, logorrhoea, accelerated and often disorganized thought, initial psychomotor agitation, occasionally anger and aggression.
III. Severe mania. It is characterized by frank psychic disorganization and behavioural oddities, increased and aimless psychomotor activity, persistent insomnia, possibly with psychotic features.

3.2.4.3 Depression

Clinical presentation of depressive episode in course of BD are quite similar to depressive episodes in MDD (see Clinical Presentation in Unipolar Depression), so that a disease onset with an MDE makes it extremely difficult to assess the probability of BD development over time.

Clinical severity of depressive episode can be assessed through the same evaluation scales used in MDD, in particular the current literature is based on HAM-D and MADRS scales.

Box 3.8: Clinical Features of Depressive Episode Suggesting a Bipolar Depression

- Early onset
- Short but frequent episodes
- Family history positive for BD
- Atypical features
- Psychomotor agitation
- Post-partum depression or psychosis
- Antidepressant-induced mood swings or irritability

3.2.4.4 Mixed Feature Episode

A mixed feature episode is characterized by the coexistence of depressive features and elements characteristic of mania. A mixed episode must be suspected in the presence of clinically significant agitation, anxiety or irritability. Mood episodes with mixed features may evolve from episodes of pure hypo-mania, depression or can be mixed from the beginning. They may last weeks-to-months and can remit or evolve into full depressive episodes; on the contrary, it is unusual for mixed feature episodes to evolve into manic episodes, although mania (especially dysphoric mania) is commonly accompanied by other mixed episode features. Mixed features occur in about 20–70% of BD patients, who are at greater risk for suicidal ideation and behaviour, comorbidity (e.g. substance use disorders and anxiety disorders) and less responsive to usual treatment.

3.2.4.5 Rapid Cycling

A rapid cycling BD is defined by the occurrence of four or more bipolar mood episodes within a year. Mood episodes (i.e. manic, hypomanic or depressive episodes) can occur in any combination and order. Clinical presentation of each episode does not differ from those that occur in non-rapid cycling pattern. Initially, these patients were supposed to be less responsive to lithium (the term originally identified them). The debate on this topic is still ongoing, regarding both predisposing factors and best treatments for this condition.

3.2.4.6 Manic Episode: Diagnostic Criteria

A manic episode is defined as a period of at least 1 week characterized by an abnormal and persistent condition of elevated, expansive or irritable mood associated with increased levels of energy or purposeful activity.

During this period, at least three of the following symptoms are requested (four if the mood is only irritable):

1. Excessive self-esteem or grandiosity
2. Decreased need for sleep
3. Increased loquacity or urge to talk
4. Escape of ideas or subjective experience that thoughts follow each other very quickly
5. Reported or observed distractibility (attention is easily diverted by irrelevant external stimuli)
6. Increased purposeful activity (social, work, school or sexual) or psychomotor agitation
7. Excessive involvement in playful activities that have a high potential for harm such as uncontrolled shopping, inappropriate sexual behaviour, reckless business investments

Mood alteration causes significant impairment of social or occupational functioning or requires hospitalization to prevent harm to self or others. The symptomatology cannot be associated with the effect of a substance or a medical condition.

3.2.4.7 Hypomanic Episode: Diagnostic Criteria

A hypomanic episode is defined as a period of at least 4 consecutive days characterized by an abnormal and persistent condition of elevated, expansive or irritable mood.

During this period, at least three of the following symptoms are requested (four if the mood is only irritable): the same symptoms as in the manic episode.

The episode is associated with a change in functioning, which is not characteristic of the individual when asymptomatic.

The mood alteration and change in functioning are observable by others (one must therefore ask the patient if someone close to him has noticed a change in mood, as he is not aware of it himself).

The episode is not severe enough to cause a marked impairment in social and work functioning or to require hospitalization. This type of criterion differentiates between manic and hypomanic episodes.

The episode must not be attributable to the use of drugs or substances of abuse.

3.2.4.8 Diagnostic Criteria for Bipolar II Disorder

There have been at least one depressive and one hypomanic episode, but there has never been a manic episode. The hypomanic and major depressive episodes are not better explained by any kind of psychotic disorder.

The symptomatology causes significant distress or impairment of social, occupational or general functioning.

3.2.4.9 Cyclothymic Disorder: Diagnostic Criteria

Numerous periods with hypomanic symptoms and numerous episodes with depressive symptoms that never met the criteria for hypomanic or major depressive episode. These periods must have been present for at least half the time in the last 2 years (1 year for children and adolescents) with symptom-free periods not longer than 2 months. The symptomatology causes a significant impairment in social, occupational or global functioning, and it cannot be explained by any kind of psychotic disorder, nor by the effect of a substance or any medical condition.

3.2.5 Treatment

Once a patient is diagnosed with BD, psychoeducational information and clinical reasoning should be given in the most comprehensible and unambiguous way, according to the subject's capabilities of comprehension, to make clinical interventions effective on the short- and long-term period. This includes making the patient aware that, at present, current available treatment for BD might reveal to be insufficient to manage acute episodes, and it might be difficult to tailor the treatment in order to prevent relapses and recurrences and suicidal behaviour, so that restoring premorbid functioning become a challenge for clinical psychiatrists and frustrating for patients themselves.

Translational research ("from bench to bed") is now giving hope as new treatment options are being developed and validated by means of clinical trials, which includes making the patient aware that, at present, current available treatment for BD might reveal to be insufficient to manage acute episodes, and it might be difficult to tailor the treatment in order to prevent relapses and recurrences and suicidal behaviour, so that restoring premorbid functioning become a challenge for clinical psychiatrists and frustrating for patients themselves. Translational research ("from bench to bed") is now giving hope as new treatment options are being developed and validated by means of clinical trials.

Box 3.9: Criteria for Hospitalization in Acute Manic Patients

Severe mania
Mania with mixed features
Persistent insomnia despite full-dose hypnotics and sedatives
Psychomotor agitation
Suicidal ideation or behaviour
Aggressive behaviour
Dangerous behaviour
Psychotic features
Difficult management at home
Lack of insight and judgment capacity
Inadequate or lack of adherence to therapies
Psychotropic substance abuse

3.2.5.1 Treatment of Manic Episodes

The first aim of mania treatment is to restore a regular and sufficient wake–sleep rhythm, even in consequent stages of the episode

- At the onset of premonitory symptoms (in particular, insomnia), the use of sedatives (e.g. benzodiazepines or antihistamines) is recommended.
- *Stage I*: The addition of lithium salts or adjustment of dosage according to plasma dosage (up to a lithium level of 0.7 mEq/L) is recommended.
- *Stage II*: Lithium salts may be combined with an antiepileptic or a major sedative if agitation is present.
- *Stage III:* In the case of full-blown mania with psychotic symptoms, it is recommended to add an antipsychotic agent.

Fundamental antimanic drugs are mood stabilizers: the first choice falls on lithium salts, but also antiepileptic compounds (valproate, carbamazepine, lamotrigine, gabapentin) are effective. Second-generation antipsychotics (aripiprazole, lurasidone, olanzapine, quetiapine, risperidone, paliperidone, asenapine, cariprazine) have antimanic properties as well. Table 3.3 summarizes the drugs that are usually administered in manic patients.

Table 3.3 CANMAT 2018 Guidelines recommended acute treatments for mania

First-line monotherapies
Lithium
Valproate
Quetiapine
Aripiprazole
Paliperidone (>6 mg)
Risperidone
First-line combination therapies
Lithium or valproate + quetiapine or risperidone or aripiprazole
Second line
Lithium + valproate
Olanzapine ± lithium or valproate
Carbamazepine
Haloperidol
Third line
Carbamazepine ± lithium or valproate
Haloperidol ± lithium or valproate
Chlorpromazine
Clozapine

Among non-pharmacologic therapies, a second-line treatment includes ECT and a third-line option includes rTMS (see Chap. 12)

Lithium salts are preferred in case of classic euphoric mania, while valproate is best addressed to dysphoric/irritable mood mania, multiple previous episodes and/or substance abuse disorder. Carbamazepine might be the treatment of choice in comorbid anxiety or substance abuse disorder patients, mood-incongruent psychotic features, no family history of BD. Combinational therapies are indicated in patients with severe mania, needing a fast control of symptoms, or with partial response to acute or maintenance monotherapy. For those patients with strong anxiety, valproate, quetiapine and olanzapine showed best efficacy. In case of mixed feature mania, combination therapy of valproate and second-generation antipsychotics are useful, preferring aripiprazole, olanzapine, asenapine and ziprasidone. Despite lack of strong evidence from meta-analyses, mood-incongruent psychotic features seem to benefit from lithium or valproate plus a second-generation antipsychotic. In case of mania in a rapid cycling BD, a double mood stabilizer therapy is necessary, but no clear evidence of superiority of one compound is available, so choice is guided by previous response to maintenance therapy.

Given that the combination of lithium and a first-generation antipsychotic (e.g. haloperidol, chlorpromazine) is still probably the most effective first-line treatment in severe or psychotic mania.

3.2.5.2 Treatment of Depressive Episodes

Antidepressant therapy in bipolar depression is controversial, and caution in dosage and timing is mandatory, because of concerns for manic switch and induction of rapid cycling course. In order to minimize these events, patients with bipolar depression who are taking an antidepressant medication, are concurrently prescribed a mood stabilizer or a second-generation antipsychotic. Among the latter category, there is evidence for the use of quetiapine, lurasidone and olanzapine in bipolar depression (while other antipsychotics, e.g. haloperidol or risperidone, are the most effective during manic phases). Among mood stabilizers, evidence supports the use of carbamazepine in rapid-cycling BD patterns; otherwise, lithium is still the first-line therapy. The additional use of antimanic drugs to antidepressants seems to reduce the risk of manic switches and to improve depressive symptoms, too.

It is a good practice to continue antidepressants for approximately 2–4 months after remission of the depressive episode and then slowly taper the dosage until discontinuation if possible (see Unipolar Depression and Pharmacologic Treatment chapters for discussion of antidepressant drugs and operational instructions). If hypomanic or manic symptoms emerge during antidepressant treatment, the antidepressants should be promptly stopped and the dosage of antimanic drugs should be optimized.

Box 3.10: Treatment-Emergent Mania

Treatment-emergent mania is observed in 20–40% of BD patients exposed to antidepressant drugs and their usage might "reveal" a bipolar diathesis in subjects previously diagnosed with MDD.

Differently from *DSM-IV*, which included treatment-emergent mania within substance-induced mood disorders, the *DSM-5* specifies that "A full manic episode that emerges during antidepressant treatment but persists at a fully syndromic level beyond the physiological effect of that treatment is sufficient evidence for a manic episode and, therefore, a bipolar I diagnosis". This widens the category of BD by recognizing that developing mania or hypomania during antidepressant treatment beyond the physiological effect unveils a subjective and still not amendable predisposition to BD.

Box 3.11: Chronobiological Therapy

Given the pivotal role of circadian alterations in patients affected by BD described above, in particular the disruption of sleep-wake cycle and sleep architecture, chrono-biological interventions have been developed as non-pharmacological treatment for BD patients. The main therapies that consist in manipulation of the intrinsic biological clock are bright light therapy, total sleep deprivation and dark therapy. Here follows a brief introduction to these treatments (see Chap. 12 for in-depth discussion).

Bright Light Therapy

Light is the main regulator of sleep–wake rhythm alternation through the stimulation of the suprachiasmatic nucleus. Although bright light therapy (BLT) was initially used for seasonal depression, it has been demonstrated that this therapy is also effective in non-seasonal depression. The best day time to administer light therapy is early in the morning, and the antidepressant effect is greater if exposure occurs 2 h earlier than the usual wake-up time.

Total Sleep Deprivation

Total sleep deprivation (TSD) is a chrono-biological therapy mainly used in the treatment of bipolar depression (sometimes also in MDD), as an alternative or enhancement of antidepressant drug treatments. TSD shows a prompt and efficacious antidepressant effect, and to gain stable improvement, three deprivation cycles (36 h awake) are consecutively repeated, alternated with one night of sleep recovery; BLT is administered in the middle of the night awake and after the night slept; a pharmacological stabilization treatment is added (e.g. lithium salts). It has been demonstrated that TSD therapy is also effective in unipolar depression.

Dark Therapy

As prolonged wake has antidepressant effects, it has been observed that forced rest has antimanic properties, counteracting the ongoing circadian rhythm alteration that take place during an acute phase of BD. Dark therapy (DT) consists of 14 h of enforced darkness (from 6 p.m. to 8 a.m.) in a separate and quiet room, depriving the manic patients of any possible activating stimulus. Given the challenging applicability of DT even in hospital settings, virtual darkness therapy (VDT), in which the patient wears specific glasses with green-light wavelengths blocking lenses, thus preventing stimulation of biological clock is used in specialized hospitals.

3.2.5.3 Long-Term Therapy

The importance of preventive therapy derives from the high risk of relapse immediately after an acute episode and from the increased cumulative risk of a new episodes after each episode, besides the need of managing residual symptoms and restore premorbid functioning. The notion of neuroprogression, that is the complex changes in regional brain volumes, neuro-circuitry and consequent long-term mood and cognitive symptoms that stem from multiple recurrent episodes, support the need for adequate preventive treatments in order to maintain cognitive functions, brain plasticity and minimize the disability burden of BD. To date, there is no single recommended treatment in the world, but there are guidelines drawn up by various associations (the most widely accepted are the British and the American ones) that provide recommendations for drug choice and duration. As a rule, as it can be evicted from the Table 3.4, antidepressants as maintenance treatment are not recommended in BD patients.

The choice of maintenance treatment should be guided by a few cues drug effective in acute phase (with dose adjustment), tolerability, safety profile, predominant polarity, individual prognostic factors, and lastly patient's preference. Even with treatment, 20–25% of BD patients will face yearly recurrences, and recurrent episodes occur in 25–40% of untreated subjects. Given that relapse rates as high as

Table 3.4 According to the CANMAT 2018 Guidelines, recommended maintenance treatments for BD

First-line mood-stabilizer monotherapies
First line: Lithium
Second line: Valproate
Third line: Lamotrigine
Fourth line: Carbamazepine
Other first-line maintenance therapies
Quetiapine
Asenapine
Second-line or adjunctive maintenance therapies
Lithium or valproate + antipsychotic (for 6 months, e.g. quetiapine or aripiprazole)
Atypical antipsychotic (olanzapine, risperidone, paliperidone, ziprasidone, lurasidone)
Third-line or adjunctive maintenance therapies
Olanzapine + fluoxetine
Aripiprazole + lamotrigine
Clozapine

50–90% within 5 months after abrupt discontinuation of mood stabilizers, general recommendation is to continue mood stabilizer prophylaxis indefinitely, although in cases of sustained stabilization (e.g. 5 years free from mood episodes) and good prognostic factors it might be tried to gradual taper prophylactic treatment. Apart from pharmacotherapy, psychoeducational interventions and psycho-social support are fundamental in helping patients reach an adequate insight and achieve a satisfactory level of functioning.

Box 3.12: Treatment-Resistant Bipolar Disorder

Scientific literature on treatment-resistant bipolar disorder (TRBD) is scarce, and a shared definition is longing. A classical definition by Sachs (1996) encoded these practical definitions: treatment refractory mania when manic symptoms do not remit after 6 weeks of at least two antimanic drugs; treatment refractory depression when no improvement is observed after two adequate antidepressant trials taken for minimum 6 weeks per agent (but one must remember that not all clinical guidelines include antidepressants in bipolar depression treatment, adding complexity to the topic). Many other definitions have come with time, focusing either on an issue or another. TRBD is a challenge for psychiatrists as it is associated with high morbidity rates, high functional impairment and disability burden, suicidal behaviour, complex drug regimens and higher utilization of healthcare services.

Factors that contribute to TRBD might be poor medication compliance or dosage, long history of BD (e.g. neuroprogression), comorbidities (e.g. anxiety, substance or benzodiazepines abuse, neuroinflammation), personality traits (e.g. self-criticism, social avoidance, low irritability), drug-metabolizing status (e.g. CYP450 genotype), true pharmaco-resistance.

The main drug options in treatment-resistant bipolar depression include adding lamotrigine to current lithium treatment, using lurasidone, ketamine or clozapine. Many other molecules have been or are currently studied as adjunctive treatments, trying to target the neuroinflammatory and endocrine alteration that concur to bipolar depression, and conclusive data are still awaited. Besides drug optimization, somatic treatments such as BLT, TSD, rTMS and ECT can be applied (see Chap. 12). In case of mania resistant to mood stabilizers (lithium, valproate, carbamazepine), these can be combined, or second-generation antipsychotics can be used (e.g. olanzapine, quetiapine, aripirazole, asenapine). Also in manic patients, somatic therapies can be used, in particular ECT can quickly improve symptomatology.

3.3 Other Mood Disorders

3.3.1 Dysthymia

Dysthymia is a common disorder with a prevalence of 3–5% in the general population. It is classified among the mood disorders and the characteristics that distinguish it from major depressive disorder are its chronicity and the lower severity of its symptoms.

According to *DSM-5* criteria, the diagnosis of dysthymia requires at least 2 years of depressed mood for most of the time and at least two symptoms among changing in appetite, changing in sleeping, loss of energy, lower self-esteem, decreased capability of concentration and decision-making, hopelessness.

In dysthymia, symptoms of the emotional-cognitive sphere prevail over vegetative and psychomotor symptoms. Furthermore, this disorder is mainly characterized by low self-esteem, anhedonia, exhaustion, irritability and poor concentration. However, it is not easy to distinguish qualitatively the symptoms of dysthymia from major depression. This disorder, compared to major depression, is more frequently associated with some psychiatric comorbidities such as anxiety disorders and substance abuse.

Dysthymia is of particular relevance as it has a high risk of developing a major depression and may be responsible for a high degree of impairment of the overall functioning of the individual. Unlike major depressive disorder, dysthymic symptoms do not resolve without pharmacological intervention, but antidepressant drugs have an easy exhaustion of their clinical effectiveness over time. It is therefore a useful combined/integrated treatment, with pharmacological intervention and psychotherapy.

3.3.2 Premenstrual Dysphoric Disorder

Premenstrual dysphoric disorder (PDD) is defined by a set of affective and somatic symptoms that occur for several days around the menstrual cycle. Patients usually complain of breast tenderness, bloating, weight gain, irritable mood with ease of crying and depressive-like symptoms such as loss of interest and ability to concentrate, pathological changes in sleep and appetite. In order to make a diagnosis of premenstrual dysphoric disorder, the symptoms must be of such magnitude as to cause an impairment in patient's job and/or social life. It is also necessary to assess if the symptomatology pattern is present for at least two menstrual cycles. PDD differs from major depression and dysthymia for its short duration, given the resolution of symptoms following the arrival of menstruation and for its typical coupling with menses. PDD begins with menarche, and the main risk factors are high weight, stressful events and family history. The prevalence of the disorder is 1.8% in females.

3.3.3 Disruptive Mood Dysregulation Disorder

Disruptive mood dysregulation disorder is a disorder characterized by manifestations of anger and irritability in occasions that do not warrant such behavioural responses. It can be diagnosed between 6 and 17 years, if the symptoms occur for at least 1 year, more than once a week, in different areas of life (school, home) and there must be an impairment of interpersonal functioning. The prevalence has not been unequivocally defined, it is thought to be between 2 and 5% with higher rates in males than females.

Further Reading

Agrawal A, Nurnberger JI Jr, Lynskey MT. Item response modeling of DSM-IV mania symptoms in two representative US epidemiological samples. Psychol Med. 2010;40(9):1549–58. https://doi.org/10.1017/S0033291709992005.

American Psychiatric Association. Diagnostic and statistical manual of mental disorders. 5th ed. Arlington: American Psychiatric Association; 2013.

Andersen BL, Rowland JH, Somerfield MR. Screening, assessment, and care of anxiety and depressive symptoms in adults with cancer: an American society of clinical oncology guideline adaptation. J Oncol Pract. 2015;11(2):133–4.

Baek JH, Eisner LR, Nierenberg AA. Epidemiology and course of unipolar mania: results from the national epidemiologic survey on alcohol and related conditions (NESARC). Depress Anxiety. 2014;31(9):746–55. https://doi.org/10.1002/da.22261.

Bauer M, Whybrow PC, Anst J, Versiani M, Moller HJ. World Federation of Societies of Biological Psychiatry (WFSBP) guidelines for biological treatment of unipolar depressive disorder, Part2: maintenance treatment of major depressive disorder and treatment of chronic depressive disorders and subthreshold depressions. World J Biol Psychiatry. 2002;3(2):69–86.

Flint J, Kendler KS. The genetics of major depression. Neuron. 2014;81(3):484–503.

Fornaro M, Carvalho AF, Fusco A, Anastasia A, Solmi M, Berk M, Sim K, Vieta E, de Bartolomeis A. The concept and management of acute episodes of treatment-resistant bipolar disorder:

a systematic review and exploratory meta-analysis of randomized controlled trials. J Affect Disord. 2020;1(276):970–83. https://doi.org/10.1016/j.jad.2020.07.109.

Gigante AD, Young LT, Yatham LN, Andreazza AC, Nery FG, Grinberg LT, Heinsen H, Lafer B. Morphometric post-mortem studies in bipolar disorder: possible association with oxidative stress and apoptosis. Int J Neuropsychopharmacol. 2011;14(8):1075–89. https://doi.org/10.1017/S146114571000146X.

Goldberg JF, Truman CJ. Antidepressant-induced mania: an overview of current controversies. Bipolar Disord. 2003;5(6):407–20. https://doi.org/10.1046/j.1399-5618.2003.00067.x.

Goodkind M, Eickhoff SB, Oathes DJ, Jiang Y, Chang A, Jones-Hagata LB, Ortega BN, Zaiko YV, Roach EL, Korgaonkar MS, Grieve SM, Galatzer-Levy I, Fox PT, Etkin A. Identification of a common neurobiological substrate for mental illness. JAMA Psychiat. 2015;72(4):305–15. https://doi.org/10.1001/jamapsychiatry.2014.2206.

Goodwin FK, Jamison KR. Manic-depressive illness: bipolar disorders and recurrent depression. 2nd ed. New York: Oxford University Press; 2007.

Hawton K, van Heeringen K. Suicide. Lancet. 2009;373(9672):1372–81.

Kessing LV, Willer I, Andersen PK, Bukh JD. Rate and predictors of conversion from unipolar to bipolar disorder: a systematic review and meta-analysis. Bipolar Disord. 2017;19:324–35. https://doi.org/10.1111/bdi.12513.

Kupfer DJ, Frank E, Grochocinski VJ, et al. Demographic and clinical characteristics of individuals in a bipolar disorder case registry. J Clin Psychiatry. 2002;63(2):120–5. https://doi.org/10.4088/jcp.v63n0206.

Le Strat Y, Dubertret C, Le Foll B. Prevalence and correlates of major depressive episode in pregnant and post-partum women in the United States. J Affect Disord. 2011;135:128–38.

Mathias de Almeida K, Nery FG, Moreno RA, Gorenstein C, Lafer B. A sib-pair analysis of impulsivity in bipolar disorder type I. Compr Psychiatry. 2013;54(8):1148.

McFarland DC, Walsh L, Napolitano S, Morita J, Jaiswal R. Suicide in patients with cancer: identifying the risk factors. Oncology (Williston Park). 2019;33(6):221–6.

McIntyre RS, Soczynska JK, Cha DS, et al. The prevalence and illness characteristics of DSM-5-defined "mixed feature specifier" in adults with major depressive disorder and bipolar disorder: results from the international mood disorders collaborative project. J Affect Disord. 2015;172:259–64. https://doi.org/10.1016/j.jad.2014.09.026.

Merikangas KR, Jin R, He JP, Kessler RC, Lee S, Sampson NA, Viana MC, Andrade LH, Hu C, Karam EG, Ladea M, Medina-Mora ME, Ono Y, Posada-Villa J, Sagar R, Wells JE, Zarkov Z. Prevalence and correlates of bipolar spectrum disorder in the world mental health survey initiative. Arch Gen Psychiatry. 2011;68(3):241–51. https://doi.org/10.1001/archgenpsychiatry.2011.12.

Mitchell PB, Loo CK, Gould BM. Diagnosis and monitoring of bipolar disorder in general practice. Med J Aust. 2010;193(S4):S10–3. https://doi.org/10.5694/j.1326-5377.2010.tb03890.x.

Modabbernia A, Taslimi S, Brietzke E, Ashrafi M. Cytokine alterations in bipolar disorder: a meta-analysis of 30 studies. Biol Psychiatry. 2013;74(1):15–25.

Munkholm K, Braüner JV, Kessing LV, Vinberg M. Cytokines in bipolar disorder vs. healthy control subjects: a systematic review and meta-analysis. J Psychiatr Res. 2013;47(9):1119–33. https://doi.org/10.1016/j.jpsychires.2013.05.018.

Murray CJ, Vos T, Lozano R, Naghavi M, et al. Disability-adjusted life years (DALYs) for 291 diseases and injuries in 21 regions, 1990–2010: a systematic analysis for the Global Burden of Disease Study 2010. Lancet. 2012;380(9859):2197–223. https://doi.org/10.1016/S0140-6736(12)61689-4.

Nurnberger J Jr, Roose SP, Dunner DL, Fieve RR. Unipolar mania: a distinct clinical entity? Am J Psychiatry. 1979;136(11):1420–3. https://doi.org/10.1176/ajp.136.11.1420.

Parker GB, Graham RK. Clinical characteristics associated with treatment-resistant bipolar disorder. J Nerv Ment Dis. 2017;205(3):188–91. https://doi.org/10.1097/NMD.0000000000000517.

Rock PL, Roiser JP, Riedel WJ, Blackwell AD. Cognitive impairment in depression: a systematic review and meta-analysis. Psychol Med. 2014;44(10):2029–40.

Rutledge T, Kenkre TS, Thompson DV, et al. Depression, dietary habits, and cardiovascular events among women with suspected myocardial ischemia. Am J Med. 2014;127:840–7.

Sachs GS. Treatment-resistant bipolar depression. Psychiatr Clin N Am. 1996;19(2):215–36. https://doi.org/10.1016/s0193-953x(05)70285-9.

Saveanu RV, Nemeroff CB. Etiology of depression: genetic and environmental factors. Psychiatr Clin N Am. 2012;35(1):51–71.

Solomon DA, Leon AC, Endicott J, Coryell WH, Mueller TI, Posternak MA, Keller MB. Unipolar mania over the course of a 20-year follow-up study. Am J Psychiatry. 2003;160(11):2049–51. https://doi.org/10.1176/appi.ajp.160.11.2049.

Strakowski SM, Adler CM, Almeida J, Altshuler LL, Blumberg HP, Chang KD, DelBello MP, Frangou S, McIntosh A, Phillips ML, Sussman JE, Townsend JD. The functional neuroanatomy of bipolar disorder: a consensus model. Bipolar Disord. 2012;14(4):313–25. https://doi.org/10.1111/j.1399-5618.2012.01022.x.

Sugaya L, Hasin DS, Olfson M, Lin KH, Grant BF, Blanco C. Child physical abuse and adult mental health: a national study. J Trauma Stress. 2012;25(4):384–92. https://doi.org/10.1002/jts.21719.

Terao T, Tanaka T. Antidepressant-induced mania or hypomania in DSM-5. Psychopharmacology (Berl). 2014;231(1):315. https://doi.org/10.1007/s00213-013-3358-4.

Turner S, Mota N, Bolton J, Sareen J. Self-medication with alcohol or drugs for mood and anxiety disorders: a narrative review of the epidemiological literature. Depress Anxiety. 2018;35:851–60.

Turvey CL, Coryell WH, Solomon DA, Leon AC, Endicott J, Keller MB, Akiskal H. Long-term prognosis of bipolar I disorder. Acta Psychiatr Scand. 1999;99(2):110–9. https://doi.org/10.1111/j.1600-0447.1999.tb07208.x.

Vieta E, Berk M, Schulze TG, Carvalho AF, Suppes T, Calabrese JR, Gao K, Miskowiak KW, Grande I. Bipolar disorders. Nat Rev Dis Primers. 2018;8(4):18008. https://doi.org/10.1038/nrdp.2018.8.

Wehr T. Sleep-loss as a possible mediator of diverse causes of mania. Br J Psychiatry. 1991;159(4):576–8. https://doi.org/10.1192/bjp.159.4.576.

Wehr TA, Sack DA. Sleep disruption: a treatment for depression and a cause of mania. Psychiatr Ann. 1987;17(10):654–63.

Yatham LN, Kennedy SH, Parikh SV, et al. Canadian Network for Mood and Anxiety Treatments (CANMAT) and International Society for Bipolar Disorders (ISBD) 2018 guidelines for the management of patients with bipolar disorder. Bipolar Disord. 2018;20(2):97–170. https://doi.org/10.1111/bdi.12609.

4 Psychotic Disorders

M. Bosia, M. Spangaro, F. Cocchi, J. Sapienza, L. Tonet, F. Martini, L. Fregna, C. Passani, and R. Cavallaro

4.1 Schizophrenia

4.1.1 Introduction

Schizophrenia is a common paradigm of mental illness; it is characterized by a constellation of psychopathological aspects clinically and biologically related in a nonlinear, but rather coherent syndromic presentation and temporal pattern. Even if the clinical presentation and temporal distribution of symptoms across the lifespan have a variety of possibilities, the life perspective is that of a chronic condition, leading to cognitive and functional deterioration, which is not common to any other disease with a juvenile onset for the complexity of its different aspects.

A severe mentally ill is not necessarily a patient suffering from schizophrenia, but, when it happens, schizophrenia is probably the most frequent severe psychiatric illness.

Schizophrenia has represented since centuries madness itself like delusions are as a psychopathological entity. In delusion, you lose the qualitative features of the sentiment and perception of reality (what you believe firmly is not understandable

M. Bosia · F. Cocchi · F. Martini (✉) · L. Fregna (✉)
San Raffaele Hospital, Milan, Italy
e-mail: bosia.marta@hsr.it; cocchi.federica@hsr.it; francesca.martini1@hsr.it; fregna.lorenzo@hsr.it

M. Spangaro · R. Cavallaro
San Raffaele Hospital, Milan, Italy

Vita-Salute San Raffaele University, Milan, Italy
e-mail: spangaro.maroc@hsr.it; cavallaro.roberto@hsr.it

J. Sapienza · L. Tonet · C. Passani
Vita-Salute San Raffaele University, Milan, Italy
e-mail: sapienza.jacopo@hsr.it; tonet.leonardo@hsr.it; passani.carolina@hsr.it

R. Cavallaro, C. Colombo (eds.), *Fundamentals of Psychiatry for Health Care Professionals*, https://doi.org/10.1007/978-3-031-07715-9_4

and not shared in the common view of your peers, even considering the sociocultural background). However, while delusion itself is not typical of schizophrenia or any other disorder and can disappear with full recovery in other illnesses (i.e., mood disorders), when it happens in the context of this diagnosis, it may attenuate, become less or minimally important for patients' inner life and behavior (after an adequate treatment), but never be canceled. In the best situations, it may be referred to a past, closed experience, but never negated as true in the time it occurred.

This means that schizophrenia alterations in reality perception and sentiment become permanent in the autonoetic experience of the subject, that is, a permanently altered functioning of cognition supporting the long-term experience of consciousness itself.

Again, delusion experience alone is neither diagnostic nor characteristic of the illness. What is peculiar and practically always appreciable in all patients is the long-term change in cognition, in a wide sense.

This change in cognition is related to the "biogenesis" of the so-called positive symptoms (pathological mental activities absent in normal functioning), in their development and maintenance, but even more to the early progressive and chronic loss of some of the main components of vitality in every person experience: volition, engagement in life and relationship, affect, features that characterize the so-called schizophrenic autism (Bleuler), the prevalence of a (poor) inner world in a person's cognitive and affective investment in opposition to that in real world, and social life he could live.

Emil Kraepelin called this illness at the beginning of the twentieth century "Dementia Praecox"; this definition is impressive and strongly related to the absence, at that time, of any treatment improving symptoms and lives of the patients. Nowadays and since 50 years ago (this changed the paradigm of treatment of schizophrenia from support and assistance to care and cure), schizophrenia treatment is available starting from drugs that are effective in reducing the most acute and disturbing symptoms, but have a limited effect on the so-called negative symptoms (normal mental activities pathologically diminished in schizophrenia) and cognitive deficits developed. Nevertheless, their effects, even if limited to the "positive symptoms" treatment, are permissive toward the nonpharmacological treatment with psychosocial, neurocognitive, and sociocognitive rehabilitation that improve negative and cognitive symptoms.

This last part of the interventions, commonly delivered among patients with schizophrenia, has reached in the last 20 years much more milestones than drugs in building the hope of treatment of schizophrenia letting possible, also for this dramatic illness, terms as "remission" and also "recovery".

Nowadays, in most cases, schizophrenia is a treatable illness: this does not correspond to full restoring, usually, but a significant proportion of patients restore their life perspective, remaining a fragile population to be supported; they can go back to work, study, and have significant relationships. When this target is reached, the remaining part of the problem of schizophrenia is stigma, harder than for other psychiatric illnesses, due to the lack in comprehension of some experiences for the nontechnicals and also to the behavioral evidence of strangeness or marginality of these lives.

This chapter will lead to a basic comprehension of phenomena appreciable in the illness, a knowledge of the main treatments and their effects, and of the main targets and instruments of treatment.

4.1.2 Definition and Epidemiology

Schizophrenia is a chronic psychotic disorder, typically deteriorating over time, which strongly affects daily functioning in a negative way, determining deficits in cognitive, emotional, and social domains. Symptoms are grouped in three main categories: positive, negative, and cognitive.

Schizophrenia affects almost 23 million people worldwide, with a lifetime prevalence in the general population that averages 0.7%, and an annual incidence of 15 per 100,000.

> Schizophrenia is among the top ten leading causes of disease-related disability in the world due to the pervasiveness of related deficits and life-long course.

Historically, the risk of developing schizophrenia over one's lifetime was thought to be similar among males and females, but recent large-scale studies revealed a slightly higher incidence among males (male/female relative risk = 1.4). The age of onset ranges between adolescence and early adulthood, with sex-related differences. Indeed, male patients show first symptoms of the disease earlier (age of onset: 20–28 years) compared to females (age of onset: 26–32), with a mean difference of onset of about 6 years. The later age of onset in women is associated with higher attainment of social role functioning before illness, which is related to a longitudinal better outcome.

The overt onset of illness is frequently anticipated by the presence of prodromal symptoms, detectable since early adolescence and consisting in one or more of the following clinical manifestations: social withdrawal and isolation, impaired social/scholastic function, bizarre thoughts, speech or behavior, neglected personal hygiene, blunted and inadequate affect, and loss of personal initiative. Delayed attainment of various developmental milestones, cognitive impairments during childhood or adolescence, neurological "soft signs" and poor social adjustment have been linked to an increased likelihood of developing schizophrenia. However, it is still unclear whether such impairments represent risk factors per se or early manifestations of the disease.

Concerning epidemiological differences in the worldwide distribution of the disease, a near-uniform global distribution suggests a relative independence from cultural and ethnical factors.

Nonetheless, ethnical and sociocultural environment may indirectly represent determinants of illness, since they can be associated with significant risk factors for the disease such as migration and urbanization.

4.1.3 Etiopathogenesis

Despite a multitude of studies, to date etiology of schizophrenia is still unclear. Schizophrenia is considered a multifactorial disease, associated with genetic and environmental factors that structure the phenotype of the illness by interacting with each other. Although no single DNA mutations or specific gene variation are able to directly cause the disease, genetic variations and gene–environment interactions together determine over 80% of the liability for developing schizophrenia. In this context, epigenetic represents the bridge between gene and environment, and a growing number of studies have been focusing on the analysis of its neurobiological effects.

> Epigenetics do not determine DNA code alterations, but modulate gene expression by modifying DNA structure.

Genome-wide association studies revealed a large set of genetic loci associated with schizophrenia, each with a minimal effect in determining disease susceptibility. Among these, the most relevant are those involved in neurotransmission and immune response/modulation. Genetic alterations such as single-nucleotide polymorphisms (SNPs) and copy number variants (CNVs) located within these loci of interest are able to significantly contribute to the development of the illness. Moreover, a growing body of evidence suggests that abnormalities in genetics-mediated brain maturation during the early teen years (and especially that of the hub areas in the brain) might be crucial to the development of the disorder.

Nowadays, the strongest etiopathogenetic theory revolves around the concept of "schizophrenia vulnerability." It is hypothesized that individual genetic liability interacts with other environmental and biological risk factors, leading together to develop schizophrenia once a critical threshold is crossed.

4.1.3.1 Heritability

A family history of schizophrenia in a first-degree relative represents one of the most widely replicated risk factors for the disorder. However, as discussed above, heritability follows a complex, multigenic, nonmendelian transmission pattern, probably involving reduced genetic penetrance (variants that do not express in every carrier as disease), gene-expression variability (variants that express differently between carriers), and gene–gene interactions. Schizophrenia shows indeed a higher incidence in certain families, despite over two-thirds of the cases occur sporadically without an aggregation in familial clusters. The concept of familiarity is a topic of main interest, since family members share both genetic heritage and life experiences (environmental aspects), and thus represents an actual demonstration of the *gene × environment* etiopathogenetic theory. Familiar studies and comparisons between twins and sibling clearly indicate a higher risk of schizophrenia in relatives of affected patients, compared with the general population. Indeed, lifetime risk in first-degree relatives of patients is 6.5%, and it rises to more than 40% in monozygotic twins of affected subjects. However, despite the same genetic substrate and family environment, the rate of concordance between homozygous twins deviates from the theoretical expected rate of concordance of 100%, evidencing the crucial role of external contributing risk factors.

4.1.3.2 Environmental Risk Factors

A multitude of epidemiological studies have highlighted the presence of a large number of environmental factors associated with an increased risk for schizophrenia. Among these, there are both biological and psychosocial risk factors that may occur from antenatal and perinatal periods until early adulthood. However, once again, it is important to underline that none of these elements are necessary or sufficient to cause schizophrenia alone but only increase the likelihood by a small percentage.

Overall, the main environmental factors that seem to have a greater impact on the susceptibility of developing the disease are urbanicity, substance use/abuse, and childhood trauma. Although the causal mechanism is still unclear, it has been hypothesized that the different environmental stressors ultimately impact on brain structure and function through a common neurobiological pathway, the dysregulation of the striatal dopaminergic neurotransmission ("stress sensitization").

Pregnancy and birth complications, together defined as "obstetric complications," have been individuated as early risk factors for schizophrenia, as different studies reported a higher number of these events among patients. Obstetric complications can be grouped in three main categories: complications of pregnancy (bleeding, preeclampsia, diabetes, maternal malnutrition, maternal infections, Rh incompatibility), abnormal fetal development (low birth weight, congenital malformations, small head circumference), and complications of delivery (asphyxia, uterine atony, emergency cesarean section). Concerning parents, among other relevant biological risk factors, is also reported higher maternal/paternal age, due to an increased likelihood of de novo mutation and aberrant epigenetic regulation correlated with age.

Adverse childhood experiences and trauma (sexual/physical/psychological abuse, neglect, parental death or separation, and bullying) have been extensively investigated and associated with increased odds of developing psychosis in adulthood. Particularly, childhood trauma has been associated with the most severe forms of positive symptomatology and affective symptoms. Life events that bring negative changes in personal circumstances and/or involve an element of threat occurring in a period ranging between 3 months and 3.6 years before the onset of the illness are the most involved in inducing psychotic transition.

Other relevant stressors occurring during early stages of life and associated with the development of schizophrenia are *urbanicity*, defined as growing up or moving to an urban environment, and *migration*. These factors are associated with a mean twofold increased risk for the disease, although the specific mechanisms underlying these data are not fully understood yet. It has been hypothesized that both urbanicity and migration are associated with a large cluster of environmental stressors (i.e., social inequality, isolation, discrimination, lack of appropriate accommodation, lower social status) that together contribute to the increased incidence of psychosis.

Besides environmental and social stressors, epidemiological studies individuated a large set of risk factors associated with schizophrenia. Among these, *substance abuse* is probably the most important determinant of the disease, after genetic liability. Schizophrenia is indeed associated with the use/abuse of several substances, such as cannabis, psychostimulants (cocaine and amphetamines), or hallucinogens, and these harmful habits have been widely reported to induce psychotic symptoms. In this context, cannabis plays a major role, representing the most frequently involved psychoactive substance in psychotic onsets. Besides being able to induce and worsen positive symptoms, several prospective epidemiological studies consistently reported an association between cannabis use and greater risk for schizophrenia. A relationship between degree of abuse (dose and frequency of consumption, THC concentration) and risk of psychosis has been extensively demonstrated as well. Moreover, frequent use of cannabis, especially at a younger age, and high potency THC not only increase the risk of schizophrenia of several times but are also associated with earlier onset, more severe positive symptoms, greater cognitive impairment, and worse long-term clinical and functional outcome.

4.1.3.3 Neurobiological Correlates

A multitude of cross-sectional studies strongly support the presence of structural and functional brain alterations in schizophrenia. Whole brain and gray matter volume reductions are largely described, particularly in frontal, temporal, and postcentral cortical regions. Moreover, increased ventricular size, conditioned by atrophy, is a frequent finding. Structural magnetic resonance imaging (MRI) has been used to investigate longitudinal volume changes, finding a progressive decline in both the first-episode psychosis and chronic patients. Notably, a large body of evidence suggests that longitudinal progression of cortical thinning of frontal and temporal areas correlates with deficits in several cognitive domains. Alterations in brain connectivity seem to play a crucial role in schizophrenia as well. Diffusion tensor imaging (DTI) studies indicate a decreased integrity of the microstructure of the white matter (WM). Interestingly, disruption of WM integrity seems to be progressive along the course of the disease. In particular, structural alterations in the myelin sheet seems to play the major role in the disruption process of the WM, hampering neuronal signaling. Taken together, these structural alterations result in impaired neurofunctional networks, as shown by positron emission tomography (PET), spectroscopy, and functional-MRI studies. Specifically, patients with schizophrenia are characterized by decreased cortical glutamatergic and dopaminergic activity, associated with cognitive and negative symptoms, and by increased dopaminergic synthesis and signaling in the striatum, underlying positive symptomatology.

Over the last decade, in vivo imaging approaches have significantly contributed to elucidate neurotransmitter abnormalities underling the disorder, strongly contributing to the development of new and integrated etiopathogenetic theories.

Overall, schizophrenia is characterized by a dysregulation of multiple neurotransmitters in multiple pathways, with main alterations involving dopaminergic and glutamatergic systems.

Historically, the "*dopamine hypothesis*" was the first proposed to explain etiopathogenesis of schizophrenia. It was based on two main observations: the efficacy of antipsychotic drugs acting by blocking dopamine mesolimbic D2 receptors in managing psychotic symptoms, and the capacity of drugs increasing dopamine, such as amphetamine, to induce psychotic symptoms. Schizophrenia was therefore hypothesized to be caused by the hyperactivity of subcortical mesolimbic D2 pathways. However, this theory was not able to explain the presence of negative and cognitive symptoms, neither the frequent lack of response to antipsychotic D2-blocking activity. Subsequent studies showed that cognitive and negative symptoms were associated with reduced prefrontal cortical activity, leading to a reconceptualization of the dopamine hypothesis implying a prefrontal hypodopaminergia as the primary alteration. However, as mentioned before, dopaminergic dysfunction does not comprehensively explain the full range of clinical features of the disorder, not including alterations of other neurotransmitter systems known to be dysregulated in schizophrenia. Among these, the *glutamatergic system* is highly implicated in the etiopathogenesis of the illness, representing the major excitatory pathway in the central nervous system involved in critical processes such as neural development, synaptic plasticity, and cognition. Glutamate dysregulation was hypothesized to be involved in the etiopathogenesis of schizophrenia as a consequence of the observed psychotogenic effects of phencyclidine and its derivative ketamine, drugs antagonizing the glutamatergic *N*-methyl-D-aspartate receptor (NMDAr). Following studies consistently reported a widespread reduction of glutamatergic signaling among patients with schizophrenia.

However, there is much more to psychosis than dopamine and glutamate, and there is much more to treatment of psychosis than D2 antagonism. For instance, serotoninergic hyperactivity (specifically at 5HT2a receptor) has been hypothesized to underlie positive symptoms. In fact, serotonin was examined early for a role in schizophrenia because of the psychomimetic actions of its agonists, such as lysergic acid diethylamide (LSD) and mescaline. Moreover, drugs with serotoninergic antagonist action seem to improve psychotic symptoms. Dysregulation of other neurotransmitters, such as GABA and choline, have been reported as well, although less consistently and with marginal effects on clinical outcomes, but rather on cognition.

More recently, an etiopathogenetic model attempted to integrate glutamatergic and dopaminergic alterations of schizophrenia. Specifically, it has been proposed that a deficient cortical glutamatergic transmission, associated with diminished prefrontal activity, may lead to increased striatal dopamine synthesis and release by hypo-activating GABAergic transmission, which physiologically inhibits striatal dopamine activity.

However, although intriguing, even this integrated model only partially elucidate the broad range of neurobiological alterations and does not take into account the high inter-individual variability observed among patients with schizophrenia.

Indeed, rather than a unique disorder characterized by high neurobiological and clinical variability, schizophrenia is being reconceptualized as a spectrum of chronic psychotic disorders (or "*schizophrenias*") sharing a multitude of biological and phenotypic features. In this view, treatment-resistant schizophrenia (TRS) has been proposed as a different subtype of schizophrenia, defined as "normodopaminergic." TRS occurs in approximately one third of patients diagnosed with schizophrenia and is characterized by persistence of positive symptoms despite adequate antipsychotic treatment, detectable from illness onset. Studies reported that TRS patients, differently from responders, show normal striatal dopaminergic synthesis and higher cortical glutamate levels. Consistently, clozapine is the only effective antipsychotic in TRS and the only antipsychotic showing a glutamatergic receptor affinity, thus further suggesting the presence of two (at least) distinct neurobiological subtypes of schizophrenia.

4.1.4 Clinical Presentation

Schizophrenia is characterized by a great heterogeneity of clinical manifestations and outcome, resulting in a merge of positive, negative, and cognitive symptoms whose severity varies across patients and through the course of illness. In the great majority of affected subjects, schizophrenia has a devastating impact on functional outcome, determining long-term disability.

Besides the structural feature of no insight on delusions and hallucinations, patients affected by schizophrenia usually lack insight on the illness itself, being unaware of it and its potential consequences, thus not recognizing the need for treatment and showing poor compliance to therapy.

For this reason, a god patient–doctor relationship and psycho educational interventions on illness, its consequences, and positive perspectives given from an adequate treatment are of extreme importance.

4.1.4.1 Positive Symptoms

Positive symptoms are more prominent during acute exacerbations and include the reality distortion symptoms of hallucinations and delusions, as well as disorganized thoughts, speech, and behavior.

Box 4.1: Positive Symptoms

- Hallucinations
- Delusions
- Disorganization of thought and/or speech and behavior

Hallucinations are perceptions that occur without an actual external stimulus. They may involve all the senses, but the most common hallucinations in schizophrenia are the auditory ones, typically complex (voices). Schizophrenia presents among individuals' different types of auditory hallucinations, sometimes coexisting: dialoguing in third person about the patient, imperative, threatening, accusatory, and commenting voices. Frequently, auditory hallucinations and delusional beliefs influence each other and together contribute to an altered interpretation of reality: as an example, the patient hears voices insulting/threatening him/her and is thus led to believe that someone is pursuing him/her or reinforces the concomitant persecutory delusion.

Delusions are false beliefs that are held despite contrary evidence on which the patient has no insight, presenting with variable content.

Some of the most frequent contents of delusions, according to their content, in schizophrenia are

- *Persecutory*: It is the most common of the subtypes. Typical themes are that of being persecuted, controlled, poisoned, the object of a conspiracy, hindered in actions.
- *Influencing and control:* Delusions of thought control, insertion, withdrawal, and broadcasting have also been historically linked to the diagnosis of schizophrenia even if not exclusive. Patients may believe that their acts or bodily functions and mental activity are not under their voluntary control or ruled by individual physiology, but under the control and influence of other people or entities by means of instruments (radio waves or similar) or magical powers; these themes are usually referred to as delusions of passivity.
- *Reference*: Beliefs that random or neutral events are addressed to the subject in a special and meaningful way. Common ideas of reference include believing that occurrences on the television or radio (certain words said or songs played) are meant to deliver a special message to the subject.
- *Somatic and body transformation*: Patient believes he is going to be transformed in an animal or that some parts of his body are changing measure, consistency, or even constitution (i.e., "my bones are made of glass" or "I'm going to be transformed in a dog" or "my inner organs have been moved from their original position") frequently associated with somatic hallucinations. Other types of somatic delusions are that of infestation with parasites, body dysmorphic, and body odor.
- *Bizarre:* Ideas are clearly implausible, unusual, and hardly understandable (i.e., being an alien or a robot or having a microchip implanted in the brain to influence his/her thoughts).
- *Erotomania*: The patient believes another person, usually someone of higher status or famous, is in love with him/her.
- *Grandiose:* The patient believes he/she has great talent, special power, knowledge, a relationship with someone famous, or has made an important discovery. Grandiose delusions may have genealogical or religious content.

Among these, persecutory delusions and ideas of reference are the most frequently reported in schizophrenia, often co-occurring and determining suspiciousness and active social withdrawal.

Nevertheless, in response to the presence of all themes of delusions, patients have important affective involvement, especially during acute phases, and can appear depressed, anxious, irritable, litigious, or even aggressive. Acting delusion or in defense of the content of the delusion is frequent (i.e., aggression of the hypothetical prosecutor, stopping to eat for fear of poisoning by a family member, resuming suddenly a work, remaining barricaded himself at home, or leaving to escape from a threaten).

Disorganization is a key feature of patients with schizophrenia, potentially affecting thought, speech, and behavior, regardless of the severity of hallucinations or delusions. Thought disorganization is identified by the lessening or total loss of connections between ideas, affecting both behavior and communication. Disorganized behavior can be directly observed and may range from simple problems sustaining goal-directed self-care behaviors such as personal hygiene, to unpredictable and bizarre socially inappropriate outbursts.

Severely disorganized thought can result in inconsistent and disorganized speech that can lose its communicative finality. The most commonly observed forms of abnormal speech are tangentiality and circumstantiality, while more severe thought disorder includes derailment, neologisms, and word salad, which consists in a total lack of logical associations with words thrown together without any sensible meaning.

4.1.4.2 Negative Symptoms

Negative symptoms are core features of schizophrenia, conceptualized as an absence or diminution of normal processes and being also referred to as deficit symptoms. This group of symptoms identifies different declinations of the same psychopathological element, that is, disinterest/detachment from the outside world. Negative symptomatology progressively grows in importance and pervasiveness during the course of the disease, representing the main feature of the residual phase (when positive symptoms are strongly attenuated or absent). Symptoms can be divided into two main clusters: diminished expression, including alogia and affective flattening, and avolition, including apathy, anhedonia, and asociality. Negative symptoms are often correlated with a marked reduction of facial mimicry and gestures and frequently result in a loss of goal-oriented behaviors, personal interests, and libido. Decreases in personal care and a failure to comply with social conventions represent common aspects as well. Negative symptoms are independent from positive symptomatology, being very resistant to treatment and negatively affecting functional outcome.

Box 4.2: Negative Symptoms

- Apathy (absence of emotions)
- Abulia or avolition (lack of will/initiative)
- Alogia (poor content of speech, blocks, latency of response)
- Anhedonia (inability to experience pleasure)
- Affective flattening

4.1.4.3 Cognitive Deficit

Cognitive deficits represent, together with negative symptoms, the most critical dimension in schizophrenia treatment, determining high disability and being almost completely resistant to antipsychotic drugs. Cognitive impairment is often detectable before the onset of illness, is scarcely related to psychotic symptoms, and remains stable or worsens along the course of illness. The severity of cognitive impairment has been significantly linked to global functional outcome and patient's quality of life and is a main target of treatment. Among patients with schizophrenia, deficits are present in several cognitive domains, making it difficult to establish a clear pattern of specific deficits associated with the disorder. Patients typically perform one to two standard deviations below healthy controls on a broad range of neurocognitive measures, with the most prominent impairments observed in working memory, executive functions and processing speed. Neurocognitive functions represent part of the structuring elements of the complex cognitive domain of social cognition, also impaired in schizophrenia. Social cognition is a multifactorial construct that includes the abilities of individuals to understand themselves and others in the context of social interactions, such as empathy and theory of mind. Social cognitive abilities have been directly linked to quality of life, frequency and significance of interpersonal relationships, work attainments, and personal achievements. Language and communication are also altered in schizophrenia, with a strong impact on social, as well as global functioning. These deficits are detectable at different levels, encompassing the syntactic structure and higher-order domains involving the integration of context, such as figurative language comprehension.

Box 4.3: Impaired Cognitive Functions
- Executive functions
- Working memory
- Processing speed
- Selective and sustained attention
- Verbal fluency
- Verbal memory
- Psychomotor coordination

4.1.4.4 Catatonic Features

Catatonia refers to a syndrome characterized by alterations of movements and behavior. It has long been considered a core dimension of schizophrenia, also previously characterizing a clinical subtype. However, catatonic symptoms have been reported to occur in more than 10% of patients with acute psychiatric illnesses.

Box 4.4: Catatonic Syndrome

- Indifference to external stimuli, negativism
- Immobility, maintenance of forced postures (catatonic postures), and waxed flexibility (postures assumed by passive mobilization of patient's arts). This might be interrupted by sudden, bursting psychomotor agitation in some cases
- Automatic obedience
- Echolalia and echopraxia
- Stereotypes and mannerisms

4.1.4.5 Functional Impairment

Every aspect of schizophrenia symptomatology contributes to the impairment in patients' daily functioning, which hampers the achievement of an adequate quality of life. A certain degree of functional decline is often detectable before the acute onset of illness, persisting and gradually worsening also during periods of symptoms' remission. Patients with schizophrenia show impairments in several functional domains, including self-care activities, psychosocial, and work functioning. Functional deterioration determines a great social and financial burden on healthcare systems and society. Moreover, severe functional impairment is also associated with a higher risk of physical morbidities and, therefore, with reduced longevity.

Functional outcome is mainly influenced by neurocognitive functions and social-cognitive skills such as empathy, theory of mind, and emotion recognition. Taken together, these deficits negatively affect social interactions, further impairing real-world functioning and patient's autonomy. Most of them are targeted by current rehabilitative programs in articulated designs addressed to all these components.

Notably, functional impairment is only marginally affected by positive symptomatology. Similarly to negative and cognitive symptomatology, antipsychotic pharmacological treatment is not yet clearly effective in improving functional outcome directly, but through the mediation of the reduction of positive symptoms.

4.1.4.6 Clinical Course

Schizophrenia is characterized by a chronic and progressive course of illness, with a relapsing/remitting pattern of acute exacerbations. Remissions are generally partial and incomplete, with the persistence of a certain degree of symptoms. Negative and cognitive symptoms tend to progress along the course of illness, increasing proportionally with the number of relapses and conditioning progressive functional impairment (educational and work performance) and social disability (social withdrawal). Nevertheless, most of primary cognitive and functional deterioration occurs during the early stages of the illness, typically within 3–5 years after symptom onset. Nowadays, with current treatment available, following these disruptive early stages, the illness stabilizes and, despite possible subsequent exacerbations, there is generally no further significative illness-driven decline in cognition and, partially, in functioning. However, the magnitude of this deterioration appears to be at some degree related to the duration of untreated psychosis, as also evidenced by patients showing treatment resistance is characterized by worse longitudinal cognitive and functional outcome and the different course in the preneuroleptic era, almost invariably leading to a severe worsening. Indeed, on the one side, persistence of active positive symptoms has been associated with brain structural and functional decline and, on the other, limit patient's ecological daily cognitive training by leading to active social withdrawal. So a secondary (to nonfunctioning due to persistent positive symptoms and negative symptoms) worsening in cognition may be seen when illness is not under control. So, despite the poor correlation of cognitive and positive symptoms seen in current research, in clinical reality the control of positive symptoms and the reduction of re-exacerbations with continuous use of antipsychotics is fundamental and underlines the need to favor compliance to treatment.

The clinical course can be divided into three major "phases" even if the illness is a chronic one (four, if we consider prodromal phase).

Premorbid Phase

A substantial proportion of patients show a significant premorbid impairment detected in terms of cognitive functioning and poor social adjustment (varying degrees of nonspecific negative and cognitive symptoms) that frequently hamper the achievement of standard scholastic results. Deficits mainly affecting motor coordination, motor sequence, and sensory integration, named neurological soft signs, are also detectable in subjects who will later develop schizophrenia, even if mainly subtle.

Box 4.5: Neurological Soft Signs (NSS)

- Altered domains
- Motor coordination: diadochokinesia, finger-thumb opposition
- Sensory integration: gait, tandem gait, two-point discrimination
- Complex motor tasks: finger to nose test, fist edge-palm test (prono-supination task)
- Spatial motor task: right-left orientation, graphesthesia, stereognosis
- Primitive reflexes

Prodromal Phase

The onset of the disease is often preceded by a prodromal phase characterized by attenuated positive symptoms (subthreshold unusual ideas or perceptions), declining of functioning in terms of educational or work performance, and progressive social withdrawal with loss of interests. The subject experiences a reduction in the quality and quantity of social and affective relationships, appearing detached from life events. This phase is frequently accompanied by a sense of alienation of the subject with respect to the surrounding environment, together with a great anguish and an internal feeling of disruption of psychic functions. This phenomenon, defined as *delusional mood,* is described as a terrifying experience in which the individual has the subjective perception that something indefinable and devastating is going to happen. Moreover, bizarre ideas, fatuous or perplexed expressions, and strange behaviors may be reported, such as inappropriate laughter, often secondary to unusual perceptions. *Delusional perceptions* (abnormal meanings attributed to real perceptions) or *delusional intuitions* (sudden and aberrant enlightenment) are also frequently described.

Acute Exacerbation

The first psychotic episode is generally recognized as the formal onset of the disease and usually occurs in late adolescence or early adulthood. It is characterized by the presence of delusions, hallucinations, and disorganization, also accompanied by behavioral problems like psychomotor agitation or acting related to the delusional and hallucinatory contents (a patient may stay weeks at home without going out with sunblinds down because he is scared by people observing him and prosecuting him). The first years of illness are often characterized by repeated episodes of acute psychosis, with variable duration of inter-episode remission or attenuation of positive symptoms. During these episodes, hospitalization is often required, aimed at clinical management and behavioral observation and control sometimes necessary for the behavioral consequences of severe delusional or hallucinatory or disorganization symptoms and the severe anguish the patient may show, and faster therapeutic optimization. Fewer episodes and lower severity are achieved with early and effective pharmacological treatment and its maintenance over time as prophylaxis. The lack of compliance with antipsychotic treatment is the main cause of relapses. Exacerbations can be triggered also by stress and substance abuse.

Chronic Phase

Throughout the course of illness, a phase of relative stability can be reached in which positive symptoms become less severe, while negative symptoms and cognitive deficits remain the more prominent features. In this phase, the functional impairment persists and should be addressed with specific rehabilitative intervention.

4.1.4.7 Diagnosis

At first, schizophrenia is generally diagnosed on the presence of positive symptoms (not necessarily continuous) and impaired functioning in the absence of significant mood symptoms, neurological illness, or substance use that can account for the psychotic symptoms. Despite the absence of reliable diagnostic tests, biomarkers, and pathognomonic symptoms, neuroimaging (to exclude organic causes in the differential diagnosis), diagnostic interviews, and cognitive testing are frequently used to support the clinical diagnosis based on symptoms.

The diagnosis of schizophrenia must be carefully evaluated due the aspecifical nature of most symptoms and symptoms combination in psychotic onsets of any kind (in particular care should be used to exclude the onset of a mood disorder or an acute psychosis driven by a substance abuse or a medical condition or a medication).

Nevertheless, the onset of a significant part of cases is insidious and not necessarily attributed to a mental illness due to the later presentation of positive symptoms and attenuates progressive losing of functioning and performance attributed to an adolescent "crisis."

Psychic examination and anamnestic collection are the main diagnostic tools for a correct clinical assessment. Anamnesis must be focused on specific issues, such as prodromal symptoms, family history, and specific risk factors (substance use/abuse). Another crucial aspect in evaluating patients is functional assessment: scholar/working functioning, social skills, personal autonomy, and daily functioning in the domestic/family environment.

The psychiatric interview allows the clinician to explore the current positive and negative symptomatology of the patient, and also to estimate the cognitive impairment.

Specific standardized rating scales are available to support clinicians in the assessment of the severity of symptoms and functional impairment. Concerning psychopathology, the Positive and Negative Syndrome Scale (PANSS) represents the gold standard for evaluation of schizophrenia, specifically investigating the presence and intensity of positive, negative, and general psychiatric symptomatology. PANSS is one of the most specific tools for schizophrenia. Concerning cognitive assessment, there are various well-validated and comprehensive neuropsychological batteries such as the Measurement and Treatment to Improve Cognition in Schizophrenia (MATRICS) Consensus Cognitive Battery (MCCB) and the Brief Assessment of Cognition in Schizophrenia (BACS). These batteries assess domains of cognition that are typically impaired and strongly correlated with functional outcome in schizophrenia. The explored functions are working memory, attention/vigilance, verbal learning and memory, visual learning and memory, and executive functions (reasoning, problem-solving) and speed of processing. Frequently, the cognitive assessment also includes an evaluation of Intelligence Quotient, through the Wechsler Adult Intelligence Scale (WAIS).

Finally, different scales have been developed in order to assess functional status and quality of life as well, evaluating both patient's and clinician's point of view. In this context, the most frequently used scales are the Global Assessment of Functioning (GAF) scale, the Specific Level of Functioning (SLOF) scale, the Quality-of-Life Scale (QLS), and the University of California, San Diego (UCSD) Performance-Based Skills Assessment (UPSA). These scales allow the evaluation of real-life behavior, offering a good perspective on global daily functioning and also providing a reliable measure of well-being.

Box 4.6: DSM-5 Diagnostic Criteria

1. Presence of at least two of the following symptoms lasting at least 1 month:
   - Delusions
   - Hallucinations
   - Disorganized speech
   - Disorganized or catatonic behavior
   - Negative symptoms
2. Level of functioning in one or more areas (school, work, social relations, self-care) is markedly low compared to the level reached before onset. In childhood or adolescence, there is the inability to reach the expected level of interpersonal, educational, or work functioning.
3. Continuous signs of the disorder must be present for at least 6 months (prodromal symptoms, negative or cognitive symptoms or attenuated psychotic symptoms).
4. Exclusion of a mood disorder or schizoaffective disorder.
5. Symptoms are not caused by a substance or a medical condition.
6. If a diagnosis of autism spectrum disorder is present, adjunctive diagnosis of schizophrenia occurs only in the presence of criteria described above.

4.1.4.8 Differential Diagnosis

The following differential diagnoses need to be considered when hypothesizing a diagnosis of schizophrenia:

- *Brief psychotic disorder*, differentiated by the duration of an episode of at least 1 day but less than 1 month, and with eventual full return to premorbid level of functioning.
- *Schizophreniform disorder*, differentiated by duration of psychotic illness. Diagnostic criteria are basically the same as for schizophrenia, but the duration of the psychotic episode ranges between 1 and 6 months. If the diagnosis is made before recovery, it should be qualified as "provisional," or changed to Schizophrenia if duration of the episode will exceed 6 months.
- *Delusional disorder*, differentiated by the rare presence of other psychotic symptoms (disorganized speech or behavior, negative symptoms, and cognitive deficit), the later age of onset.
- *Bipolar disorder* with psychotic features, differentiated by the presence of manic/hypomanic features, usually characterized by grandiose or persecutory delusions and/or hallucinations, or depressive episodes with delusions and hallucinations, usually characterized by themes of guilt and ruin, cyclic pattern of manifestations with restoring of functioning to premorbid level in interepisode periods.
- *Depression with psychotic features*, differentiated by the presence of severe depressive symptoms and delusions/hallucinations with contents of guilty and ruin and return to basal functioning at the end of the episode.

- *Schizoaffective disorder*, differentiated by the presence of mood episodes overlying chronic psychotic symptoms which remain after the manic or depressive episode is closed, and better functional outcome.
- *Personality disorders* such as schizotypal, schizoid, and paranoid personality disorders.
- *Drug and medication-induced psychosis.*
- *Psychosis secondary to organic causes.*

4.1.4.9 Comorbidities

Frequently, comorbidity with substance or alcohol abuse occurs. For this reason, differential diagnosis with substance-induced psychotic disorder results sometimes challenging. Moreover, nicotine dependence or smoking is highly prevalent in patients with schizophrenia compared to the general population.

Schizophrenia is characterized by a decreased longevity, with a life expectancy about 20% lower than the general population (61 vs. 76 years). Increased mortality is due to both psychiatric and medical conditions. Indeed, on the one side, schizophrenia is associated with higher rates of suicide, in particular at onset, with one-third of individuals attempting suicide at least one time over the entire course of illness and 5% of patients actually dying. On the other side, people with schizophrenia show elevated metabolic risk factors and incidence of cardiovascular morbidity and related mortality, compared with the general population. Indeed, the presence of obesity in patients with schizophrenia is two times higher than in the general population. Metabolic disorders are a consequence of rapid weight gain leading to obesity, dyslipidemia, and glucose intolerance which may develop into type 2 diabetes. It is difficult to determine whether the high prevalence of metabolic disorders in this population is independent of antipsychotic treatment or is a consequence of medication, especially some atypical antipsychotics such as olanzapine and clozapine. Besides pharmacological treatment, a number of explanations have been proposed, such as unhealthy lifestyle and dietary habits (related to the dimensions of negative symptoms and disorganization) that facilitate the development of obesity among patients with schizophrenia, genetic predisposition to altered glucose and lipid metabolism, and alterations of the hypothalamic–pituitary–adrenal axis linked to hypercortisolemia. Moreover, a crucial factor that further worsens the severity of medical and metabolic comorbidities is the lower capacity of patients with schizophrenia to seek medical assistance and to implement lifestyle interventions. Together with social isolation and stigma, reduced help-seeking behavior leads to a systematic under-recognition and undertreatment of cardiovascular disease of people with schizophrenia within primary care, which might contribute to the substantial cardiovascular-related morbidity and premature mortality observed in this patient group.

Metabolic alterations do not only negatively affect physical health but can lead to a worse psychopathological outcome as well. Indeed, recent findings pointed out that metabolic syndrome is associated with greater cognitive impairment, and, thus, also with poorer longitudinal functional status and clinical outcome.

4.1.4.10 Prognosis

Although positive symptom remission is achieved in almost 50% of patients within 5 years following treatment initiation, schizophrenia can lead to severe long-term disability, negatively affecting quality of life of both patients and their families. However, as previously evidenced discussing neurobiological and clinical features of the disorder, high heterogeneity of longitudinal outcome and prognosis is observed among patients, according to the severity of their symptomatology. Indeed, while many patients with schizophrenia have a lifelong vulnerability to recurrent episodes of re-exacerbations, a large proportion will have few relapses and get a good functional recovery. When adequately treated with both pharmacological and rehabilitative interventions, patients can achieve even high levels of autonomy, although often needing formal or informal financial and daily living support. The best possible personal autonomy, restored relationships, and ability to get and maintain a work are the targets toward which to look at as "the" target when treating a patient affected by schizophrenia, but the results may be quite variable in response in relation to personal premorbid resources social-familiar support and quality of medical support from services. Nevertheless, more specific factors and illness presentation and course features have been individuated as potential predictor of longitudinal outcome. Among positive prognostic predictors are reported acute onset, good premorbid cognitive functioning, short duration of untreated psychosis, early treatment response, absence of substance abuse, later age of onset, and female gender. On the other side, high number of relapses, impaired premorbid adjustment, prominent negative symptomatology, poor insight and low adherence to antipsychotic treatment represent the stronger predictors of poor longitudinal outcome.

4.1.5 Treatment

Treatment should address all the dimensions of the disease, therefore including positive, negative, and cognitive symptoms. Despite pharmacologic advances, the treatment of schizophrenia remains a challenge, and poor or suboptimal outcomes are still frequently observed. Ultimate treatment goals are remission and functional recovery. In order to achieve these targets, antipsychotic pharmacotherapy should always be integrated with cognitive-behavioral, social, and neurocognitive rehabilitative interventions. Pharmacological guidelines are treatment benchmarks. However, for each patient, it is necessary to adopt a personalized therapeutic approach taking into account current symptoms, comorbid conditions, past therapeutic response, adverse effects, and environmental influencing factors that could affect compliance.

The first and mandatory therapeutic approach for schizophrenia is pharmacological, based on antipsychotic therapy, aimed to reduce acute psychotic symptoms such as hallucinations and delusions and possibly induce remission (see Box 4.7). Although large-scale and meta-analytic studies indicate a similar efficacy of the first- (typical) and second-generation (atypical) antipsychotics during acute phases of the disease, the advantage of a lower rate of extrapyramidal side effects may favor the second-generation antipsychotics for the need of chronic treatments.

Anyway, antipsychotic response is individual, and the results of effectiveness found in randomized controlled studies and meta-analyses do not necessary apply to the individual; antipsychotic response is now considered to have high interindividual variability and a high concordance of response in the same individual across time, i.e., after drug withdrawal due to noncompliance. There is no "winner" drug on average, but the winner drug for that patient. Resistant schizophrenia is a clear example: a rigid assessment of this response phenotype through sequential treatment with typical and possibly more than one atypical antipsychotic treatment leads to indication to clozapine, indicated specifically for resistant patients, with rates of response up to 70% in comparison with no response to the previously tried treatments.

Sometimes, especially after years of treatment, a good response may decay and a change in antipsychotic required to maintain stability, or a resistance may develop, leading to the indication to clozapine.

Different antipsychotics may fit the neurobiology of each subject, and this is the major advantage for clinical efficacy of having nowadays a wide number of available antipsychotics, including the old ones: the higher number allows to treat at the best a higher number of patients in comparison to the time where only neuroleptics (strongly homogeneous in the main pharmacodynamic profile, with shared high affinity to D2 receptors prevailing at clinically used dose over the other, minor properties) were available.

Especially during the acute phase, concomitant treatment, addressing specific symptoms such as insomnia and/or agitation, is often indicated. Benzodiazepines are typically the first choice for this purpose.

After the onset phase, with a diagnosis clearly confirmed, antipsychotic treatment should be continued throughout the course of the disease. This is crucial to prevent relapses and maintain clinical stabilization, which, in turn, is important to improve social withdrawal and functional impairment and to make possible rehabilitation. Indeed, a good response to antipsychotics, together with rehabilitative interventions, allows the recovery of functional capacities and the improvement of quality of life.

The discontinuation of antipsychotic therapy during stable phases of the illness is associated with poorer cognitive and functional outcomes, higher rates of relapse, and incomplete inter-episodic remission (persistence of residual symptoms). If the patient shows poor compliance to pharmacological treatment, antipsychotics can be administered intramuscularly once or twice a month, and, recently even every 3 months (depending on of the pharmacokinetic properties) as long-acting injectable (LAI) formulations in order to avoid the risk of self-discontinuation.

During antipsychotic treatment, a primary goal is to limit side effects to improve compliance and thus to prevent relapses. Prompt detection and treatment of adverse effects of antipsychotics are crucial in the overall management of schizophrenia. Side effects may undermine treatment adherence and affect functional capacity, subjective well-being, quality of life, and life expectancy. The adverse effects that should be regularly monitored include sedation, sexual and reproductive system dysfunctions, extrapyramidal symptoms, weight change, alterations in blood pressure, lipid profile, and fasting blood glucose levels. Recent data have shown that the assessment of potentially problematic adverse effects, particularly metabolic, such as increases in glucose and lipid levels, is still suboptimal in clinical practice and need to be significantly implemented.

Concerning cognitive symptoms, antipsychotic agents have no or minimal effects on neuropsychological performance, and despite extensive economical and research efforts, to date, no effective newly developed procognitive drugs are available yet. On the other hand, different nonpharmacological rehabilitative approaches showed promising results and are starting to be routinely implemented in clinical practice. Among these, cognitive remediation techniques were proved effective to the point that they are currently considered the best treatment option, being able to induce significant changes not only at the behavioral but also at the biological level, leading to a global improvement in neural system dysfunction. However, cognitive remediation has been shown to be strongly effective when delivered together with cognitive-behavioral and social skills rehabilitation, leading to persisting cognitive improvements. However, it is important to remind that a requirement for a good outcome of rehabilitative interventions is the presence of clinical stabilization, particularly concerning positive symptomatology. Response to pharmacological treatment is thus a necessary requirement for rehabilitation.

Although schizophrenia was long considered a disease characterized by progressive clinical and functional deterioration, nowadays the goal of treatment is not just remission, but also recovery (see Box 4.7). Evidence supports the idea that the course of the illness is not inevitably deteriorating and that functional recovery can be achieved.

Box 4.7: Remission and Recovery

Remission

Remission in schizophrenia, as for other chronic psychiatric and nonpsychiatric illnesses with relapsing and remitting courses, does not require symptoms to be completely absent and can be associated with some residual symptomatology.

Remission has been defined as "a state in which patients experience an improvement in core symptoms to the extent that any remaining symptoms are of such low intensity that they no longer interfere significantly with behavior and are below the threshold typically utilized in justifying an initial diagnosis of schizophrenia."

A univocal definition of remission for schizophrenia is lacking. The Remission in Schizophrenia Working Group (RSWG) developed a consensus definition of symptomatic remission that consists of two elements: a severity criterion covering the core symptoms of schizophrenia (low scores on eight diagnostically relevant symptoms in the PANSS) coupled with a time criterion (symptom severity criteria should be achieved for a minimum period of

6 months). Symptomatic remission is a challenging but achievable objective for a significant proportion of patients with schizophrenia, given the ongoing efforts to better understand the condition and the ongoing search for better treatments.

Symptomatic remission of positive symptoms does not necessarily mean that the patient is functioning well, because other illness components may persist, leading to functional impairment.

Recovery

Remission is a necessary, but not sufficient, step toward recovery, which is a longer-term and more complex goal to achieve. Recovery is less precisely defined than remission. In general, recovery can be defined as sustained symptom remission, accompanied by functional rehabilitation (e.g., cognitive, social, and vocational) and reduced use of medical health services. It is generally assumed that recovery comprises both objective and subjective components: the objective domain refers to clinical outcomes which are evaluated through operationally defined criteria; the subjective domain refers to the process of positive changes in an individual's subjective experience of themselves as human beings. Not all patients with schizophrenia can achieve recovery. In literature, the lack of consensus on its definition gives rise to heterogeneous data, with the percentage of patients with schizophrenia achieving recovery varying from 13.5 to 50%. Recovery may be influenced by multiple factors such as adherence and response to treatment, support from family and society, adjustment coping, and reappraisal. Factors that have negative effects on recovery are higher levels of cognitive deficits and negative symptoms, stigmatization (both self-stigma and social stigma), poor service engagement, side effects of medications, and lack of self-awareness.

4.1.5.1 Pharmacological Treatment

Pharmacological treatment is based on antipsychotics, both typical (first generation) and atypical (second generation), each of which shows unique and different receptor activity and affinity (see Chap. 11). It is recommended to use a single antipsychotic (preferably atypical) at the lowest effective dose. There is a time latency in response since initiation of antipsychotic, which is typically of 4–6 weeks. However, an early response can be observed during the second week of treatment. In case of lack of response and/or tolerability issues despite adequate doses, a switch to another antipsychotic is recommended. In case of treatment resistance, as reported above, clozapine is the only indicated and evidence-based drug for TRS. Augmentation strategies (adding a second antipsychotic drug) should be contemplated only in case of incomplete response to clozapine.

Treatment-Resistant Schizophrenia

Poor response to at least two antipsychotic drugs at adequate dosage and duration of treatment.

Antipsychotic treatment is chronic over time with the aim to prevent exacerbations. The drug to which the patient has clearly responded in the exacerbation must be continued, but doses used in maintenance phase tend to be lower to minimize side effects and overall drug exposure. For some antipsychotics with high interindividual plasma-level variability, like clozapine, periodical plasma level evaluation may improve efficacy and minimize side effects avoiding under- and over-dosing and letting compliance to be controlled definitely.

When assessing resistance, a crucial aspect is, in fact, is to exclude nonresponse due to impaired or absent compliance; as previously mentioned, low adherence to treatment is one of the stronger predictors of poor longitudinal outcome, leading to greater risk of relapse, hospitalization, and eventually suicide. Large-scale studies reported that almost 75% of patients discontinue antipsychotic medication within 18 months due to insufficient efficacy, intolerable side effects, or for other reasons. Key drivers of low compliance include lack of insight, attitude toward medication (frequently a premorbid feature related to the subject's psychology and cultural background), side effects, and substance abuse. Factors positively related to adherence are a good therapeutic relationship with physician and perception of benefits of medication that must be underlined and strengthened by clinicians, together with obtaining the lower possible rate and severity of side effects.

4.1.5.2 Noninvasive Neurostimulation Techniques

To date, transcranial direct current stimulation (TDCS) and transcranial magnetic stimulation (TMS) have low levels of evidence to be effective in improving poor pharmacological response in schizophrenia. Possible applications of these noninvasive brain stimulation concern reduction of cognitive, positive, and negative symptoms, depending on the targeted cortical area.

4.1.5.3 Electroconvulsive Therapy (ECT)

Data on ECT are limited; ECT may be useful in treating catatonic symptoms in schizophrenia and as adjunctive treatment in certain resistant patients unresponsive even to clozapine.

4.1.5.4 Psychiatric Rehabilitation

Remission of positive symptomatology can be achieved through pharmacological treatment. However, pharmacotherapy's effect on the negative-cognitive dimension is poor. Rehabilitation programs aim to improve impaired cognitive, emotional, and social skills through cognitive training programs, group activities, and individually targeted interventions. The theoretical goal is to allow the patient to have a role, relationship, and a work for self-sustainment.

Rehabilitative interventions have been developed to complement psychopharmacological treatments and aim to assist patients in attaining their highest level of functioning, the better degree of symptom control, and the greatest level of subjective life satisfaction. Indeed, antipsychotic treatment has only limited efficacy on cognitive impairment, insight, and social skills, whereas rehabilitation interventions specifically target these aspects of the disease. Moreover, rehabilitation interventions also aim to promote the recovery process in its subjective component, by encouraging self-determination and active empowerment. According to literature, many rehabilitative programs have proven their effectiveness in favoring functional recovery.

Main rehabilitative programs include cognitive remediation therapy (see Box 4.8), social cognitive interventions, psychoeducational intervention concerning disease, cognitive-behavioral therapy, and social skills training.

Box 4.8: Cognitive Remediation Therapy (CRT)

CRT for schizophrenia is a behavioral training-based intervention that aims to enhance through exercise, pen and paper or computerized, several cognitive functions such as attention, memory, executive function, social cognition, or metacognition durably. The ultimate goal of CRT is to improve functioning, limiting the impact of cognitive impairment on everyday life and supporting the effects of concomitant cognitive-behavioral and social rehabilitation. Its efficacy has definitely been demonstrated in patients with schizophrenia.

4.1.5.5 Clinical Case

U. is a 43-year-old engineer who stopped working 15 years ago. He is not married and lives with his mother. He spends most of his time alone, inactive, and his functioning is limited to a few daily home tasks that the patient performs only if stimulated. He rarely leaves the house, and, if he does, he is always accompanied by his mother or brother.

U. came to the attention of the psychiatrist nearly 15 years ago when he had troubles at work. At the time he claimed to clearly hear the voices of his chief and colleagues insulting and speaking ill of him among themselves by reason of his incredible skills. One day, he ran away from the workplace thinking they wanted to kill him, and he went to the police station to report everything. Finally, the police officer called the psychiatrist.

Nowadays, these contents of thought, despite being referred to the past, are pointed out with less emotional participation through a digressive and sometimes illogical speech. Despite suspiciousness is inferred by his glances, U. trusts his psychiatrist, probably because he knows him for a long time. During interviews with him, sometimes he described a sort of "psychological violence" and "mental dialogues" with deceased relatives. In the last period, U. has become very religious, often feeling very close to god and believing in a sort of interconnection with Him. Sometimes he gets frustrated, and irritability emerges relatively to his performance difficulties: since the onset of the disease, he was no longer able to work. Moreover, difficulties in maintaining concentration and attention are frequently reported. U. is indeed often distracted, as if he is listening to something/someone.

4.2 Other Psychotic Disorders

The spectrum of psychotic disorders includes different conditions, both primary and associated with substance use or medical pathologies. They show similarities to schizophrenia in clinical manifestation and only partially share etiopathogenetic mechanisms and risk factors. According to DSM-5, these disorders are defined by abnormalities in one or more of the following five domains: delusions, hallucinations, disorganized thinking, grossly disorganized or abnormal motor behavior, and negative symptoms. Each of them exhibits different combinations of symptoms and deficits. We will summarize the main clinical features, differential diagnoses, and treatment indications for delusional disorder, schizoaffective disorder, brief psychotic disorder, and substance-induced psychotic disorder.

4.2.1 Delusional Disorder

4.2.1.1 Definition

Delusional disorder, or paranoid disorder in the previous nomenclature, is usually a chronic condition mainly characterized by the presence of one or more delusions related in themes to the main delusional idea, typically in the absence of other psychotic features. In addition, globally, functioning is relatively preserved without evolution to defective states.

4.2.1.2 Epidemiology

Delusional disorder is less frequent than other conditions, such as schizophrenia and mood disorders, and its lifetime prevalence in the general population is estimated at around 0.02%, without gender predominance. The relatively low prevalence may be in part due to underreporting. People with delusional disorder may not seek mental health help unless taken to attention by family or friends, because of their poor/absent insight of illness, but they may maintain a good general functioning.

It has a later age of onset as compared to schizophrenia. The mean age of onset is about 40 years, but it can occur at any age in life. Despite this, the appearance of delusions in patients older than 60 years is more likely attributable to an organic condition, such as dementia.

4.2.1.3 Clinical Presentation

The clinical picture is mainly characterized by the presence of one, less commonly more, but related, delusion. According to DSM-5 criteria, to be diagnosed, the delusion must persist for at least over a month, but it is typically a chronic disorder. The delusion is usually systematized and pervasive and revolves around one main topic. The patient displays high emotional involvement and behaves according to the delusional idea. Although the disorder may manifest with any type of delusional content, typically the contents are plausible and believable, while bizarre delusions are rarer in this condition than in schizophrenia. Delusional disorder, differently from schizophrenia, does not globally compromise psychic functions, nor lead to functional impairment. Moreover, insight is generally absent only in the field of events and experiences related the delusional idea.

In addition to delusions, hallucinations may rarely be present, but they are not prominent and are closely related to delusional content.

Mood may be altered secondary to and congruent with delusional content. For example, a patient with paranoid delusion may be depressed and anxious due to the preoccupation of being injured or betrayed by a friend or a spouse.

In delusional disorder, general behavior is not much altered, and global functioning is relatively preserved. However, patients may engage in abnormal behaviors related to the delusional idea that can result in occupational, marital, and social difficulties, and, not infrequently, legal problems.

The course of delusional disorder, although chronic, is usually better than that of schizophrenia. A remission period is achievable, but, even in the attenuation or absence of active delusions, the insight on previous ones remains poor or absent.

Box 4.9: DSM-5 Diagnostic Criteria of Delusional Disorder

1. One (or more) delusion lasting 1 month or longer.
2. Diagnostic criterion 1 of schizophrenia has never been met.
3. Hallucinations, if present, are not prominent and related to the delusional theme.
4. Functioning is not markedly impaired, and behavior is not bizarre.
5. Symptoms are not caused by substances or a medical condition.

4.2.1.4 Main Differential Diagnoses

- *Schizophrenia:* It can be differentiated from delusional disorder by the simultaneous presence of other psychotic symptoms (disorganized speech or behavior, negative symptoms, and cognitive deficit), the earlier age of onset, the substantial functional impairment, the worse course of illness.
- *Depressive and bipolar disorders and schizoaffective disorder*: In delusional disorder, mood alteration, if present, is not prominent but secondary to delusional content. If delusions occur exclusively during mood episodes, the diagnosis should be a depressive or bipolar episode with psychotic features. Nevertheless, a comorbidity between delusional disorder and a mood disorder in comorbidity should be considered.
- *Delirium, major neurocognitive disorder, psychotic disorder due to another medical condition, and substance/medication-induced psychotic disorder:* Delusions may occur in all these disorders, and the distinction should be made considering the coexistence of other symptoms, the age of onset, the course, and the presence of comorbidities. Delirium is characterized by abnormalities in consciousness, which is instead clear in delusional disorder. Psychotic disorder due to medical condition or medication induced can be considered in older patients, affected by other medical diseases and/or treated for them. Imaging or laboratory tests should be performed to rule out any organic cause.

- *Illness anxiety disorder:* In this disorder, the patient is worried about a physical illness, but his/her concerns, differently from delusional disorder somatic type, can be at least transiently reassured by medical exams. Also, some people with delusional disorder of this type may seek multiple medical assessments, but with their delusional idea not changing in front of clear evidence.
- *Body dysmorphic disorder* previously known as *dysmorphophobia*: This disorder is characterized by excessive concern about the appearance of one's body, often associated with behavioral abnormalities such as requests of medical procedures (i.e., plastic surgery). At onset, ideas that these patients have about their bodies are not delusional, but only extremely overvalued. Insight varies, although it is mostly poor. Onset is usually earlier than delusional disorder, tending to begin during the first decades of life. However, over time these preoccupations can develop a delusional quality, with almost 50% meeting criteria for a delusional disorder (somatic type).

4.2.1.5 Treatment

Antipsychotic medications are considered the mainstay of treatment, and FGAs appear to be slightly superior to SGAs. However, SGAs are more frequently used due to their more acceptable side effect profiles. A positive response to antipsychotic treatment occurs in nearly 50% of the cases. Due to the chronic course of the disorder, treatment should be continued indefinitely.

However, the treatment of delusional disorder is particularly challenging due to the patients' lack of insight and hypersensitivity to side effects, compared to patients with schizophrenia, both undermining adherence to the treatment regimen.

4.2.1.6 Clinical Case

P. is a 40-year-old history teacher. She has been married for many years. P. came to psychiatric attention as her husband brought her to the emergency room because she has completely refused to eat for 2 days. About 5 months ago, she started complaining of heartburn and digestive difficulties, only partially improved after a pharmacological treatment prescribed by her general practitioner. Since then, her husband says he noticed a change. P. is often silent, detached, and sometimes grumpy toward him. She appears worried, restless, and struggles to fall asleep. She spends a lot of time away from home, often stays at work late for overtime, and is evasive when he asks for explanations. Although she has always been careful about her physical fitness, she is described by her husband as a "good eater." However, for the last few months, she has been complaining of loss of appetite and has been eating irregularly. In fact, she eats mainly packaged foods, refusing anything cooked by her husband, and drinks only bottled water that she buys herself. In the last month, she has lost about 5 kg.

At the interview with the doctor, P. asks her husband not to attend. She appears suspicious and in a state of alertness. At first, she is not very talkative, but gradually she expresses her anxiety about the certainty that her husband wants to "get rid of her." She says that 4 months ago, she found out that her husband had an extramarital affair with the neighbor. She reports that, at first, she noticed strange coincidences and then she became certain when her husband turned on the radio while the song "Arsenic" was playing. It was precisely in those days that the gastric problems started, according to her.

P. denies having ever heard voices or unusual noises. After her initial hesitation, her speech is fluid and coherent, and her mood appears depressed in consequence of the delusional beliefs.

4.2.2 Schizoaffective Disorder

4.2.2.1 Definition

As suggested by its denomination, schizoaffective disorder (SAD) is a disorder whose hallmark is the presence of affective symptoms, either depressive or manic, co-occurring with characteristic symptoms of schizophrenia, such as hallucinations and/or delusions. Researchers debate whether the diagnosis of SAD represents a comorbidity of schizophrenia and a mood disorder or rather actually exists as the mid-point on a continuum between schizophrenia and bipolar disorder. Indeed, since it was first defined in literature, its actual clinical distinction and validity as an independent nosological category have been largely discussed with conflicting points of view, leading to a large number of misdiagnosis.

4.2.2.2 Epidemiology

SAD has a lifetime prevalence of 0.3%, about one-third less common than schizophrenia.

The depressive subtype is more frequent in females (2:1), whereas no significant sex-related differences are reported for the bipolar subtype.

Similarly to schizophrenia, the onset of SAD is typically in early adulthood, although possible at any age, and later for women. The depressive subtype of SAD seems to be more frequent in the elderly, whereas the bipolar subtype seems to prevail in young people.

The risk for SAD is increased among individuals who have a first-degree relative with schizophrenia, bipolar disorder, or SAD.

4.2.2.3 Clinical Presentation

DSM-5 diagnostic criteria for SAD require the occurrence of an uninterrupted period of illness in which the characteristic symptoms of schizophrenia (criterion A for schizophrenia) are present simultaneously with affective symptoms meeting the criteria for a major mood episode, either depressive or manic. Moreover, in order to distinguish a SAD from a mood disorder with psychotic features, more than 2 weeks of psychotic symptoms in the absence of symptoms of a major mood episode are required. Lastly, affective symptoms need to be present for a significant period during active or remission psychotic phases of illness, in order to differentiate SAD from schizophrenia.

DSM-5 separates bipolar and depressive subtypes: in bipolar type, at least a manic episode has occurred in the lifetime, whereas in depressive type only major depressive episodes have occurred.

Great heterogeneity characterizes clinical presentations of subject with SAD: some experience predominantly mood symptoms, other mainly psychotic symptoms. Psychotic symptoms usually occur before the onset of a prominent major mood episode, persisting after its remission. Similar to bipolar disorder, major mood episodes are recurrent, with variable durations typically longer in the depressive subtype.

Patients with SAD exhibit higher suicidal risk and higher rates of hospitalization and substance abuse than patients with schizophrenia or mood disorders.

Most patients with SAD tend to have a nondeteriorating course of illness, with an intermediate prognosis between schizophrenia and bipolar disorder. However, approximately 20–30% of patients show a deteriorating course with persistent psychotic symptoms, closely resembling longitudinal clinical course and deteriorating functional outcome of schizophrenia.

Predictors of poor outcome in SAD include poor premorbid functioning, insidious onset, absence of a precipitating factor, predominance of psychotic symptoms, early age at onset, poor inter-episode remission, SAD depressive subtype, and a family history of schizophrenia.

Box 4.10: DSM-5 Diagnostic Criteria of Schizoaffective Disorder

1. Diagnostic criterion 1 of schizophrenia + concomitant diagnosis of major depressive episode or manic episode.
2. Along the course of the illness, delusions or hallucinations, lasting at least 2 weeks, are present without the concomitant diagnosis of a depressive or manic episode.
3. Symptoms that meet criteria for a major mood episode are present for the majority of the total duration of the illness.
4. Symptoms are not caused by substances or medical conditions.

4.2.2.4 Differential Diagnosis

- *Bipolar and major depressive disorder:* The presence of psychotic symptoms out of a mood episode is crucial to distinguish SAD from a mood disorder with psychotic features. Patients with a mood disorder only experience psychotic features during a manic or depressive episode, whereas patients with SAD have psychotic symptoms during and out of mood episodes.
- *Schizophrenia:* Depressive symptoms may occur in schizophrenia, especially during the prodromal or residual phases, but they are less well characterized and often associated with negative symptoms. In patients with SAD, mood symptoms are prominent, whereas once the psychotic symptoms predominate for most of the course of illness, the diagnosis leans toward schizophrenia. In SAD,

global functioning is usually less impaired than in schizophrenia (and impairment is not a defining criterion in contrast to schizophrenia).

- *Delusional disorder:* Mood alterations, especially depression, commonly develop during its course. However, such presentations do not meet the criteria for SAD because the psychotic symptoms in delusional disorder are mainly delusions (they do not meet Criterion A for schizophrenia).
- *General medical and substance-induced conditions, delirium, and major neurocognitive disorder:* They may present with psychotic and mood symptoms, even combined. These conditions can be distinguished from SAD thanks to evidence from the clinical history, physical examination, laboratory findings, and any other investigations that the symptoms are pathophysiologically a consequence of these conditions.

4.2.2.5 Treatment

As for schizophrenia and bipolar disorder, long-term complex pharmacological therapies are required, targeting both psychotic and affective symptoms. Antipsychotics, mood stabilizers (lithium, anticonvulsants), and antidepressants are commonly used in combination, according to patient's stage of illness.

Since most patients present with prominent psychotic symptoms over time, treatment usually requires chronic treatment with antipsychotics. The SAD subtype may guide the choice of the add-on therapy. Mood stabilizers are indicated in patients with the bipolar subtype, whereas antidepressants in those with the depressive subtype, at least during the episodes. Patients with SAD tend to respond to lithium worse than patients with bipolar disorder, but atypical antipsychotics have shown mood-stabilizing properties in these patients and are thus often prescribed also for this purpose.

Similarly to other disorders, treatment resistance may occur in a minority of patients that can be thus prescribed with clozapine as second-level pharmacological treatment.

4.2.3 Substance-Induced Psychotic Disorder

4.2.3.1 Definition

Substance-induced psychotic disorders (SIPD) are brief psychotic syndromes occurring during or soon after the use of a substance or its withdrawal. This condition may persist for days or weeks after substance intoxication has resolved, or even longer if patients continue using the substance (see also Chap. 9).

4.2.3.2 Epidemiology

Although it is difficult to precisely evaluate the prevalence of these disorders, as they may be transient and patients may not seek medical help, SIPDs are common disorders with estimates of incidence ranging from 1.52 to 6.53 per 100,000 person-years. Up to 25% of first-episode psychosis is represented by SIPDs, and approximately 25% of these can progress to schizophrenia.

4.2.3.3 Clinical Presentation

Many substances with psychotomimetic properties are known to cause (or exacerbate) psychotic reactions resembling a primary psychotic disorder. The psychotogenic effect of these substances is related to their specific activities in different brain circuits (see Chap. 9).

The clinical presentation of an individual diagnosed with SIPD is quite similar to those presenting psychosis in the absence of substance use. It may include psychotic symptoms, such as delusions, hallucinations, disorganized thinking, grossly disorganized behavior, psychomotor disturbances (excitement or stupor), and an abnormal affect, which may range from intense fear to ecstasy. However, substance-induced psychotic symptoms are usually short-lived, remitting after sustained abstinence.

The risk of developing psychotic symptoms can vary according to the substance of abuse, the frequency and duration of abuse, and individual vulnerability to develop psychosis. Virtually all substances of abuse may induce acute psychotic syndromes, although cannabis, cocaine, amphetamines, and hallucinogens are associated with a greater risk for psychosis. Moreover, for some substances, in particular cannabis, there is a dose–response relationship, with psychosis occurring especially in those individuals who have been using high doses and over a lengthy period.

Cannabis use can cause a range of psychotic symptoms of variable severity, from acute short-term symptoms, related to intoxication, to severe and persistent symptoms leading to a diagnosis of cannabis-induced psychotic disorder (CIPD). Cannabis-induced toxic psychosis is characterized by mild alteration of consciousness, thought disorganization, hallucinations, distortion of time perception, and "dream-like" euphoria, usually resolving within a week. Absence of confusion and derealization distinguish cannabis-induced functional psychosis, associated with heavy and long-lasting THC assumption, clinically resembling an acute episode of schizophrenia and showing a rapid and good response to antipsychotic medications. If symptoms persist after discontinuation, functional psychosis can progress to a cannabis-induced chronic psychosis, hardly distinguishable form a chronic schizophrenia triggered by cannabis use and requiring long-term treatment.

The diagnosis of SIPD is not stable over time: a significant proportion of people with substance-induced psychosis later have a transition to a diagnosis of a schizophrenia-spectrum disorder or, less frequently, bipolar disorder, usually within the first 3–5 years after substance-induced psychosis.

Cannabis-induced psychosis and amphetamine-induced psychosis are associated with the highest risk for transition. Among cannabinoids, high-potency cannabis and synthetic cannabinoids carry the greatest risk of psychosis. Young age is associated with a higher risk of conversion to schizophrenia, with the highest risk in the range of 16–25 years.

Alcohol-induced psychosis/hallucinosis is a rare condition characterized by acute onset of auditory hallucinations (characteristically in the form of derogatory voices), often associated with persecutory delusions. These symptoms usually occur in clear consciousness and absence of thought process disorder in individuals with heavy alcohol consumption. These symptoms usually improve quickly, within a week, although they can become chronic due to ongoing alcohol abuse. Alcohol-induced psychosis should be differentiated from psychotic symptoms that may occur in alcohol withdrawal delirium ("delirium tremens"), wherein hallucinations are more likely visual than auditory, and associated with alterations in consciousness, neurovegetative symptoms, and psychomotor agitation. Chronic use of alcohol can also more frequently determine a jealousy delusional disorder.

Box 4.11: DSM-5 Diagnostic Criteria of Substance-Induced Psychotic Disorder

1. Presence of delusions and/or hallucinations.
2. Symptoms of criterion 1 start during or soon after substance intoxication or withdrawal or after exposure to a medication.
3. The substance/medication is capable of inducing the symptoms observed.
4. The disturbance is not better explained by other psychotic disorders.
5. The disturbance does not occur exclusively during a delirium.
6. The disturbance causes significant distress or functional impairment.

4.2.3.4 Differential Diagnosis

The differentiation of substance-induced psychotic disorder (SIPD) from primary psychotic disorder is particularly difficult.

Clinically, the most straightforward way to make an appropriated differential diagnosis is by assessing the temporal relationship between onset of psychosis and substance assumption.

According to DSM-5, the distinction should be based on the persistence of psychotic symptoms for a prolonged period (about 1 month) after the cessation of substance use. However, this criterion may be difficult to apply because patients may continue using the substance. Some clinical characteristics may assist differential diagnosis during the acute episode. Onset before substance use, personal/family history of other psychotic disorders, poor premorbid functioning, prevalence of auditory hallucinations, poor insight into symptoms, emotional detachment, and apathy suggest the presence of primary psychotic disorders rather than SPID. On the other side, some clinical features may help direct diagnosis to SIPD: a later age of onset, better insight into symptoms, prevalence of altered state of consciousness (confusion, disorientation), prevalence of visual or tactile hallucinations, fewer or less severe negative symptoms, higher levels of impulsivity, aggression and suicidal thoughts, overexcitement and excessive emotional involvement, and personal or family history of substance use disorders.

4.2.3.5 Treatment

Management of patients with SIPD is complex, requiring treatment of both psychosis and substance-related organic symptoms. Moreover, patients with SIPD are usually less compliant with treatment and have a generally poorer response to antipsychotic treatment.

Antipsychotic treatment is indicated to manage acute psychotic symptoms and agitation and may be discontinued after symptoms remission and with careful observation.

However, for individuals with persisting symptoms or who are not able to establish periods of abstinence, antipsychotic medications should be indicated as long-term treatment to prevent relapses. Moreover, pharmacotherapy should be accompanied by psychoeducational and psychosocial interventions targeting the substance use disorder (i.e., CBT). Some studies have suggested that treatment with atypical antipsychotic medications is also associated with decreased craving and substance use.

4.2.4 Brief Psychotic Disorder

4.2.4.1 Definition

Brief psychotic disorder (BPD) is an acute and transient psychotic syndrome that lasts at least 1 day but less than 1 month and is followed by complete remission, with possible future relapses. Severe psychological stressors may trigger BPD.

Over time, such clinical syndromes have been renamed as "bouffée délirante," "cycloid psychosis," "reactive psychosis," "emotional psychoses," or "atypical psychosis," in an attempt to differentiate them from other psychotic disorders.

4.2.4.2 Epidemiology

Reliable data on its incidence and prevalence are not available, but the disorder is considered uncommon, accounting for 4–10% of first-episode psychosis.

The average age of onset is the mid 20–30s, but onset can occur across all the lifespan.

BPD is more common in women (2:1), especially when related to a stressor. Moreover, it is more frequent in individuals with personality disorders, most commonly histrionic, narcissistic, paranoid, schizotypal, and borderline.

Studies reported a higher incidence of brief psychotic disorder in developing countries and in populations known to be under high stress, such as refugees, immigrants, earthquake victims, etc.

4.2.4.3 Clinical Presentation

According to DSM-5, diagnosis of BPD requires the presence of at one least positive psychotic symptom (delusions, hallucinations, or disorganized speech). Other than psychotic symptoms, there may be other symptoms such as labile mood, confusion, and impaired attention and memory, which are usually more common at the onset of brief psychotic disorder than at the onset of other chronic psychotic disorders. Occasionally, depressive symptoms may follow the resolution of the psychotic symptoms. BPD is also associated with an increased risk of suicidal behavior.

A BPD may occur in response to a traumatic or stressful life event, such as delivery, bereavement, as well as to large-scale tragedies environmental disaster, war, or pandemic, but also personal events with peculiar emotional impact for the subject.

Regardless of the objective entity of the traumatic event, the stressor usually has major significance and strong emotional impact for that subject.

Although during an acute episode the level of functioning results significantly impaired, the prognosis is generally very good with complete remission of symptoms and return to premorbid functioning, unlike other psychotic disorders. Some individuals, however, may experience relapses in their lifetime, especially in the setting of a stressful psychosocial milieu.

Furthermore, it is to note that patients with BPD have low diagnostic stability and a high transition rate (around 50%) to long-lasting psychiatric disorders, mainly schizophrenia or mood disorders.

Some positive prognostic indicators for the brief psychotic disorder are sudden disease onset, presence of stressful triggers, short duration of symptoms, good premorbid adjustment, no familiar history for psychosis.

Box 4.12: DSM-5 Diagnostic Criteria of Brief Psychotic Disorder

1. Presence of one (or more) of the following symptoms. At least one of these must be A, B, or C:
   (A) Delusions.
   (B) Hallucinations.ssss
   (C) Disorganized speech (e.g., derailment or incoherence).
   (D) Grossly disorganized or catatonic behavior.
2. The duration of symptoms is at least 1 day but less than 1 month.
3. The disturbance is not better explained by another psychotic or mood disorder, and it is not induced by substances or medical conditions.

4.2.4.4 Differential Diagnosis

It is essential to consider other possible conditions before determining a final diagnosis of brief psychotic disorder. Differential diagnoses include any disorder manifesting with psychotic symptoms: mood disorders with psychotic symptoms, schizophrenia-spectrum disorders, delusional disorder, substance-induced psychotic disorders, and psychotic symptoms secondary to medical conditions (i.e., delirium). Also, personality disorders should be taken into account, especially those included in cluster A (schizoid, schizotypal, and paranoid), often characterized by mild positive symptoms, and the borderline personality disorder, which frequently presents dissociative symptoms that need to be distinguished from actual hallucinations.

A diagnosis of brief psychotic disorder can only be made after the symptoms have remitted, as the symptomatology may otherwise be an early manifestation of another disorder with a psychotic component.

4.2.4.5 Treatment

Due to the limited evidence on the treatment of brief psychotic disorder, current treatment recommendations are based on antipsychotic pharmacological interventions known to be effective in other psychotic disorders. Indeed, the treatment of brief psychotic disorder is similar to the treatment of an acute exacerbation of schizophrenia and consists of the use of two main classes of drugs, which are antipsychotics and benzodiazepines. Although BPD usually shows a complete resolution of symptoms within 1 month after onset, maintenance treatment with antipsychotics for some months after symptoms remission is indicated.

Further Reading

APA. Diagnostic and statistical manual of mental disorders. 5th ed. Washington, DC: American Psychiatric Association; 2013.

Buonocore M, Spangaro M, Bechi M, Baraldi MA, Cocchi F, Guglielmino C, Bianchi L, Mastromatteo A, Bosia M, Cavallaro R. Integrated cognitive remediation and standard rehabilitation therapy in patients of schizophrenia: persistence after 5 years. Schizophr Res. 2018;192:335–9.

Fatemi SH, Folsom TD. Schizophrenia. In: Fatemi S, Clayton P, editors. The medical basis of psychiatry. Berlin: Springer; 2016.

Fett AK, Viechtbauer W, Dominguez MD, Penn DL, van Os J, Krabbendam L. The relationship between neurocognition and social cognition with functional outcomes in schizophrenia: a meta-analysis. Neurosci Biobehav Rev. 2011;35(3):573–88.

Galderisi S, Mucci A, Buchanan RW, Arango C. Negative symptoms of schizophrenia: new developments and unanswered research questions. Lancet Psychiatry. 2018;5(8):664–77.

Green MF, Horan WP, Lee J. Nonsocial and social cognition in schizophrenia: current evidence and future directions. World Psychiatry. 2019;18(2):146–61.

Howes OD, McCutcheon R, Owen MJ, Murray RM. The role of genes, stress, and dopamine in the development of schizophrenia. Biol Psychiatry. 2017;81(1):9–20.

Jongsma HE, Gayer-Anderson C, Lasalvia A, Quattrone D, Mule A, Szoke A, et al. Treated incidence of psychotic disorders in the multinational EU-GEI study. JAMA Psychiatry. 2018;75(1):36–46.

Kahn RS, Keefe RS. Schizophrenia is a cognitive illness: time for a change in focus. JAMA Psychiatry. 2013;70(10):1107–12.

Keepers GA, Fochtmann LJ, Anzia JM, Benjamin S, Lyness JM, Mojtabai R, Servis M, Walaszek A, Buckley P, Lenzenweger MF, Young AS, Degenhardt A, Hong SH. The American Psychiatric Association practice guideline for the treatment of patients with schizophrenia. Am J Psychiatry. 2020;177(9):868–72.

Keshavan MS, Tandon R, Boutros NN, Nasrallah HA. Schizophrenia, "just the facts": what we know in 2008 Part 3: neurobiology. Schizophr Res. 2008;106(2–3):89–107.

Kraguljac NV, McDonald WM, Widge AS, Rodriguez CI, Tohen M, Nemeroff CB. Neuroimaging biomarkers in schizophrenia. Am J Psychiatry. 2021;178(6):509–21.

Lieberman JA, First MB. Psychotic disorders. N Engl J Med. 2018;379(3):270–80.

Marder SR, Cannon TD. Schizophrenia. N Engl J Med. 2019;381(18):1753–61.

McCutcheon RA, Reis Marques T, Howes OD. Schizophrenia—an overview. JAMA Psychiatry. 2020;77(2):201–10.

Mitchell AJ, Vancampfort D, Sweers K, van Winkel R, Yu W, De Hert M. Prevalence of metabolic syndrome and metabolic abnormalities in schizophrenia and related disorders—a systematic review and meta-analysis. Schizophr Bull. 2013;39(2):306–18.

Morin L, Franck N. Rehabilitation interventions to promote recovery from schizophrenia: a systematic review. Front Psychiatry. 2017;12(8):100.

Murray RM, Englund A, Abi-Dargham A, Lewis DA, Di Forti M, Davies C, Sherif M, McGuire P, D'Souza DC. Cannabis-associated psychosis: neural substrate and clinical impact. Neuropharmacology. 2017;15(124):89–104.

Murrie B, Lappin J, Large M, Sara G. Transition of substance-induced, brief, and atypical psychosis to schizophrenia: a systematic review and meta-analysis. Schizophr Bull. 2020;46(3):505–16.

Potkin SG, Kane JM, Correll CU, Lindenmayer JP, Agid O, Marder SR, Olfson M, Howes OD. The neurobiology of treatment-resistant schizophrenia: paths to antipsychotic resistance and a roadmap for future research. NPJ Schizophr. 2020;6(1):1. https://doi.org/10.1038/s41537-019-0090-z.

Reinhardt M, Ghoneim D, Huggins B, Mani A, Murante T, Cohen C. Schizophrenia spectrum and other psychotic disorders. In: Tampi R, Tampi D, Boyle L, editors. Psychiatric disorders late in life. Berlin: Springer; 2018.

Sadock BJ, Ruiz P, Sadock VA. Kaplan and Sadock's comprehensive textbook of psychiatry. 10th ed. Philadelphia: Lippincott Williams and Wilkins; 2017.

Schizophrenia Working Group of the Psychiatric Genomics Consortium. Biological insights from 108 schizophrenia-associated genetic loci. Nature. 2014;511(7510):421–7.

Taylor DM, Barnes TRE, Young AH. Prescribing guidelines in psychiatry. 13th ed. New York: Wiley Blackwell; 2018.

Vita A, Barlati S. Recovery from schizophrenia: is it possible? Curr Opin Psychiatry. 2018;31(3):246–2.

Anxiety-Related Disorders

5

M. Locatelli, I. Vanelli, L. Fregna, F. Martini, E. Manfredi,
F. Pacchioni, G. Bigai, L. Tonet, C. Passani, and C. Colombo

5.1 Anxiety Disorders

5.1.1 Definition

Commonly speaking, anxiety is perceived as a negative feeling or an obstacle to the person's overall functioning. In origin, anxiety is an innate process of the organism, selected by evolution and present in all animals, whose function is to trigger reactive and adaptive responses aimed at survival and self-preservation.

The concept of physiological anxiety comes close to the concept of fear that we have in common with other animals and is the emotional response to an imminent threat, while anxiety is a typically human emotion, connected to the sense of time, which refers to the anticipation of a future threat that is not always precisely identifiable. The two states can of course coexist but they differ because while the first is more oriented to a defence reaction through the fight or flight, that is to move away from danger (attack–flight reaction), anxiety has the function of predisposing the

M. Locatelli · I. Vanelli · L. Fregna (✉) · F. Martini (✉)
IRCCS San Raffaele Scientific Institute, Milan, Italy
e-mail: locatelli.marco@hsr.it; vanelli.irenelucia@hsr.it; fregna.lorenzo@hsr.it; francesca.martini1@hsr.it

E. Manfredi · F. Pacchioni · G. Bigai · L. Tonet · C. Passani
Vita-Salute San Raffaele University, Milan, Italy
e-mail: manfredi.elena@hsr.it; pacchioni.federico@hsr.it; bigai.giorgia@hsr.it; tonet.leonardo@hsr.it; passani.carolina@hsr.it

C. Colombo
IRCCS San Raffaele Scientific Institute, Milan, Italy

Vita-Salute San Raffaele University, Milan, Italy
e-mail: colombo.cristina@hsr.it

R. Cavallaro, C. Colombo (eds.), *Fundamentals of Psychiatry for Health Care Professionals*, https://doi.org/10.1007/978-3-031-07715-9_5

subject to a behavioural response. The imminent, perceived or real danger, through a state of vigilance or alert that leads to prudent or avoidance behaviour.

We therefore speak of "physiological anxiety" for a state of psychic alert in which, in front of a stimulus perceived as threatening or dangerous, the organism puts in place a series of mechanisms aimed at improving performance and an adaptive response to the threat. This alarm reaction allows the organism to recruit all the resources to defend its integrity. The result is an increase of vigilance, attention, and ability to react against a perceived danger.

This process is innate and shared by all individuals, although it varies in its expression and intensity based on the subjective component—that is the personal background and experience—and determines how a given stimulus is perceived. From this also derives the importance of environmental factors and life events in determining the threshold for activation of the alert state and therefore the extent of the anxious response. In addition to inter-individual differences, there can also be intra-individual differences: the anxious reaction to a given circumstance can vary throughout the life of a subject based on previous experiences, outcomes, and feedback. Physiological anxiety should therefore be considered advantageous, as it is functional and adaptive to the integrity of the individual.

On the contrary, "pathological anxiety" occurs when the reaction to an anxiogenic stimulus is dysfunctional or disadvantageous, for example it compromises the individual's ability to react, up to a condition of inhibition on a behavioural and mental level which can result in clumsy movement and loss of lucidity, instead of an adequate psychophysical reaction, resulting therefore in maladaptive behaviour.

Anxiety can also be defined as "pathological" in relation to the stimulus: a person can have an anxious reaction in the absence of a real stimulus or an anxious reaction that is disproportionately excessive or prolonged for the stimulus itself.

In psychiatry, the term "anxiety" refers to a pathological condition that is connoted as an emotional state with a markedly unpleasant content and associated with a condition of disproportionate alarm: it is characterized by a sense of apprehension, uncertainty, fear and alertness, with anticipation of negative events towards which the person feels hopeless and helpless. It is often linked to hypothetical dangers, which are anticipated at the level of thought and feelings, imagined, and projected into the future. In pathological anxiety, the subjective component is more important than the extent of the dangerous event itself.

Characteristics of anxiety are that it is anachronistic (erroneous anticipation of the damage: a threat is expected in the future but it is perceived as already present now and catastrophically), phantasmatic (representation of imagined dangers that are not really present), and stereotyped (it is often repetitive because it is linked to the patient's cognitive distortion).

Lastly, anguish is an anxious mood state of even greater intensity, with a marked sense of oppression, restlessness, and the perception of despair and even physical pain or tightness.

The figure shown here represents the Yerkes–Dodson curve, which relates the intensity of anxiety with the efficiency of the performance: it allows for depicting the continuum that exists between physiological (adaptive) and pathological (maladaptive) anxiety. A normal degree of anxiety is required for a satisfactory performance: in the first phase, as the level of anxiety increases, the individual's performance improves up to an optimal level. If the level of anxiety continues to increase, the physical and cognitive performances decrease and become dysfunctional up to a total inability to react.

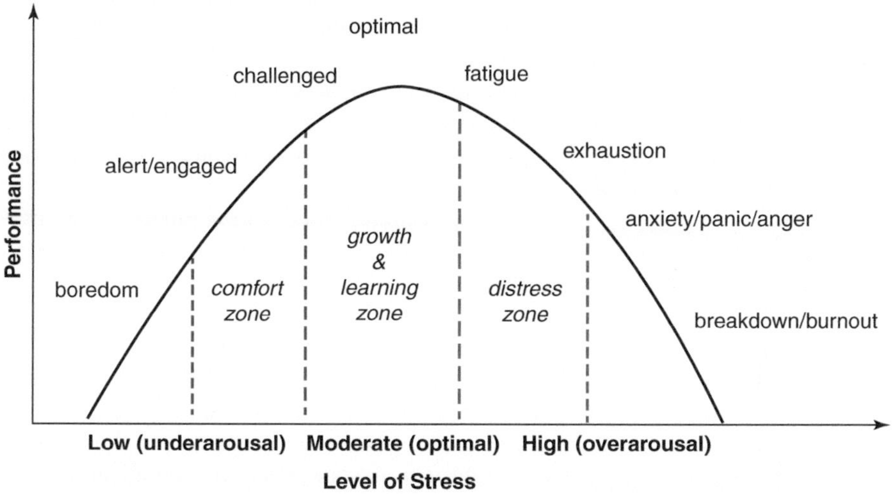

By the term "adaptive anxiety", we indicate those anxiety reactions that lead to the improvement of performance; planning and adaptation to the environment and that require an adequate level of vigilance. The optimal level of the Yerkes–Dodson curve is that corresponding to a state of activation in which psychic and somatic systems prepare the individual for the most appropriate reaction for that specific situation—that can be altogether referred to the "flight or fight" response.

An important characteristic of adaptive anxiety concerns its duration, which must be limited in time and end with the suspension of the stimulus.

If the level of anxiety continues for prolonged times, this results in a consumption of resources until they are totally exhausted and one would slip into a state of maladaptive anxiety.

"Maladaptive anxiety" is a reaction that is disproportionate in terms of intensity and/or duration with respect to the inducing stimulus. This involves a progressive consumption of resources, which is followed by a decrease in performance, as shown in the right part of the Yerkes–Dodson curve. Anxiety is maladaptive even when the alarm state is activated in response to a stimulus that is not actually dangerous or in the absence of stimulus at all.

5.1.2 Aetiopathogenesis

5.1.2.1 Physiological Mechanisms of Stress Response

A sensory system (e.g. sight, hearing) perceives the danger and transmits the message through the ascending reticular formation to the thalamus and the limbic system. The thalamus stimulates the hypothalamus and activates the endocrine system and the sympathetic nervous system, initiating the body's response to stress. The thalamus performs a primary relay function between the exteroceptive sensory systems and the primary sensory areas of the cerebral cortex, which then project to the associative areas for an integrated processing of the stimulus. From the cortex, the information is then transferred to subcortical structures involved in affective, behavioural and somatic responses. The main of these subcortical areas is the amygdala, a subcortical nucleus responsible for the acquisition and expression of conditioned fear, with a wide spectrum of reciprocal connections with the cortical and limbic structures involved in the multisystemic response to stress. From these cortical and subcortical areas, signals are sent to the brainstem and hypothalamus, resulting in the autonomic, endocrine and behavioural response of fear.

The anxious response consists of a cascade of finely tuned physiological processes that result in various modifications aimed to cope with the threatening stimulus. Three phases can be distinguished.

- *First phase*: alarm. The phase of shock, of fright, when a threatening stimulus is presented, and during which energy resources are mobilized through biochemical and hormonal responses. It has a latency of a few seconds and prepares the body to deal with the dangerous situation. The organism perceives, consciously or unconsciously, a stress factor (physical, psychological or biological), that is something unexpected, new or unusual, able to represent a difficulty or a potential danger and therefore implements a cascade of lightning-fast responses to try to survive the threat. Whatever the cause, the biochemical process of the stress reaction is the same. The hypothalamus, which, in addition to other brain areas, is closely connected with the endocrine and immune system, causes numerous chemical and electrical changes in the body. In particular, this activation causes the secretion of specific hormones such as cortisol and, through a direct brain–adrenal gland connection of the orthosympathetic nervous system, adrenaline and noradrenaline, which will recruit various organs important for the response; production of beta-endorphins, the body's own innate painkillers that raise the pain threshold. The effect is an increase in metabolism. Blood from peripheral areas (peripheral vasoconstriction and coagulation facilitation, important in case of wounds) and secondary organs are directed to the most necessary ones (heart with increased range, lungs with tachypnoea, muscles whose tone increases) to maximize efficiency. The skin becomes pale, sweaty and cold. Digestive function tends to stop, causing nausea. Blood circulation also decreases in areas of the brain that specialize in processing information and solving problems: therefore, anxiety increases and concentration decreases.
- *Second phase*: resistance or adaptation. The duration of each stress reaction depends above all on this phase, which lasts as long as special promptness and capacity for action is required. It is the phase in which one adapts to the new circumstances, the body reacts to the threatening stimulus by using its resources to cope with the danger as long as it persists. In this phase, the hypothalamus–pituitary–adrenal axis plays a fundamental role. The fundamental event is the overproduction of cortisol, which results in the downregulation of the immune system

(reduction in the absolute number of NK cells, alteration of the CD4/CD8 ratio, increase in circulating neutrophils, reduced proliferation of lymphocytes and phagocytosis of neutrophils). There is an increase in the expression of IL-1, IL-6, TNF-α; reduction of IL-2, INF-γ, and MHC-II; reduced activity of NK cells): this is not worrying in the short term but becomes a problem in case of chronic stress.
- *Third stage*: exhaustion. Once the anxious stimulus stops, the physiological mechanisms gradually die out. It is fundamental as a prolonged stressful condition creates the situation of pathological maladaptive anxiety, which is associated with the consumption of energy resources (catabolic state) of the body. It starts in a few minutes and can last a long time. When the danger is overcome or when the energy is totally consumed, the final phase of the stress response begins, which aims to ensure the body the necessary rest period. The exhaustion phase is felt like a significant drop in energy associated with relief or pleasant numbness. Biochemically, the exhaustion phase begins with a rapid decrease in adrenal hormones (adrenaline, noradrenaline, and cortisol) and energy reserves. The parasympathetic system takes over the control of the vegetative system: normal blood flow in the digestive system, brain and skin is restored.

Neuroimaging studies have shown that anxiety disorders have been related to both hyperactivity of the amygdala and decreased activity of the hippocampus, a condition that leads to excessive activation of the hypothalamus–pituitary–adrenal system (HPA).

5.1.2.2 Chemical Basis of Anxiety

The neurotransmitters mainly involved in anxiety disorders are serotonin and noradrenaline; in particular, studies have demonstrated the following.

- An increase in serotonin turnover at the level of the prefrontal cortex, amygdala and lateral hippocampus;
- Central hyperactivity of noradrenaline, with downregulation of postsynaptic α-2 adrenergic receptors.

Norepinephrine dysregulation is associated with the onset of panic attacks, in Klein's theory of the "false suffocation alarm". Around the 1990s, it was seen that inhaling CO_2 in subjects predisposed to anxiety (increasing the partial pressure of CO_2 as it occurs with tachypnoea, a frequent phenomenon in anxious subjects) led to hyperactivation of respiratory chemoreceptors, with metabolic alkalosis, direct stimulation of the locus coeruleus and hypoxia. Subjects have hypersensitivity to these suffocation signals such that, following a false alarm signal, they develop a panic attack.

Other neurotransmitters involved are as follows

- GABA: the role of the GABAergic system is highlighted by the action of BDZ in anxiety. The system is widely distributed in the hippocampus, amygdala, frontal and occipital cortex.

- Dopamine (DA): subjects with high levels of trait anxiety, particularly predisposed to developing panic attacks, seem to show an increase in the central activity of the dopaminergic pathways (in particular, frontal cortex).
- Some neuropeptides also participate in the anxiety response, with collateral functions: cholecystokinin (CCK), corticotropin-releasing factor (CRR), and tachykinins as substance P, neurokinin A and B are abundantly expressed in the circuits connected to anxiety.

5.1.3 Clinical Presentation

Anxiety disorders are the most common mental disorders, with a prevalence greater than 28%; they often develop in adolescence or early adulthood and are twice as frequent in women. They can often be accompanied by depressive symptoms, particularly when the anxiety problem continues over time, and substance abuse/addiction (substances, medications, and alcohol).

A characteristic of anxiety is the presence of fluctuating levels of intensity and duration. It has a tendency to pervasiveness, markedly conditioning the life of those affected. It is perceived as a threat to one's integrity that generates a condition of apprehension and expectation of damage, which is associated with the cognitive distortion of pejorative amplification of reality and therefore a feeling of helplessness and despair.

A symptom frequently encountered in patients is insomnia, with difficulty falling asleep and with a sleep fragmented by numerous awakenings.

As anticipated above, there are many systems and organs involved in the anxious response. Their involvement explains the somatic equivalents that accompany psychic anxiety and that are frequently found in anxiety disorders.

- Cardiovascular system
  - Increased heart rate and output
  - Increased blood volume and blood pressure
  - Peripheral vasoconstriction
  - Coronary dilation
  - Positive cardiac inotropic effect
  - Increase in blood levels of glucose, free fatty acids, cholesterol
- Respiratory system
  - Increased oxygen exchange
  - Tachypnoea
- Digestive system
  - Xerostomia
  - Contractions of the oesophagus
  - Increase in gastric acid secretion
  - Changes in peristalsis
- Metabolic effects of catecholamines
  - Activation of glycogenolysis and lipolysis
  - Increase in free fatty acids, glucose, and lactate

- Muscles, skin, immune system
  - Increased muscle tone
  - Increased smooth muscle contractions
  - Increased perspiration
  - Decreased skin temperature
  - Reduction of immune activity

5.1.4 Treatment

The therapeutic approach of anxiety disorders involves the integration of psycho-educational interventions, psychotherapeutic interventions, and somatic therapy (mainly pharmacological).

Psycho-educational interventions consist of explaining the mechanisms underlying the anxious response to the patient; for example, it is normal to have a fast heartbeat, shallow breathing, and tightness in the chest as the body prepares for an attack and flight reaction.

- Psychotherapy aims at a cognitive restructuring of the patient's way of facing the perceived dangerous situation; the most used are cognitive-behavioural therapy, desensitization with imaginative and in vivo exposure, autogenic training, social skills training, various relaxation techniques, and biofeedback.
- Somatic pharmacological therapy is similar for all anxiety disorders and uses drugs with a purely symptomatic purpose, with a rapid but short-term effect and drugs with a delayed but long-term effect. Among the somatic non-pharmacological therapies, the most recent is transcranial magnetic stimulation (TMS), which can be used as an enhancement of drug therapy.

5.1.4.1 Pharmacotherapy

The degree of discomfort of the patient is often very high and requires a pharmacological intervention that is often effective to reduce the intensity of anxiety while taking into account possible side effects. In addition to reducing anxiety levels, one of the first objectives to be achieved with drug therapy is the restoration of good sleep, often compromised in anxiety disorders.

The ideal is personalized therapy, which is a therapy tailored as much as possible to the patient's profile and needs. Before setting up a drug therapy, an accurate mental state examination is required, accompanied by a thorough medical history interview with particular attention to the psychopharmacological history. A careful medical history must also be conducted, concerning the presence of possible organic pathologies that may affect the choice of the drug. Another principle is to give a therapy that is as simple as possible, with the least possible number of drugs at the minimum effective dosage.

The choice of the drug, with the expected results, the latency times to be able to observe the first benefits and their possible side effects must be shared with the patient. Sometimes, and only with the patient's permission, it may be appropriate to provide adequate information also to family members accompanying the patient to the visit regarding drug therapy, in order to optimize adherence to treatment by the patient who in such way feels supported by family members.

During the entire duration of the drug treatment, it is important to monitor the occurrence of possible or probable side effects and any interactions with other concomitant therapies, especially in elderly patients.

Pharmacological therapy used are as follows

- Antidepressants
  - Selective serotonin reuptake inhibitors (SSRIs)
  - Serotonin/norepinephrine reuptake inhibitors (SNRIs)
  - Tricyclics antidepressants (TCAs)
  - Atypical antidepressants
- Benzodiazepines
- Beta-blockers
- Anticonvulsants

The class of drugs of first choice is that of antidepressants and in particular selective serotonin reuptake inhibitors (SSRIs). Benzodiazepines (or other symptomatic drugs such as beta-blockers and some antiepileptic drugs with anxiolytic action such as gabapentin and pregabalin) are often used in combination, at least in a first phase and for a limited time, especially when acute symptoms require rapid action (e.g. high levels of anxiety with its somatic correlates, panic attacks, insomnia). In any case, it will be appropriate to reduce their use, as soon as possible, since benzodiazepines give tolerance and dependence and have a consequent potential of abuse.

5.1.5 Classification

Anxiety disorders are clinical conditions characterized by pathological anxiety that can be distinguished from each other based on the type of objects or situations that determine it, which are associated with cognitive distortions and avoidance behaviours. Clinical evaluation to determine whether fear or anxiety is excessive or disproportionate must also take into account the patient's cultural context.

Anxiety disorders in the category of Anxiety Disorders according to DSM-5 are as follows

- Generalized anxiety disorder
- Panic disorder
- Specific phobia
- Social phobia
- Selective mutism
- Separation anxiety
- Anxiety induced by substances/drugs
- Anxiety due to another medical condition

- Anxiety disorder with another specification
- Anxiety disorder without any other specification

5.1.5.1 Generalized Anxiety Disorder (GAD)

General anxiety disorder (GAD) is characterized by a chronic state of apprehension, with concerns of a diffuse and multifocal nature. Patients present various physical symptoms (headache, lower back pain, gastrointestinal disorders) and they frequently seek internists and other speciality physicians in the belief that such symptoms are the expression of an organic problem. Only 33% of subjects with GAD point their distress back to a psychological origin from the very beginning and refer directly to a psychiatrist.

Epidemiology

This form of disorder is the most frequently encountered. In the USA, it has a prevalence of 5–11%, in Europe of 4.3–5.9%, with a double prevalence in females. It has a typical onset in late adolescence or early adulthood.

GAD generally has a chronic course but can have fluctuations in terms of intensity.

Comorbidities

50–90% of patients have comorbidity with other disorders, in particular social phobia, specific phobia, panic attacks, or even depression. It is common to find abusive behaviours that can be substances or drugs (anxiolytics).

Clinical Presentation

The patient has a pervasive state of frequent concern that persists and appears to be disproportionate to the extent of the threat over which the patient has no control.

The symptomatology of GAD is very varied and changeable

- Cognitive level: feeling nervous or on a tightrope, having exaggerated alarm responses, difficulty concentrating, light-headedness, inability to relax, difficulty falling asleep, irritability, apprehensive attitude, fear of being able to deal with situations,
- Somatic level: dyspnoea and a feeling of suffocation, palpitations, sweating or cold and wet hands, dry mouth, dizziness or a feeling of dizziness, nausea, diarrhoea or other abdominal discomforts, hot flushes, or chills, urinary discomfort, dysphagia or "knot in the throat", tremors, muscle twitching, muscle tension or pain, easy fatigue.

DSM-5 Diagnostic Criteria

1. The patient reports an excessive state of anxiety and apprehensive anticipation when confronted with different events and activities. These symptoms occur on most days for a total of at least 6 months.
2. Difficulty to control the worries.

3. The worries must be associated with at least three of these psychophysical equivalents of anxiety:
   (a) Sense of tension and tendency to feel restless
   (b) Fatigability
   (c) Reduced ability to maintain concentration for a long time
   (d) Irritableness
   (e) Physical tension
   (f) Insomnia

   In children, only one of the above symptoms is necessary.
4. This clinical condition leads to a reduction in overall functioning.
5. Symptoms must not be due to the use of drugs, medicines or organic pathologies.
6. Symptoms are not better explained by another mental illness.

Differential Diagnosis

It must always be borne in mind that a clinical picture with anxiety symptoms and physical symptoms may in the first place depend on an underlying organic pathology, which must, therefore, always be considered. In particular, the following organic and other psychiatric pathologies must be considered in the differential diagnosis.

- Hyperthyroidism
- Paroxysmal supraventricular tachycardia
- Pheochromocytoma
- Hypoglycaemic crisis
- Complex partial seizure
- Mitral prolapse
- Respiratory pathologies
- Dizzy syndromes
- Anxiety disorder due to another medical condition
- Induced anxiety disorder
- Depression
- Phobias
- Conversion disorder

Treatment

GAD therapy includes antidepressants, in particular SSRIs as drugs of first choice. Antidepressant therapy generally lasts 6–12 months; however, there is often the need to continue treatment in the long term given that GAD is chronic in nature. There is a relapse in 25% of patients 1 month after stopping treatment and in 60–80% within 1 year after stopping drug therapy.

For symptomatic purposes, especially in the early periods of treatment, it may be necessary to combine therapy with anxiolytics such as benzodiazepines. Mid-life benzodiazepines are preferred, for example, delorazepam, in divided doses (usually twice a day) to limit both the side effects of long half-life benzodiazepines (somnolence) and the plasma peak effect (which produces rapid but transient effect).

5.1.5.2 Panic Disorder

A panic attack is a massive and acute crisis that arises quickly and lasts a few minutes. It is an episode of intense alertness, anguish, fear or discomfort during which a variable number of physical and cognitive symptoms can occur suddenly and reach a peak within a few minutes. Panic attacks can be unexpected when they appear suddenly and for no apparent reason, even at night thus awakening the patient, or situational, when they occur in correlation to situations that the patient particularly fears and in which he experiences growing fear of being sick until the end of the panic attack. About half of patients have both types of panic attacks. Situational or expected ones are characterized by the presence of anticipatory anxiety concerning being able to find oneself in a certain situation or place: anticipatory anxiety is the "fear of fear", a state in which negativity, fear and worries related to a specific event are anticipated with the thought and the person already experiences them as if it were present at the moment.

A single panic attack is not enough to make a diagnosis of panic disorder. Panic Attack Disorder (PAD) is when panic attack episodes recur over time, causing fear of following panic attacks and interference with normal functioning.

Epidemiology

10–15% of the general population experiences an unexpected panic attack at some point in their life. Of these, 3–4% develop a PAD with clinical relevance.

PAD has a prevalence of 1–4% in the general population, with the ratio F:M = 3:1. The typical onset is in young adulthood (20–30 years).

The symptoms are varied and are neurological (44%), cardiac (39%) and gastrointestinal (33%). The trend of symptoms is varied as well as the frequency of attacks, which can be daily or interspersed with weeks or months. The trend is therefore fluctuating with phases of remission alternating with phases of recrudescence in periods of particular stress.

Pathogenic Hypotheses

- Genetic predisposition: a higher prevalence of PAD is observed in first degree relatives (7–21%) compared to the general population; concordance between monozygotic twins (22–73%) is greater than that between dizygotic twins (0–11%).
- Respiratory hypothesis: Klein's false alarm theory of suffocation (noradrenergic and locus coeruleus hyperactivity in response to false hypoxia alarm, which determines the sensation of suffocation).
- Dysregulation of the brainstem and the cardiac–respiratory–vestibular system.

Clinical Presentation

Criterion B in DSM-5 (at least one of the attacks is followed by a period of at least 1 month of persistent concern about the consequences of the attacks (loss of control, heart attack, fear of going crazy) and behavioural changes (need for companionship in certain situations, avoidance, for example for embarrassment or fear of being negatively judged by others precisely because of the symptoms of panic evident on the outside) illustrates the two key factors of the so-called panic loop. The panic loop is the conditioning process of the person concerning the association of situations/events as triggers the panic attack: by associating a certain situation with the panic attack, the patient begins to feel discomfort and fear for that situation and will decide to avoid it, or he will change his lifestyle and habits, for fear of having another panic attack.

DSM-5 Diagnostic Criteria

1. The panic disorder is characterized by recurrent and unexpected panic attacks. A panic attack is a sudden episode characterized by a cognitive and physical set of symptoms that cause in the subject an intensive fear and discomfort. Particularly, four (or more) of the following symptoms must occur:
   (a) Tachycardia.
   (b) Diaphoresis.
   (c) Trembling or shaking.
   (d) Dyspnoea.
   (e) Feelings of choking.
   (f) Chest tightness.
   (g) Nausea.
   (h) Dizziness.
   (i) Altered sensation of body temperature.
   (j) Paraesthesia.
   (k) Dissociation.
   (l) Fear of going out of control.
   (m) Worry of dying.
2. A complete panic attack is defined when at least 4 of these symptoms are present.
3. At least one of the attacks has been followed by 1 month (or more) of one or both of the following.
   (a) Constant worry that the panic attack will recur.
   (b) A significant maladaptive change in attack-related behaviour with an important conditioning in the subject's life.
4. Symptoms are not caused by a substance abuse or an organic clinical condition.
   Symptoms are not better explained by another mental illness.

Complications of the panic loop are as follows

- Agoraphobia: development of avoidance behaviours for places or situations from which it would then be difficult to move away or in which they may not receive help in the event of a panic attack. In these cases the patient tends to identify the phobic partner in a family member or a person of reference, that is, the one who must be present to face the feared situations and who, being aware of his problem, can intervene and provide assistance if necessary, resulting in a significant limitation of the patient's autonomy and freedom of movement. Up to 50% of individuals with PAD develop agoraphobia.

- Abuse of substances with an anxiolytic effect (alcohol, benzodiazepines, cannabinoids): they are used as self-medication for anticipatory anxiety.
- Hypochondria: patients with panic attacks often go repeatedly to the emergency room fearing they have a serious physical illness.
- Secondary depressive syndrome.

Differential Diagnosis

- Hyper/hypothyroidism.
- Hypoglycaemic crisis.
- Pheochromocytoma.
- Heart attack or angina.
- Alterations of perception: déjà vu, déjà vécu, hallucinations.
- Asthmatic attack.
- Abuse of stimulants (caffeine, amphetamines, cocaine).
- Alcohol and benzodiazepine withdrawal.
- Temporal lobe epilepsy.

Treatment

Treatment is mainly based on antidepressants, of which the first choice falls on paroxetine. Therapy should be prolonged for at least 6–12 months before going to evaluate a possible reduction of drugs, after which the maintenance dosage is half the attack dose. In the case of panic disorder, there may be a resolution of the disorder, so once the patient is cured and is stable, they may no longer have panic attacks in the course of their life. This happens in 35–50% of patients, with improved response rates particularly seen in patients with agoraphobia. However, the therapy is ineffective in 20–30% of cases.

In case of acute panic attack or in the case of anticipatory anxiety, benzodiazepines with a short half-life can be prescribed for symptomatic purposes, for example, alprazolam 0.5–1 mg. Who assists the patient during a panic attack can reassure them, stimulate a diaphragmatic breathing that allows better oxygenation of the blood, and provide a paper bag to help them in ventilation.

5.1.5.3 Agoraphobia

Clinical Definition and Presentation

It is the anxiety related to being in embarrassing places or situations or from which it would be difficult to get away or in which help may not be available in the event of a panic attack.

Characteristic agoraphobic triggers are crowded or closed places (stadiums, shopping malls, cinemas, means of transport queuing in traffic, tunnels), wide-open spaces (freeways, bridges, wide streets, squares). These situations are avoided, reduced or endured with much discomfort and with the constant anxiety of having a panic attack, severely limit the patient's autonomy and quality of life.

Epidemiology

Agoraphobia is mostly found in association with a clinical history of panic attacks (with a prevalence of 1.1%); however, the presence of agoraphobia alone without panic attacks is possible (prevalence of 0.8%). It has a chronic course and a prevalence of 2–6% in the general population.

Comorbidities

Agoraphobia can also be associated with other anxiety disorders (49–64%) and depressive disorders (33–52%). In one-third of cases, there is concomitant substance abuse.

DSM-5 Diagnostic Criteria

1. Pronounced fear or anxiety for at least two of the following situations:
   (a) Using public transportation.
   (b) Being in open spaces.
   (c) Being in enclosed locations.
   (d) Being in a crowd or standing in line.
   (e) Being outside of the home alone.
2. Patients feel that escape might be difficult or that help might not be available in the event of developing panic-like symptoms.
3. The agoraphobic situations almost always trigger fear or anxiety.
4. Patients actively avoid agoraphobic situations and/or need a companion to face them.
5. The fear or anxiety is out of proportion to the actual danger posed by the agoraphobic situations and to the socio-cultural context.
6. The fear, anxiety, or avoidance typically lasts for 6 months or more.
7. The fear, anxiety, or avoidance causes clinically significant distress or impairment in social and/or occupational functioning.
8. The fear or anxiety is out of proportion to that generated by a medical disorder that may be present.
9. The fear, anxiety, or avoidance is not better explained by the symptoms of another mental disorder.

Note: A patient may be both diagnosed with panic disorder and agoraphobia at the same time.

Treatment

In case of acute anxiety symptoms or concomitant panic attack, benzodiazepines with a short or short-medium half-life can be used, for example, alprazolam, clonazepam, lorazepam.

For long-term treatment, antidepressants are used. Although SSRIs are considered the first choice drugs in PAD and agoraphobia, TCAs (in particular clomipramine and imipramine) are even more effective, but they are used as a second option for they have greater side effects. For both categories, what has been said for anxiety disorders in general applies: start with a low dose checking for tolerability and then gradually increase to the effective dose in order to minimize paradoxical effects. In addition to drug therapy, psychotherapy can be set up, whose orientation generally has a cognitive-behavioural approach.

5.1.5.4 Specific (Simple) Phobia

In psychiatry, "phobia" means that a normal stimulus, even harmless in itself, is experienced as a danger capable of triggering an exaggerated somatic, psychic and behavioural reaction in some individual. The fear that the individual feels is actual and present when they are exposed to the phobic stimulus and is at the same time mixed with feelings of repulsion and disgust.

The characteristics of the phobia are as follows

- Disproportion of fear compared to the actual situation.
- Invasive nature of fear thoughts.
- Inability to dismantle them with rational arguments, awareness of the unreasonableness or exaggeration of fears.
- Intense anxious reaction induced by exposure to the phobic stimulus.
- Tendency to establish avoidance behaviours.

Epidemiology

Specific phobia has a lifetime prevalence of 7–12%, with a F:M ratio of 2:1 and with variability between countries. It typically has two peaks: the first in childhood, in most cases before the age of 12 and another in early adulthood. Phobias that arise in childhood tend to spontaneous remission; it tends to have a more prolonged course in females. For some phobias (e.g. dark, blood, thunderstorms) spontaneous remissions can occur.

Clinical Presentation

Despite being some fear understandable and shared by many people around the world, phobias become pathological when the fear leads to avoidance behaviours that negatively affect the subject's quality of life. Avoidance behaviours will be more or less disabling, in relation to the diffusion of the feared objects and situations.

The mechanism of phobias consists in identifying a stimulus, which is loaded in cognitive terms with negative values and therefore with anticipatory anxiety, then when the patient is in the specific situation anxiety reaches a peak level (similar to panic loop described above).

Some examples of phobias are as follows

- Acarophobia: insects
- Acrophobia: heights
- Agoraphobia: open spaces
- Brontophobia: thunder
- Claustrophobia: closed spaces
- Ereuthophobia: blushing

- Ochlophobia: crowd
- Pathophobia: diseases
- Rupophobia: dirty
- Thanatophobia: death
- Sitophobia: food
- Xenophobia: foreigners
- Zoophobia: animals

DSM-5 Diagnostic Criteria

A. Marked fear or anxiety about a specific situation or object.
B. The phobic object or situation nearly always provokes immediate fear or anxiety.
C. Patients actively avoid phobic object or situation.
D. The fear or anxiety is out of proportion to the actual threat posed by the phobic object or situation.
E. The fear, anxiety, or avoidance typically lasts for 6 months or more.
F. The fear, anxiety, or avoidance causes clinically significant distress or impairment in social and/or occupational functioning.
G. The fear, anxiety, or avoidance is not better explained by the symptoms of another mental disorder.

Treatment

In order to control symptoms, anxiolytics can be administered in the case of acute anxiety. On some occasions, beta-blockers may be indicated too.

However, the principal therapy is not pharmacological: exposures according to the principles of cognitive-behavioural therapy with the cognitive restructuring of the meanings pathologically attributed to the phobic stimulus and gradual exposures to progressively reduce the intensity of anxiety and gradually let the patient recover.

5.1.5.5 Social Anxiety Disorder (Social Phobia)

Social anxiety disorder: It is characterized by a marked and persistent fear of social situations of interaction or in which a service must be provided, in which the subject is exposed to possible (negative) judgment by others.

Shyness is not synonymous with social phobia: shyness consists of being too self-conscious, it is a condition limited to the present moment but once the situation is overcome, the subject does not preclude subsequent situations. It can be present in some periods of childhood and is quite common in adolescence, then in most people, shyness decreases over time.

Epidemiology

It is a condition that markedly interferes with social relationships, it has a prevalence of 3–13%. It has an early onset, between 15 and 30 years. The incidence is higher in women than in men.

Risk factors seem to be the female gender, positive family history for social phobia, shyness or a tendency to behavioural inhibition in early childhood.

Comorbidities
Social phobia frequently occurs with alcohol, substance abuse, and depression.

Clinical Presentation
Individuals with social phobia feel embarrassed during social situations and worry that anxiety symptoms (e.g. flushing, sweating, and tremor) may reveal their state of discomfort to others. All this leads the patient to create impediments and obstacles in their daily life, isolating themselves from social situations, implementing avoidance strategies or delegating to others. During exposure to feared situations, there are marked neurovegetative manifestations such as palpitations, dizziness, redness, tremors, sweating, and hot flashes.

The most frequently feared social and performance situations in social phobia: conversation, public speaking, participating in small groups, eating, drinking or writing in public, talking to people who have an authority role, attending a party.

Specific subtypes of social phobia can be distinguished
- Specific social phobia: it is an anxiety limited to one or two isolated situations; it is characterized by anticipatory anxiety.
- Generalized social phobia: anxiety is extended to almost all interpersonal situations, interpersonal difficulties and behavioural inhibition are constant. The subjects implement behaviours of referral, avoidance, renunciation and withdrawal towards interpersonal relationships. It is associated with paranoid and schizotypal personality structures (Cluster A).

Complications include the following
- Social withdrawal and impairment of school or work performance, with subsequent development of feelings of inadequacy and inferiority.
- Substance abuse: alcohol (in 10–20%), anxiolytics and cannabinoids.
- Depressive syndrome: secondary demoralization, risk of suicide.

DSM-5 Diagnostic Criteria
A. Marked fear or anxiety about one or more social situations in which patients may be scrutinized by others.
B. Patients are concerned about being evaluated negatively.
C. Patients actively avoid social situations.
D. The social situations are avoided or endured with intense fear or anxiety.
E. The fear or anxiety is out of proportion to the actual danger posed by the social situation.
F. The fear, anxiety, or avoidance lasts for 6 months or more.
G. The fear, anxiety, or avoidance causes clinically significant distress or impairment in social and/or occupational functioning.
H. The physiological effects of a substance or another medical condition cannot explain the fear, anxiety, or avoidance.

I. The symptoms of another mental disorder cannot explain the fear, anxiety, or avoidance.
J. The fear or anxiety is out of proportion to that generated by a medical disorder that may be present.

Specify If
Performance only: If the fear is restricted to speaking or performing in public.

Differential diagnoses
- Agoraphobia: avoidance of situations in which it would be difficult to receive help;
- Depression: social withdrawal and loss of confidence, but there was normal premorbid social functioning;
- Body dysmorphism;
- Cluster A Personality Disorder (where there is voluntary social withdrawal with egosynthony for interpretation and self-reference).

Treatment
The best treatment of social phobia involves a multidisciplinary approach, in which both pharmacotherapy and psychotherapy are used. Symptomatic drugs used are short half-life benzodiazepines (e.g. alprazolam, clonazepam) while drugs with curative purposes are antidepressants (mainly SSRIs, venlafaxine). In case of performance anxiety, beta-blockers can be used just before the phobic situation, for example, atenolol 50–100 mg or propranolol 20–40 mg. Cognitive-behavioural psychotherapy, relaxation techniques, and exposure exercise complete the treatment.

5.1.5.6 Other Anxiety Disorders

Separation Anxiety Disorder
Separation anxiety disorder is a typical condition of the child, burdened by excessive anxiety evoked by the separation from a parental figure. This condition is characterized by excessive sensitivity to separation with an intense anxious component, which interferes in the activities of daily life and with the normal development of the child.

Selective Mutism
Selective mutism is characterized by a persistent inability of the patient, typically a child, to communicate verbally in certain social life contexts that are selectively perceived as threatening (e.g. school). Conversely, the child normally talks at home with family members, such as siblings. Sometimes patients do not even speak in the family context if there are friends or relatives from outside the close family unit (e.g. grandparents, uncles, cousins).

Substance/Medication-Induced Anxiety Disorder

Substances such as anxiolytics, cannabinoids, alcohol which can give, after abrupt suspension, an anxious condition. So it is essential to adequately investigate the patient's complete history.

5.2 Somatic Disorders

5.2.1 Introduction

The process of somatization is defined as the tendency to experience, conceptualize, or communicate psychological conditions or content through bodily sensations, functional modifications or somatic metaphors. It relates to all those situations in which the body occupies the whole space of communication and becomes the sole vehicle for transferring psychic messages. Body language not only participates and amplifies emotions but also can completely replace them.

This definition encompasses the complex and varied mechanisms that can be brought into play by patients who complain of physical complaints that cannot be explained from an organic point of view.

Before the patient can be included in one of the disorders of the somatization phenomenon the following criteria have to be fulfilled

- The presence of an underlying medical condition must be excluded.
- Even if a related medical condition exists, the intensity of the symptoms must not be proportionate to it.
- There are no demonstrable physiopathological mechanisms capable of explaining the symptoms.

5.2.2 Somatic Symptom Disorder (SSD)

5.2.2.1 Definition

Somatic Symptom Disorder is characterized by multiple and recurrent somatic complaints, lasting for years, for which medical intervention is sought; it has a chronic course and can lead to drug abuse, disability, and iatrogenic illness.

The somatic symptoms must be associated with at least one of these psychological or behavioural responses

- Excessive thoughts regarding the severity of the symptoms.
- High or persistent levels of anxiety.
- Excessive time and energy spent on concerns about health status.

5.2.2.2 Epidemiology

SSD is more frequent in the female population where it reaches a lifetime prevalence of 2%, whereas men's prevalence is less than 0.2%. There is a tendency towards familiarity (10–20% of female relatives of individuals with the disease are affected).

5.2.2.3 Diagnostic and Clinical Criteria

SSD is similar in its basic clinical features to the so-called Briquet syndrome, a homogenous form of hysteria characterized by multiple somatic symptoms and a chronic course. All patients with SSD have complicated medical histories. Any bodily organ or disorder can become a target of the somatization process (abdominal pain, menstrual problems, and headache are the most frequently occurring symptoms).

The most peculiar aspect in the SSD clinic is the dramatic, exaggerated, and emotionally charged way in which the patient presents his or her history and the discomforts related to his or her symptoms.

The patient appears vindictive and accusatory towards the health care organization that has not been able to find a solution to the suffering produced by all her physical symptoms. The patient is often a woman who has recently given up her job and declares that she is unable to carry out her domestic activities. Marital relationships are described as highly unsatisfactory and the illness makes emotional relationships even worse. Sexuality is described as lacking in gratification. The DSM-5 definition requires only one body symptom that is distressing or disruptive to daily life and lasts at least 6 months. One of the following psychological or behavioural responses is also required (Box 5.1).

- Disproportionate thoughts about the severity of symptoms.
- Persistent high level of anxiety about the symptoms.
- Too much time and energy spent worrying about health.

In DSM-IV a total of 8 physically unexplained symptoms taken from four different symptom groups was required, of which at least 4 had pain and 2 had gastrointestinal symptoms.

Box 5.1: Diagnostic Criteria for Somatic Symptom Disorder According to DSM-5

A. One or more somatic symptoms that cause deep concern.
B. At least one of the following.
   1. Exaggerated and persistent thoughts about the severity of the symptoms.
   2. Persistently high anxiety about symptoms or health.
   3. Excessive time and energy spent on these symptoms or health concerns.
C. The state of being symptomatic lasts more than 6 months, but the specific symptom may change during that time period.

5.2.2.4 Aetiology

Socio-cultural, biological, and psychodynamic factors are involved in the aetiology of SSD that interact or promote the establishment of a pathological personality profile and the development of other disorders, especially depressive and anxiety disorders. As regards the first factor, an association is recognized with the following.

- Low socio-cultural level.
- Childhood development in a family with subjects prone to somatization.
- Relationship with parents who were only capable of providing emotional care during states of illness.
- Health culture and practice that tends to privilege illness behaviour rather than the psychological expression of illness and therefore selects symptoms of a somatic nature.

From a biological point of view, SSD is interpreted as neurophysiologically determined by an abnormal lowering of the general threshold level to various stimuli. Thus, it is the experience of somatosensory amplification that determines a cognitive alteration of signal analysis.

5.2.3 Illness Anxiety Disorder

5.2.3.1 Definition

Illness anxiety disorder (IAD) takes the form of the pervasive belief and worry that one has or will have a serious illness whose origin is not discovered or excessive concern about one's health status. Somatic symptoms are not present, or if present are of mild intensity. This belief and/or fear is manifested through an unrealistic interpretation of the presumed physical symptoms.

5.2.3.2 Epidemiology

Prevalence estimates of IAD are based on those of DSM-3 and DSM-IV for the diagnosis of hypochondria. IAD is frequently observed in general practitioners' offices, where the prevalence of IAD in the patient population is about 10%. There is no difference between men and women; onset is more common during the third to fourth decade of life and in the elderly.

5.2.3.3 Clinical and Diagnostic Criteria

As in other SSD, the basic diagnostic condition is the absence of any organic pathology underlying the symptomatology that can fully justify the patient's state of mind, worries or beliefs (Box 5.2).

Hypochondriacal symptoms may accompany other psychiatric disorders (major depression, schizophrenia) or be a temporary response to conditions of existential stress. The patient with IAD complains of physical symptoms, which, although caused by some organic disorder, are often exaggerated in their consequences, but above all are interpreted as caused by a serious illness that no doctor has yet been able to diagnose.

The symptoms affect the whole body and all organs and are expressed through pain and complaints in the heart or gastrointestinal tract. Headaches suggest tumours or aneurysms, mild arrhythmias suggest an impending heart attack, asthenia suggests AIDS. The symptomatology is chronic and the patient moves to the health care environment in the belief that some investigation has been overlooked or that he will finally find the doctor who can cure him.

Box 5.2: Diagnostic Criteria for Illness Anxiety Disorder According to DSM-5

A. Concern about having or acquiring a serious illness.
B. Absence of somatic symptoms or, if present, there are only mild symptoms.
C. High anxiety about health.
D. The patient frequently monitors health status or maladaptively avoids hospital and doctor appointments.
E. Although the specific illness feared may change, illness preoccupation lasts more than 6 months.
F. Another mental disorder cannot explain the illness-related preoccupation.

5.2.3.4 Aetiology

The expression of the somatic symptom is the expression of a mental conflict, the thinking in IAD is more pervasive and pathological than in other forms of somatization. The theatre where the hypochondriac plot unfolds is the entire internal world. Freud regarded hypochondria as a disinvestment of interest and vital energy from objects to the external world. The preoccupation with diseased organs represents the equivalent of anxiety in relation to strongly aggressive and destructive internal experiences. It is as if the patient were saying, “I am afraid that someone in my body wants to harm and destroy me”.

5.2.4 Conversion Disorder

5.2.4.1 Definition

Conversion disorder corresponds to what used to be called hysterical neurosis. The core of this disorder is the patient’s use of a physical symptom-forming mechanism, which consists in transferring unacceptable drives or instincts, desires or affections to the body (conversion) via voluntary nerve pathways. In this way, the unpleasant psychic charge of the conflict is removed.

5.2.4.2 Epidemiology

Conversion disorder is on the decline compared with the nineteenth and early twentieth centuries. The prevalence among general hospital patients is high (ranging from 5 to 10%); in psychiatric wards this percentage drops considerably. Conversion disorder appears in both adolescence and early adulthood. It is more frequent in the female population, but there are two populations in which the prevalence in males is high: among victims of accidents at work and in the military. A low socio-cultural level and ethno-cultural factors also play a role.

5.2.4.3 Clinical and Diagnostic Criteria

The symptoms presented by the patient usually mimic physical neurological diseases, that is, of the sensory motor area. The diagnosis is complicated by the fact that the conversion disorder overlaps with an objective organic condition. This determines two phenomena: the first is a difficulty in differentiating the psychic quota present in the individual; the second is that often the presence of the theatrical and histrionic psychological characteristics of the patient with a conversion disorder can lead to an underestimation of possible organic aetiologies (Box 5.3).

The symptomatology may have subacute characteristics, with a chronic course, or acute and dramatic (hysterical crises) with resolution of the symptoms themselves. In the latter case, we refer above all to epileptic crises of hysterical origin or to crises that are famous, but nowadays have become very rare, such as Charcot's *arc de cercle*.

It is necessary to remember that in conversion disorder, it is not possible to find a correspondence between an anatomical distribution of the venous pathways and the symptoms complained of by the patient; in fact, the latter are the expression of the idea of the disorder that the patient has constructed fantastically or through a process of imitation.

A feature that was thought to play a diagnostic role in the past was Janette's so-called *belle indifference*: in these cases, the patient appears unconcerned about the drama and severity of the symptoms.

Box 5.3: Diagnostic Criteria for Conversion Disorder According to DSM-5

A. At least one symptom of impaired voluntary motor or sensory function.
B. Medical examination and tests rule out neurological or medical conditions.
C. Another medical or mental disorder cannot explain the symptom.
D. The symptom or deficit causes clinically significant distress or impairment in social and/or occupational functioning or warrants medical evaluation.

5.2.4.4 Aetiology

On the study of hysteria, Freud built the basis of the theory and practice of psychoanalysis and the term conversion was introduced by Freud to explain the mysterious leap from the psychic to the somatic. The organic symptom becomes the representation of different types of drives, affections or instincts that are removed and rendered unconscious, as censorial or inhibitory forces (superego) determine a conflictual situation and make them incompatible with the individual's consciousness. Through somatization (primary advantage) the subject gets rid of the conflict and therefore of the anguish produced by it and satisfies in a metaphorical way the removed aggressive, sexual drives or conflict (secondary advantage), through a physical representation.

5.2.5 Factitious Disorder

Factitious disorders are characterized by physical or psychic symptoms that are intentionally produced or simulated in order to assume the role of a sick person. The main feature of this disorder is the intentional production of behaviours or symptoms of a physical or psychic disorder. The subject may adopt subjective complaints (abdominal pain in the absence of specific symptoms or falsification of objective signs such as manipulation of a thermometer to create the illusion of fever).

Fictitious disorders must be distinguished from acts of simulation in which the subject produces the symptoms intentionally, but is seeking a secondary benefit (avoiding work obligations or avoiding being selected for stressful tasks or roles).

5.2.5.1 Epidemiology

Factitious diseases are very frequent; it has been calculated that more than 5% of contacts between doctor and patient occur for these reasons.

5.2.5.2 Clinical and Diagnostic Criteria

Individuals with a bogus disorder may seek treatment for themselves or another person after deliberately causing the injury or illness. Diagnosis requires demonstration that the individual is deliberately faking-causing signs or symptoms of illness or injury in the absence of obvious external benefit (Box 5.4). Factitious disorder has similarities to substance-related disorders, nutrition and eating disorders and impulse control disorders, as well as to other disorders related to both the persistence of disturbed behaviour and deliberate efforts to conceal such behaviour by deception. The diagnosis of factitious disorder emphasizes the objective finding of a falsification drawing symptoms of illness, without drawing inferences about the possible international underlying motivation.

Patients seem to be resistant to undergo a psychiatric examination, which does not yield encouraging results. Affected persons often involve family members, health and social workers. The clinical history invented by the patient is usually credible and plausible, although the details are often vague and inconsistent. A common aspect of the factitious disorder is that patients undergo endless examinations and even invasive and disturbing investigations, denoting a kind of self-harm.

Box 5.4: Diagnostic Criteria for Factitious Disorder (Imposed on Self) According to DSM-5

A. Simulation of physical or psychological symptoms or signs, or induction of injury or illness.
B. The patient presents himself or herself to others as sick, impaired, or hurt.
C. The deceiving behavior is clear even in the absence of external rewards.
D. Another mental disorder cannot explain the behavior.

5.2.5.3 Course and Prognosis of Somatic Symptom Disorders and Related Disorders

The disorders that appear in the group of SDS and related disorders tend to have a chronic course, whether the symptoms are long-lasting or present periods of remission of varying lengths and then reappear. In conversion disorder, crises (usually of short duration) become chronic in only 10% of cases. In general, it is possible to say that acute onset is linked to a better prognosis, whereas subtle onset is a slow and complex establishment of the various somatization symptoms that expose the patient to very serious risks.

The prognosis is linked to the patient's personality traits and the type of psychological conflict that characterizes him or her; equally important is the presence of a concomitant organic pathology. Linked to this picture of illness are affective disorders, which are often associated not only with the physical symptoms but also with the psychological symptoms of the patient.

5.2.5.4 Factitious Disorder by Proxy

A specific form of Factitious Disorder is the Factitious Disorder by proxy (Box 5.5).

The essential characteristic is "the deliberate production or simulation of physical and/or psychic signs or symptoms in another person who is in the care of the subject". Typically, the victim is a young child (usually up to 5 years old) and in 90% of cases, the perpetrator is the mother.

They appear very distant from the image of an abusive mother, on the contrary, they seem to be caring, anxious for the health of their children, very cooperative and grateful towards the doctors, which encourages the latter to investigate more and more the causes of these symptoms. The figure of the father is generally marginal; he is usually absent or mostly passive in family life.

Symptoms are usually not characteristic of known illnesses and this confuses paediatricians and other clinicians and prompts them to investigate further. It is usually a long time before clinicians start to consider the idea that the patient's illness is caused by the caregiver.

The methods used to create symptoms in victim patients are heterogeneous and often cruel.

The way in which the syndrome manifests itself varies greatly. In some cases, the caregiver may make false accusations of physical and/or sexual abuse of their child, causing the child to suffer the pain of being subjected to detailed questioning, and sometimes the caregiver may alter the results of their child's tests, for example by contaminating urine samples with poisons, herbicides or other toxic substances. Physical attacks included pinpricks on the face and body, facial injuries from tools or nails, and suffocation by pressing a hand or pillow to the face. Other equally dangerous physical attacks were voluntary undernutrition and a dirty and neglected home environment, induction of epileptic seizures or loss of consciousness.

Box 5.5: Diagnostic Criteria for Factitious Disorder by Proxy According to DSM-5

A. Intentional falsification or induction of physical or psychological symptoms or signs in another.
B. The patient presents himself/herself to others as sick, impaired or hurt.
C. The deceiving behavior is clear even in the absence of external rewards.
D. Another mental disorder cannot explain the behavior.

5.2.5.5 Treatment of Somatic Symptom Disorders and Related Disorders

The basic assumption to refer to in formulating an action strategy is that all patients with SSD and related disorders have the belief that a physical disorder is at the origin of the complained symptoms and therefore lack the ability to desomatize the removed effects and conflicts. It is for this reason that the patient appears dissatisfied when he/she is told that he/she has nothing, in fact he/she gets irritated and thinks that he/she has not been taken into consideration. This is where the psychic defences in somatization and the series of medical pathways begin to strengthen. The doctor, therefore, must:

- Establish a sincere relationship of pure understanding. In the diagnostic period, it is fundamental to build a relationship of trust both through a precise execution of the clinical investigations and through an attitude of understanding and not of disqualification.
- Collection of an existential anamnesis with the aim of placing the appearance of physical symptoms in relation to very important existential and stressful events (bereavements, painful experiences, sentimental disappointments, etc.).
- Highlighting, whenever possible, psychological symptoms or discomforts that were already present before or even concomitantly with the appearance of the physical symptoms.
- Restitution of the medical diagnosis with hypotheses of objective connections. This is the most delicate phase. It should not be communicated "you have noth-

ing" but, for example, "everything is quite well and it is difficult to explain all your complaints with the results of these tests". The doctor will respond with some examples from common experience that can show how fears, anxiety, depression and anger are always accompanied by physical experiences. If the heart beats fast during unexpected events, this is because there are nerve endings in the heart that start in the brain. The answer that will often be given will be "but I am not anxious or depressed", this is the time to move on to the next point.

- Returning hypotheses of subjective connections. Now one has to use the information gathered earlier. It is important to offer these connections to the patient as hypotheses. Therefore, it is necessary to use "you are right, he does not seem to me at this moment anxious or depressed, however it is curious that the disorders started shortly after his mother died. I think this is very important emotionally". It is important to remember not to force the patient's defences.
- Referral to psychiatric specialist. If the doctor has succeeded in gaining the patient's trust, a sentence such as, "I think it might be worthwhile to try the psychological route as well" will be readily accepted by the patient.
- Psychiatric treatment. Pharmacotherapy is necessary in the presence of co-morbidity with other psychiatric pathologies such as mood and anxiety disorders: however, it is psychotherapy that is the preferred option. Its aim is to clarify the dynamics and the psychic facts that have determined the somatization process, to build with the patient the mental equipment with which he can face conflicts and life events in a different way.

5.3 Obsessive-Compulsive and Related Disorders

Obsessive-Compulsive Disorder (OCD) has been considered as part of the Anxiety Disorders chapter up to the fourth edition of the DSM (DSM IV-TR), while from the fifth edition (DSM-5) it is placed in a separate chapter. The nosographic autonomy of this disorder depends on the fact that the anxious manifestations, even if present, would be secondary and dependent on the obsessive contents. A group of conditions related to it from a clinical, epidemiological, and sometimes aetiopathological perspective is included in the DSM chapter about obsessive-compulsive disorder.

The DSM-5 "obsessive-compulsive and related disorders" chapter currently includes the following

- Obsessive-Compulsive Disorder
- Obsessive-Compulsive and Related Disorder Due to Another Medical Condition
- Substance/Medication-Induced Obsessive-Compulsive and Related Disorder
- Other Specified Obsessive-Compulsive and Related Disorder
- Body Dysmorphic Disorder
- Hoarding Disorder
- Trichotillomania (Hair-Pulling Disorder)
- Excoriation (Skin-Picking Disorder)

5.3.1 Obsessive-Compulsive Disorder

5.3.1.1 Definition

The double etymology of the term "obsessive-compulsive" portrays the nature of the disorder immediately: obsession derives from the Latin *obsidere* which means to besiege, while "compulsive" derives from the Latin verb *compellere* or to force. The subject affected by this disorder is on one hand surrounded, pursued, by his obsessions and on the other, compelled, forced to carry out compulsions.

5.3.1.2 Epidemiology

Until the 1990s, obsessive-compulsive disorder was considered a rare disease. Instead, recent epidemiological studies have significantly revised previous estimates. Today, this disorder is in fourth place regarding psychiatric pathologies, preceded only by phobias, substance-related disorders, and major depression.

The lifetime prevalence of OCD is between 0.3 and 3.5%, with a mean prevalence of approximately 2%.

The distribution in adults is the same in males and females, while it seems to affect boys more than girls in adolescence. The average age of onset is about 20 years (66% of cases), slightly earlier in males (19) than in females (22). A second peak of onset occurs at 35 years (15% of cases). Late onset appears rarer, and any underlying organic pathologies must be carefully investigated in such cases. Obsessive thoughts are described in a high percentage (57%) of women with postpartum depression.

5.3.1.3 Aetiopathogenesis

Obsessive-compulsive disorder appears to be a multifactorial disease. Biological and psychosocial factors contribute to the pathogenesis of the disorder. Studies in different research habits support this hypothesis.

- *Genetic studies:* First-degree relatives have a 12% overall risk of developing OCD, six times greater than the general population. The concordance between non-twin siblings and dizygotic twins is approximately 50%, reaching 85% in homozygous twins.
- Neuroanatomy: brain imaging studies have shown an increase in metabolic activity in some cortical regions (orbitofrontal cortex) and subcortical regions (caudate nucleus, dorsomedial thalamus). This increase in activity appears to be reversible after drug or behavioural therapy.
- *Serotonin and Noradrenaline*: the finding of reduced levels of serotonin metabolites in the cerebrospinal fluid of patients with OCD combined with the evidence of clinical efficacy from drugs that act on the serotonin system (SSRI, SNRI, TCA), seems to indicate a role of this neurotransmitter in the pathogenesis of the disorder. The high frequency of OCD symptoms in patients with basal ganglia alterations and the antipsychotic response of some forms of OCD resistant to antidepressants seem to indicate the involvement of the neurotransmitter dopamine.

- *Behavioural factors*: from a behavioural point of view, obsessions can be considered as conditioned stimuli. A neutral stimulus is associated, by classical conditioning, with an event that produces anxiety or suffering. From that moment, the thought alone will be able to cause anxiety or discomfort until it becomes a real obsession. Always according to a behavioural model, the compulsion is established when the subject identifies an action or thought that is able to reduce an unpleasant sensation (i.e. tension, anxiety). The subject will repeat the action/thought as a strategy to avoid the unpleasant sensation and gradually, this action/thought will take on the characteristics of an indispensable ritual, a compulsion (Box 5.6).

Box 5.6 The OCD "loop"

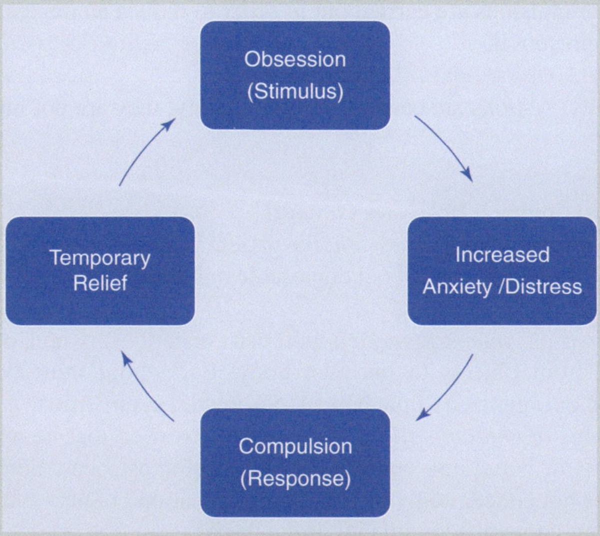

5.3.1.4 Clinical Presentation

Diagnostic criteria include persistent and recurring thoughts/images (obsessions) or repetitive behaviours (compulsions). Commonly, the patient manifests both; however, in some cases only obsessions are be present. Compulsions alone are much rarer, typical of protracted illness pictures, in which the patient automatically carries out behaviours unrelated to the content of thought (Box 5.9).

Obsession: it can be an idea, an image, a fear, or an impulse. It is a mental content; it cannot be objectified clinically. Obsessions can have different contents (Box 5.7) and can be described by some characteristics.

- Persistence/recurrence of the obsessive content and degree of impairment of the patient's functioning.
- Egodystonia: obsession produces discomfort, is experienced as intrusive and extraneous.
- Invincibility: does not depend on or respond to the will of the patient.
- Criticism/insight: the patient tends to recognize it as a product of his own mind.

Compulsion: repetition of an action or mental action that the subject carries out in response to an obsession following a rigid scheme (Box 5.8). Although sometimes they may only be mental, in most cases they are actions, behavioural manifestations that can be described by the following.

- Repetitiveness: the actions are stereotyped and repeated according to a precise pattern.
- Purpose: compulsions are carried out in order to reduce an unpleasant feeling or in order to prevent it.

Intentionality: Actions are carried out deliberately; they are not involuntary.

Box 5.7: Most Common Obsessive Contents

- *Aggressive and sexual obsession*: the patient is afraid of being able to carry out or have already carried out deplorable or harmful actions for himself or for others.
- *Obsessions of symmetry/perfection*: it can concern every area of life of the subject, from objects to his own body. Everything must be precisely arranged or organized (alphabetical, chromatic, symmetrical, etc.).
- *Obsessions of contamination*: the patient is worried that he may contract infections or come into contact with harmful substances. Sometimes the fear does not concern only physical contamination but also symbolic (fear of coming into contact with evil or the devil).
- *Obsessions related to doubt/control*: the patient fears that, in the absence of repeated checks, situations or behaviors could lead to unpleasant consequences.
- *Unrealistic, magical, and superstitious beliefs*: the patient is convinced that specific actions or thoughts can influence the outcome of events even when it is clear that there is no causal relationship between the two.

Box 5.8 Most Common Types of Compulsion

- *Control Compulsions*: repetitive and excessive control of actions, objects or situations. Typically related to doubt obsession.
- *Compulsions of washing and cleaning*: frequent and excessive washing of parts of the body, objects or environments. Related to contamination obsession.
- *Compulsions of order*: reorganization and arrangement of objects/environments according to rigid pre-established schemes. In response to obsessions of order/symmetry.
- *Counting Compulsions*: the patient carries out counts, lists, or mental operations. Usually in response to superstitious obsessions.

Box 5.9: Diagnostic criteria for obsessive-compulsive disorder according to DSM-5

1. The patient has obsessions, compulsions, or both:

   Obsessions are characterized by:

   - Thoughts, images or urges that occur repeatedly, that are experienced as intrusive and unwanted and that in most patients cause great distress or anxiety.
   - The patient tries to suppress such thoughts, images, or urges, or to neutralize them by performing a compulsion.

   Compulsions are characterized by the following

   - Mental acts (e.g. counting, praying) or repetitive behaviors (e.g. washing or cleaning, ordering, checking) that the patient feels compelled to perform in response to an obsession.
   - The mental acts or behaviors are intended to prevent or reduce anxiety.
2. The obsessions or compulsions take up a lot of time in the patient's day. In association or alternatively, they cause clinically significant distress or interfere with social and/or occupational functioning.
3. The physiological effects of a substance or another medical condition cannot explain the obsessive-compulsive symptoms.
4. Another mental disorder cannot explain the obsessive-compulsive symptoms.

Specify if:

- *With good insight*: The patient recognizes that the obsessions and compulsions are not reasonable.
- *With poor insight*: The patient believes that the obsessions and compulsions are probably reasonable.
- *With absent insight*: The patient is not self-critical about obsessions and compulsions.

Specify if:

- *Tic-related*: The patient has a positive history of a tic disorder.

Course

The onset of symptoms occurs suddenly in most patients, sometimes preceded by a stressful event (i.e. bereavement, pregnancy, brain injury). In a smaller number of patients, the onset appears more subtle, insidious, and gradual. In these cases, the patient will tend to hide the symptoms more frequently, deferring medical help requests even for several years.

The course of symptoms tends to have two main patterns: episodic (25% of cases), in which phases of well-being alternate with active phases of the disorder and chronic (75% of cases), in which the symptoms never entirely recede.

Obsessive-compulsive symptoms, especially in the chronic evolution, can have a stable or fluctuating course, with periods of partial remission alternating with a recurrence of symptoms.

Finally, in 10% of cases, the trend is progressive; this is the most severe clinical form, characterized by worsening of symptoms over time.

The impact of obsessive-compulsive disorder on the lives of those afflicted can be dramatic. Obsessions and compulsions can occupy a large part of the day, interfering with work, social, school and relational activities. In the most severe forms, the patient is wholly absorbed by the fears and rituals he puts in place to calm the resulting discomfort. In chronic states with a worsening trend, the patient can face a real cognitive impairment.

The onset of suicidal ideation is described in about half of OCD patients, while actual suicide attempts in about 25% of patients. The risk of suicide increases when depressive symptoms are associated with OCD.

Comorbidities

OCD can have several comorbidities, and these are the most frequent:

- Mood disorders: major depression and bipolar disorder, up to 60% of cases. Mood disorders can precede OCD or be a consequence.
- Anxiety disorders: in particular, panic disorder and phobias, up to 70% of cases.
- Tourette's syndrome: up to 50% of sufferers also develop OCD.

- Tic disorder: involuntary, afinalistic, intermittent movements can be present in up to 30% of patients with OCD.

5.3.1.5 Treatment

Treatment of obsessive-compulsive disorder is based on two main interventions: drug therapy and cognitive-behavioural therapy (CBT). The latter is to be preferred to the psychodynamic approach and psychoanalysis, towards which patients with this disorder seem to be refractory. It should be remembered that generally better results are obtained with the combination of drug therapy and behavioural therapy.

5.3.1.6 Pharmacological Therapy

SSRI

The first line for the treatment of OCD is usually monotherapy with antidepressant drugs belonging to the category of SSRIs. The dosage range is comparable to that for the treatment of depressive episodes; however, the beneficial effect of these drugs on obsessive-compulsive symptoms is generally delayed. It usually takes at least 4–6 weeks of treatment to get the first results, with the maximum therapeutic benefit being achieved on average after 12 weeks of treatment. When there is a response to treatment, therapy should be continued for at least 2 years, possibly reducing the dosage to approximately 50–60% of the effective attack dose.

Other Drugs

Among the non-SSRI antidepressants, clomipramine, an antidepressant drug belonging to the class of tricyclics (TCA), is the one with the greatest selectivity on serotonin re-uptake. It was the first drug approved for the treatment of OCD; however, with the advent of SSRIs with fewer side effects, it became a second choice in treating this disorder.

If drug treatment with SSRIs or clomipramine is ineffective, combination with other drug categories can be evaluated. Generally, drugs used to augment SSRI/clomipramine therapy include stabilizers (valproate, carbamazepine), second-generation antipsychotics (risperidone, olanzapine, aripiprazole), benzodiazepines, and beta-blockers (pindolol).

Venlafaxine, an antidepressant drug belonging to the SNRI category, has been shown to be effective in the treatment of OCD.

5.3.1.7 Non-Pharmacological Therapy

Cognitive Behavioural Therapy (CBT)

Among the psychological therapies, the one that has proved most effective in obsessive-compulsive disorder is cognitive-behavioural therapy. CBT has been shown to be at least as effective as drug therapy in treating OCD. The main approach of this technique is exposure and response prevention (ERP). A treatment lasting 15–20 weeks, once a week, is generally recommended. The outpatient meetings are supplemented by exercises that the patient carries out daily and independently. On the other hand, psychodynamic therapy is not recommended, while family therapy can help manage the conflict that often arises between the patient and the closest relatives due to the peculiar symptoms of the disorder.

5.3.2 Body Dysmorphic Disorder

The disorder is characterized by the patient's concern for one or more physical defects. These defects can only be imagined by the patient or, when present, are of modest entity. The patient's concern is, in both cases, excessive and can become invalidating.

5.3.2.1 Epidemiology

There is no clear epidemiological description of the disease. Patients affected by this disorder, in fact, tend to turn to various health professionals, in particular dermatologists and plastic surgeons, thus making diagnosis and epidemiological investigation difficult.

The prevalence of the disorder appears to be around 2.4% of the population, with a slight predominance of women. The number tends to increase within specific groups of patients, such as dermatological patients (9–15%) or patients who turn to cosmetic surgery (up to 16%).

5.3.2.2 Aetiology

It is not known. Comorbidity with mood disorders and OCD and the good chance of responding to antidepressant drugs suggest a role for serotonin in the disorder's pathophysiology.

The family, work, and social context seem to play a role, being the disorder more frequent in contexts in which a certain beauty stereotype is emphasized.

5.3.2.3 Clinical Presentation

Patients may complain of concerns about defects in every possible location in the body. However, several studies have identified the skin, hair and nose as the most frequent locations, which in any case can vary over time. The worry can be more or less specific and intense, ranging from "less-than-perfect" to "unsightly" (Boxes 5.10 and 5.11).

The idea of having a physical defect produces a sense of discomfort and concern such as to induce the patient to carry out repetitive behaviours to control their appearance (directly or in the mirror) or attempts to conceal the alleged defect through make-up, clothes or accessories. In some patients, avoidance behaviours may develop and can lead to real social isolation. Up to 20% of patients attempt suicide. Concerning the course, the onset is more frequently placed in adolescence while the progress is generally chronic, with periods of greater intensity of worries alternating with periods of almost complete well-being.

Box 5.10: DSM-5 Criteria for Diagnosing Body Dysmorphic Disorder

1. Concern about one or more imperfections or defects perceived in the physical appearance by the subject and that to other people are unobservable or appear minimal.
2. In response to such concerns, the subject engages in a series of repetitive physical and mental acts such as excessively looking in the mirror and comparing his or her own physical appearance with other people.
3. The worries cause impairment in social and/or occupational functioning.
4. In a patient with an eating disorder, the concern about the appearance is not better explained by preoccupations for weight or body fat.

Specify if

With muscle dysmorphia: The patient is concerned with the idea of being under-muscled.

Specify if:

With good insight: The patient recognizes that the defects in physical appearance are not true or appear mild.
With poor insight: The patient believes that defects in physical appearance are probably true.
With absent insight: The patient is convinced that defects in physical appearance are true.

Box 5.11: Muscle Dysmorphia

It is a specific form of dysmorphic disorder, which almost exclusively affects males. The patient is concerned that their body is not big and muscular enough. This worry can lead to strenuous workouts, weight lifting, or improper diets. Generally, the patient's awareness is low and to the possible consequences described for dysmorphic disorder, we can add the effects on the body of extreme training and particular nutrition.

5.3.2.4 Therapy

The most effective pharmacological approach sees antidepressant drugs belonging to the class of SSRIs and TCAs at the forefront. Anecdotally, some antipsychotics (pimozide) and monoamine oxidase inhibitors (MAOIs) have been shown to be effective. Since comorbidity with mood disorders and anxiety disorders is common, treatment should include specific therapy for these conditions as well. The timing of maintenance of the effective drug after the remission of symptoms is not known.

Surgical treatment of the real or perceived bodily anomaly is almost always doomed to failure and is accompanied by a high rate of litigation and complaints by the patient, dissatisfied with the outcome of the procedure.

5.3.3 Hoarding Disorder

Originally, this disorder was considered a category of OCD; currently it is placed in a separate category, falling within the OCD-related disorders.

In hoarding disorder the patient buys or comes into possession of objects, generally of little or no value and cannot get rid of them. This behaviour can result in a progressive accumulation of objects, up to completely occupying the patient's living and living spaces.

The fear of getting rid of such objects is motivated by the belief that they may prove useful in some way in the future or by an excessive affective attachment to them.

5.3.3.1 Epidemiology

Hoarding Disorder affects approximately 4% of the general population. The patients who reach the doctor's attention are mostly women although, according to epidemiological studies, the problem is more common in males. The disorder is ubiquitous; no predisposing cultural factors have emerged. The accumulation of objects often begins in adolescence (11–15 years) and increasingly interferes in the following years, when the subject generally becomes more autonomous in the care of living environments.

5.3.3.2 Aetiology

Factors related to temperament and personality seem to play a role in the genesis of the disorder. Temperamental characteristics of indecision or dependent personality traits are described more frequently among the accumulators. From the genetic point of view, familiarity for accumulation disorders are present in 50% of the affected population. Specific markers on the long arm of chromosomes 4, 5, and 17 have also been described, as well as polymorphisms in the COMT gene.

From a biological point of view, metabolic alterations in the occipital and posterior cingulate cortex have also been described.

5.3.3.3 Clinical Presentation

The patient is convinced that the objects he accumulates may have some function in the future and for this reason, he cannot get rid of them. In other cases, the accumulator describes an extreme emotional bond to the object that cannot be broken. In both cases, there is an overestimation of the function or of the affective/effective value of what accumulates (Box 5.12).

Objects are often accumulated in an unorganized and passive way. In fact, the patient sometimes simply avoids choosing whether to get rid of the object or not.

Anything can be accumulated: books, clothes, cards, lists, magazines, letters, and so on.

Accumulation can go to extreme levels. Patients can literally fill their homes with materials of any kind, with serious health and safety consequences for themselves and others. Often this disorder results in social or legal problems. It is frequent that patients arrive at a progressive social isolation. Other times they are forced by the owners to abandon their homes for reasons of safety or decorum. Awareness of one's problem is generally low. Sometimes completely absent. In such cases there may be a delusional ideation concerning the usefulness or necessity of the accumulated objects. In three out of four patients, an anxiety or mood disorder is present in addition to hoarding disorder. The most common are major depression and social anxiety disorder. Often, it is precisely the possible comorbidities that bring the patient to the doctor's attention.

Box 5.12: DSM-5 Criteria Box for Diagnosis of Hoarding Disorder

1. The subject experiences a persistent difficulty in parting with his or her objects, independently of their actual value.
2. Separation from objects produces distress.
3. Patient accumulates objects in active living spaces in an abnormous measure.
4. The hoarding causes impairment in social or occupational functioning.
5. Another medical condition cannot explain the hoarding.
6. The symptoms of another mental disorder cannot explain the hoarding.

Specify if:

With excessive acquisition: If difficulty discarding objects is associated with excessive gain of items.

Specify if:

With good insight: The patient thinks that hoarding-related beliefs and behaviours are problematic.
With poor insight: The patient is mainly convinced that hoarding-related beliefs and behaviours are not problematic.
With absent insight/delusional beliefs: The patient is completely convinced that hoarding-related are not problematic.

5.3.3.4 Treatment

From a pharmacological point of view, there is no standard approach. SSRIs, first-line drugs in the treatment of obsessive-compulsive disorder, instead, had very low response rates in hoarding disorder (<20%); in other cases, they had even a negative response.

The approach that has been most effective to date is based on a cognitive-behavioural model; however, the CBT protocols used for OCD have proved ineffective.

The protocols currently used are therefore specific and include three main points: the acquisition of problem solving skills, decision-making and organization, gradual exposure (imaginative or live) to stressful stimuli with gradual prevention of the response and finally restructuring cognitive dysfunctional beliefs about the disorder.

5.3.4 Hair-Pulling Disorder (Trichotillomania)

Hair pulling disorder, also known as trichotillomania, is a disorder related to OCD. It is a chronic condition that often leads to hair loss. The subject affected by this disorder repeatedly performs the gesture of pulling hair or body hair in order to relieve a state of increasing tension or for the sense of gratification that derives from it.

5.3.4.1 Epidemiology

Evaluating the real epidemiology of the disorder is not easy; the patient often feels ashamed and may not require the doctor's attention. About one third of patients wait more than a year before seeking medical attention.

Current estimates evaluate a distribution between males and females in a ratio of approximately 1–10 and a lifetime prevalence between 0.5 and 3.5% of the general population.

5.3.4.2 Aetiology

The aetiology appears to be multifactorial. Some polymorphisms in genes coding for serotonin receptors as well as morphological variations in the basal ganglia (putamen, lenticulate nucleus) have been investigated. Depressive symptoms, conditions of emotional stress, as well as childhood trauma have always been considered as important factors for the development of the disorder albeit there is no uniformity of thought in this regard.

5.3.4.3 Clinical Presentation

The symptoms essentially consist of repeated pulling of hairs or body hairs that can affect all parts of the body with a preference for the scalp and face (eyelashes, eyebrows, and beard). The act of tearing is often preceded by a feeling of increasing tension, while subsequently the patient feels a certain sense of relief or satisfaction (Box 5.13).

There are two main types of tearing

- *Focus pulling:* the act is intentional and is carried out in response to the appearance of an unpleasant experience, such as a thought, an impulse or a bodily sensation.
- *Automatic pulling*: the act is unconscious, it occurs in the course of another activity, typically sedentary.

Often the two types of pulling alternate and can be succeeded by ingestion or chewing of the hair. Ingestion can complicate the situation leading to the formation of bezoars and intestinal obstruction/malnutrition.

The patient will often see areas of the scalp with sparse, broken hair at different stages of growth. The same can be said for eyelashes, eyebrows, or other hairy areas.

The course appears to be highly variable. The disorder can have a chronic course or phases of temporary remission. The forms with the best prognosis are those that begin at a very young age (<6 years) while the adolescent forms tend to become chronic more frequently.

Box 5.13: DSM-5 Criteria for Diagnosing Hair-Pulling Disorder

1. The patient tears off enough hair to cause hair loss.
2. The subject tries to decrease or stop hair pulling.
3. The hair pulling causes impairment in social and/or occupational functioning.
4. Another medical condition cannot explain the hair pulling or hair loss.
5. The symptoms of another mental disorder cannot explain the hair pulling.

5.3.4.4 Therapy

Although there is no univocal consensus within the scientific community, the two most effective lines of treatment seem to be pharmacological and psychotherapeutic. From a pharmacological point of view, the first line of treatment includes antidepressants belonging to the class of SSRIs (fluvoxamine, citalopram) and SNRIs (venlafaxine), the use of local steroids, and antihistamines (hydroxyzine). Other proposed pharmacological strategies include the use of pimozide lithium salts, a first generation antipsychotic and opioid receptor antagonists such as naltrexone.

As far as psychotherapy is concerned, the literature data are scarce; the greatest effectiveness seems to derive from the psychodynamic approach.

5.3.5 Excoriation (Skin-Picking) Disorder

This condition has been described since the late nineteenth century. Long called "dermatillomania", it found a precise definition only recently when it was placed in the DSM-5 among the obsessive-compulsive spectrum disorders.

The key feature that describes this disorder is recurrent picking of the subject's skin. This skin picking has the characteristics of a compulsion and can lead, in the long term, to serious tissue damage with the need for specific pharmacological treatments.

5.3.5.1 Epidemiology

The disorder has a prevalence of 1.4% in the general population, increasing to 12% in the adolescent population. In 75% of cases, the female sex is affected.

5.3.5.2 Aetiology

The aetiology is unclear. From a biological point of view, some form of neurochemical alteration has been hypothesized, in particular in the metabolism of serotonin, dopamine, and glutamate. From a psychological point of view, several hypotheses have been formulated: picking in adolescents could be a manifestation of anger towards parental figures. From a psychoanalytic point of view, skin picking would be seen as a form of autoeroticism. Finally, there is the hypothesis that excoriation is a mechanism for relieving stress. It should also be noted that often the onset of the disease coincides with the onset of dermatological conditions, such as acne. In this case, however, the teasing often continues even when the favouring dermatological condition resolves.

5.3.5.3 Clinical Presentation

As with hair-pulling disorder, excoriation is often preceded by a sense of increasing tension and followed by a sense of relief. The most commonly affected site is that of the face. In other cases, the limbs, torso, fingers, scalp may be involved. Often, the patient turns over different areas to give the skin time to heal. Patients are generally highly embarrassed by the disorder and its aesthetic consequences. This discomfort can be accompanied by avoidance and social withdrawal or masking of the areas involved with makeup, clothes, and sometimes bandages (Box 5.14).

The onset of the disorder can be placed in two phases, the first around the age of 14 while the second in adulthood, between 30 and 45 years. The diagnosis can be delayed: often, the affected person is not aware of this disorder and the possibilities of treatment and can turn to the doctor's attention when skin damage is irreversible. The course of the disorder appears to be fluctuating, with periods of remission and recrudescence.

Box 5.14: DSM-5 Criteria for Diagnosing Excoriation (Skin-Picking) Disorder

1. The patient causes visible skin lesions by picking.
2. The patient makes repeated attempts to decrease or stop skin picking.
3. The skin picking causes impairment in social and/or occupational functioning.
4. The physiological effects of a substance or another medical condition cannot explain the skin picking.
5. The symptoms of another mental disorder cannot explain the skin picking.

Excoriation disorder can be disabling; most of those affected feel intense discomfort and embarrassment, implementing social avoidance behaviours. It can lead to the abandonment of school or of the workplace. The disorder can also lead to major tissue damage with the development of infections and scars that often require medical intervention, antibiotics, and even surgery.

5.3.5.4 Therapy

Although there are no real guidelines, the treatment of this condition mainly uses pharmacotherapy and psychotherapy, preferably in combination.

From a drug perspective, those found most effective in treating the disorder include SSRIs, opioid antagonists (naltrexone), and stabilizers, particularly lamotrigine.

As for psychotherapy, it appears that brief cognitive-behavioural therapy (CBT) may be an effective solution.

In some circumstances, mechanical prevention of picking utilizing aids or protective measures may help.

5.4 Trauma and Stress-Related Disorders

The disorders grouped in this chapter have in common the exposure to a traumatic or stressful event as the foundational event in disorder development.

The psychological stress that follows a traumatic or stressful event varies from person to person. In some cases, some symptoms can be understood in the context of anxiety or fear; however, many subjects, following exposure to a particularly stressful event, also manifest anhedonia, dysphoria, externalized anger and increased aggression or dissociative symptoms. The diverse range of symptoms reported depends on various factors, including genetic/temperamental predisposition and learned coping patterns to stress.

5.4.1 Adjustment Disorders

Adjustment disorders are characterized by depressive symptoms, anxiety and/or behavioural changes, without fully satisfying the criteria for a proper depressive or anxiety disorder, following an event experienced as stressful and with marked difficulty in implementing an adaptive response to the increased demand for cognitive and emotional resources as coping with such stressor would require.

Any life event can be stressful, and the ability to cope with such events varies between different individuals, so there is no defined list of stressors. To cite a few examples, there may be marriage or divorce, relocation, financial difficulties, health problems.

5.4.1.1 Epidemiology and Course

The prevalence of adjustment disorders in outpatient psychiatric populations, depending on the studies and the sample, varies between 5 and 20%, reaching peaks of 50% in psychiatric inpatients.

ADs are ubiquitous and can appear at any age, with concerns of differential diagnosis in children and adolescents, as clinical presentation in these populations is very similar to neurodevelopmental disorders.

Risk Factors

- Personality disorders.
- Cognitive disturbances.
- Substance abuse.

Comorbidity

Theoretically, an AD can occur in any subject regardless of its already known psychiatric or physical medical diagnosis, with which they are often associated. ADs are associated with high suicidal risk and a high rate of successful suicides. The AD can have an acute course if it lasts less than 6 months or persistent/chronic if it lasts more than 6 months (in the persistence of the triggering factor and in any case no later than 6 months from the interruption of the trigger).

5.4.1.2 Clinical Presentation

The characteristic element of AD is the identification of an external and stressful life event after which the disorder occurs. AD usually begins within 3 months of the onset of the stressor and generally lasts up to 6 months after the end of the stressor itself. The triggering life event can be of various types, it can be single (e.g. parenthood, relationship interruption) or multiple (e.g. work, marital, and health difficulties), repeated (e.g. frequent moving, work), or continuous (e.g. chronic illness). The subjective discomfort is manifested with loss of function in work/school settings and difficulties in social relations. It can also complicate the course of the underlying medical condition (e.g. noncompliance to check-ups or medical therapies). The AD may primarily present with a low mood, with anxiety, with anxious-depressive symptoms, with alterations in emotion and behaviour (e.g. disorganized behaviour, self-harm, substance abuse for self-medication purposes).

DSM-5 Diagnostic Criteria

1. The patient develops emotional and behavioural symptoms after being exposed to one or more stressful events. These symptoms occur within 3 months from the exposure.
2. This symptomatology has a significant clinical impact, as evidenced by one or both of the following:
   - Pronounced distress that is disproportionate to the stressor.
   - Significant impairment of functioning in various areas of life (e.g. social and working).
3. The symptoms of another mental disorder cannot explain the disturbance.
4. The symptoms do not represent normal grief.
5. Once the patient is no longer exposed to the stressor, the symptoms do not persist for more than additional 6 months.

5.4.1.3 Treatment

The first treatment choice in AD is psychotherapy, with various approaches (e.g. group therapy, individual psychotherapy). The aim is to provide the patient with new strategies for managing stressful events, improving and strengthening individual resources and with a focus also on the prevention of future episodes.

Pharmacotherapy aims to reduce symptoms. Depending on the current distress, drugs of choice are anxiolytics (e.g. mid half-life benzodiazepines, administered for a limited period in order to reduce tolerance and dependence) and/or antidepressants (the most tolerable are SSRIs, but SNRIs and TCAs can be used too).

5.4.2 Post-Traumatic Stress Disorder (PTSD)

5.4.2.1 Definition

Post-traumatic stress disorder (PTSD) is a disease that can occur in people who have suffered or witnessed a traumatic or violent event, or if a traumatic experience happened to a loved one.

The experience of a traumatic event of various kinds (e.g. wars, robberies, child abuse, muggings, kidnapping, terroristic attack, torture, natural disasters or due to the action of man, serious car accidents), determines the onset, at variable times (in the first weeks or after several months), of a symptomatology, that includes emotional and cognitive symptoms, and neuro-vegetative alterations. Characteristic is re-experiencing the traumatic event in a recurring, involuntary and intrusive way detail of the traumatic event, whose memories are very vivid and are associated with sensory, emotional, physical and behavioural components. Therefore, the clinical presentation of PTSD is very variable: in some subjects the dysphoric component prevails, in others, the fear of reliving the trauma becomes central, in still others dissociative symptoms prevail.

5.4.2.2 Epidemiology

The incidence of PTSD is about 8–15% in the general population, with some geographic differences (e.g. higher prevalence in the USA than in Europe, Asia, Africa), according to other meta-analyses and statistics on adult subjects the global annual prevalence is 1–6%. In North America (namely, the USA and Canada), lifetime prevalence rates are higher (6.1–9.2%) than in other high- and middle-income countries (2.3%) according to WHO data.

The difference in prevalence between sexes (10% females versus 4% males) could be explained by the greater global spread of sexual assaults and rapes against females, which are far more common in all societies, rather than wars and physical violence for men.

Up to 30–50% of people who have suffered violence, genocide, war veterans but also medical providers, police officers and firefighters, develop symptoms that lead to the diagnosis of PTSD.

Prognosis

Symptoms can have a fluctuating course, with exacerbations during periods of greater stress. Without treatment, about half of the patients achieved remission of symptoms after 1 year.

Positive prognostic factors are the rapid onset, duration less than 6 months, good previous functioning, good social support, absence of other comorbidities.

The course of the disease is typically chronic: only 1/3 of the patients recover within a year, while another 1/3 remain symptomatic even 10 years after the traumatic event. The presence of PTSD is associated with numerous consequences, for example, employment difficulties, less insertion and social support, difficulties in interpersonal relationships and affections, a greater degree of disability, increased risk of death from other causes (e.g. cardiovascular disease, diabetes) but also suicide (suicide attempts and actual deaths from suicide, in particular in the case of past childhood abuse).

Comorbidity

In adults, comorbidity is common with the following:

- Depression.
- Bipolar disorder.
- Anxiety disorder.
- Substance abuse.
- Conduct disorder.
- Physical illnesses, for example, autoimmune, endocrine, pulmonary, dementia, high cardiovascular risk, dementia.

In children, comorbidity is more common with the following:

- Separation anxiety disorder.
- Oppositional defiant disorder.
- Major neurocognitive disorder.

5.4.2.3 Aetiopathogenesis

The factors that contribute to the development of the disorder are as follows:

- Stressor: a stressor alone is not sufficient but is necessary to cause the disorder. Except for factors relatively common to daily life, however stressful (e.g. bereavement, illness, divorce), possible triggering stressors are very varied, and not only the experience itself but also the emotional/cultural connotation that the subject associates with it are important;
- Individual risk factors: since not all traumatic experiences generate PTSD in all subjects, risk factors have been studied that increase the likelihood that a person will develop PTSD at the same stressor.
  - Physical or sexual abuse in childhood
  - Female gender

- Youthful age
- Repeated exposure to traumatic events
- Objective severity of the event or degree of direct exposure
- Personality disorder
- Belonging to ethnic minorities, low socioeconomic status, poor psychosocial resources
- Being a widower, separated, divorced
- History of psychiatric disorders or positive psychiatric family history
- Excessive use of alcohol recently

Aetiologic hypotheses have been proposed for the development of PTSD

- Psychoanalytic model: the trauma brings a quiescent but unresolved psychological conflict to light, determining in the subject a state of repression, denial, and immobility.
- Cognitive-behavioural model: the subject is unable to rationalize the trauma and for this reason, they relive it continuously, implementing avoidance behaviours to the factors that reactivate the memory itself. There are two distinct phases in this model.
  - In the first phase, the trauma (unconditioned stimulus) produces a fear that is elicited by specific stimuli (physical or mental, such as sight, sounds, and smells) causing conditioning.
  - In the second phase, the repetition of the conditioned stimulus evokes the fear response, even in the absence of the original unconditional stimulus (the trauma). To avoid this state of malaise, the subject takes avoidances concerning both the unconditioned stimulus and the conditioned stimuli/stimuli associated with it. Furthermore, the potential secondary advantage of this condition should not be underestimated, typically assuming a condition of protection, compassion, and care by third parties.
- Biological basis: the paradigm of "learned behaviour" has highlighted the role of various neurotransmitters and somatic systems in the onset of PTSD symptoms.
  - Noradrenergic
  - Dopaminergic
  - GABAergic
  - Endogenous opioids: a low concentration of plasma β-endorphins and an analgesic response to opioid antagonists (e.g. naltrexone) have been observed in veterans with PTSD
  - Hypothalamic–pituitary–adrenal axis: low levels of plasma and urinary cortisol, increased glucocorticoid receptors in lymphocytes, and poor ACTH response after CRF stimulation have been observed in subjects with PTSD; cortisol hypersuppression could predict those subjects who will develop the disorder and those who will not, given the exposure to same traumatic events; the hyperactivation of this axis, though, differs from other mental disorders
  - Autonomic nervous system: an increase in orthosympathetic tone leads to an increase in heart rate and blood pressure, tremors, sweating and palpitations,

sleep disturbance (fragmentation and increased sleep latency); desensitization of the α2- and β-adrenergic receptors was observed per chronic downregulation; at the level of laboratory tests, increased concentrations of urinary catecholamines are observed.

5.4.2.4 Clinical Presentation

After the traumatic event, acute stress-induced disturbances may develop with dissociative characteristics (e.g. derealization, depersonalization) and sleep disturbances. During the first month of these symptoms, the diagnosis of acute stress disorder is initially made, and only if symptoms last over a month the diagnosis of PTSD can be formulated (criterion F). Such symptoms tend to disappear after a few months, in cases with a favourable prognosis (about 1/3 of cases), thanks to the activation of resilience and stress management mechanisms. On the contrary, when the prognosis is less favourable (1/3 of cases), the course can be chronic (3–6 months after the trauma) with a progressive structuring of the psychopathological picture. There are also late-onset forms, which develop at least 6 months after the event and are a slight harder to properly recognize.

The characteristic symptoms can be grouped into four clusters: intrusive symptoms, avoidance, cognitive and mood changes, and arousal and hypervigilance.

- Intrusive symptoms occur independently of the patient's will and are characterized by recurrent and intrusive unpleasant memories or dreams of the event, the sensation of reliving the experience, with illusions, hallucinations and flashbacks. The individual may act or feel as if the event was recurring if exposed to factors that recall the event (conditioned stimulus), and intense mental and physical discomfort follow. The subject may experience feelings of shame, guilt, anger, sadness, vulnerability, fear/terror with sense of unreality ("like in a dream"), strangeness ("like in a movie"), perception of one's body as different, abnormal or extraneous, confusion, absence of space-time references. To these elements is added a set of autonomic symptoms, analogous to the acute anxiety crisis or panic attack, which consists of profuse sweating, dyspnoea, sudden crying, tachycardia, nausea, diarrhoea, tremors, hypervigilance and cognitive activation.
- The great malaise that follows the re-enactment leads patients to avoid all conditioned stimuli (e.g. thoughts, sensations, conversations, activities, places, or people) that evoke the trauma.
- A peculiar aspect of PTSD is psychogenic amnesia: it is real lacunar amnesia relating to the traumatic event, different from the more frequent dissociative amnesia, in which the subject has a distorted memory of the trauma, with an inability to recall the timeline or details. There is also psychic numbing, affective flattening, apathy and anhedonia. Affectivity is reduced, with feelings of diminishing future prospects. The loss of future prospects, the freezing in a hopeless present, the so-called guilt of the survivor, the negative vision of the world, the feeling of indelible change and the irreversibility of what was before are typical symptoms of the pathological reaction to exposure to an extreme event. Cogni-

tive and mood alteration, as well as intrusive symptoms, can lead to self-injury and impulsive behaviour, socio-occupational impairment, alcohol, or substance abuse. The maladaptive aspects resulting from the disorder present a particular gender difference: in women, they mainly concern self-care, while for men they concern the intake of alcohol and drugs, unregulated behaviour and an increase in suicidal attempts.

- Another typical symptom is hyperarousal (state of hyperactivation) which is characterized by the following.
  - Difficulty falling asleep or staying asleep
  - Irritability or outbursts of anger
  - Difficulty concentrating
  - Hypervigilance (feeling of "tense nerves", inability to "let your guard down")
  - Exaggerated alarm responses (sudden jerks for minimal stimuli)

The diagnosis of PTSD requires the presence of a series of psychic and autonomic symptoms, but there are also "sub-threshold" forms (especially in the elderly, as stated above) in which not all symptoms are present but where the post-traumatic nature of the disorder is the same. Despite the syndromic incompleteness, these forms have a suicide risk equal to the full expression forms.

Clinical presentation changes depending on the traumatic stimulus and the age of the subject:

- Children change their behaviour and mood, reduce social interactions, decrease school performance.
- In adolescents, violent behaviours prevail, with a high risk to health; they lose ambition, become irritable and aggressive.
- In adults, the disorder manifests as hyperarousal, avoidance, sleep problems, worsening health, and suicidal ideation.

DSM-5 Diagnostic Criteria

1. Direct or indirect exposure to actual or threatened death, serious injury, or sexual violence.
2. At least one of the following intrusion symptoms associated with the traumatic event.
   (a) Recurrence of memories related to the traumatic event on an involuntary basis and perceived as intrusive.
   (b) Recurrent nightmares related to the traumatic event.
   (c) Experience of dissociative reactions in which the patient feels as if the traumatic event was recurring.
   (d) Psychological suffering as a result of exposure to factors that symbolize and recall the traumatic event.
   (e) Marked physiological reactions as a result of exposure to factors that symbolize and recall the traumatic event.

3. Avoidance distressing feeling, thoughts or memories associated with the traumatic event and/or avoidance of external reminders that arouse distressing feeling, thoughts or memories associated with the traumatic event.
4. Negative changes in cognitions and mood related to the traumatic event, as evidenced by at least two of the following.
   (a) Lack of ability to recall relevant aspects of the traumatic event.
   (b) Persistence of exaggerated negative beliefs and expectations about oneself, others and the world.
   (c) Persistent, distorted thoughts about the cause or consequences of the traumatic event that lead the patient to blame himself/herself or others.
   (d) Persistent negative emotional state.
   (e) Markedly diminished interest in several activities.
   (f) Feelings of detachment or estrangement from others.
   (g) Persistent incapability of experiencing positive emotions.
5. Important alterations in arousal and reactivity associated with the traumatic event, as evidenced by at least two of the following.
   (a) Irritability.
   (b) Reckless or self-destructive behaviour.
   (c) Hypervigilance.
   (d) Incremented startle response.
   (e) Difficulty in concentrating.
   (f) Difficulty in sleeping.
6. Symptoms last more than 1 month.

5.4.2.5 Treatment

The most effective treatment for PTSD seems to be the combination of drugs (that treat comorbidities) and psychotherapy (that treat the typical symptoms of the disorder).

SSRIs (e.g. sertraline, paroxetine) and SNRIs are currently the first lines of treatment, reducing the symptomatology with good tolerability and safety profile. It is also possible to use TCAs (e.g. imipramine, amitriptyline) using dosages equivalent to those of depressive disorder (range 50–100 mg per day), and continuing the treatment for at least 1 year. Other drugs that can be used are MAOIs, mood stabilizers (e.g. carbamazepine, valproate) and trazodone. Benzodiazepines are also used, albeit with caution given the frequent comorbidity with alcohol and substance abuse.

To regulate sleep at night, hypnotics and sedatives can be used. Adjunctive preventive interventions can be the administration of $\alpha 1$-antagonists in the evening to reduce the occurrence of nightmares and the administration of β-blockers in response to the hyperactivity of the noradrenergic system.

Antipsychotics (e.g. haloperidol) are reserved for the management of major agitation.

Psychotherapy must help the patient cope with the rejection of trauma, eliminate the source of stress, metabolize all the emotions associated with the event and restore the correct sleep–wake rhythm. As a psychotherapeutic approach, one can opt for both psychodynamic therapy and cognitive-behavioural therapy. Two different cognitive-behavioural modalities can be used: one involves the use of both gradual and implosive/intense exposures and can be performed in vivo or through images or films; the other provides the patient with tools to manage anxiety and stress through cognitive and/or relaxation techniques so that they can be "skilled" when a trigger or conditioned stimulus occurs during the day.

A relatively recent psychotherapeutic technique is Eye Movement Desensitization and Reprocessing (EMDR): during sessions, a sensory "bilateral stimulation" (auditory, ocular, tactile) is performed while the patient evokes the traumatic event. This technique allows the patient to activate the brain system responsible for processing the traumatic memory in order to make the associated emotions conscious and integrate them with the other information available. The subject then through a "catharsis", in a state of deep relaxation, relives and removes the traumatic event without being overwhelmed. Finally, mindfulness has found good efficacy in reducing post-traumatic symptoms.

5.4.3 Dissociative Disorders

5.4.3.1 Definition

In this type of disorder, the various systems that underlie complex mental activity (consciousness, memory, identity, emotions and perception), which are normally experienced as a continuum and in constant integration with each other, lose mutual integration. Disconnection between the various systems can produce positive or negative symptoms. The term "positive" indicates something more than the normal state of consciousness: unwanted intrusions into consciousness and behaviour, with loss of the continuity of subjective experience. On the contrary, the term "negative" indicates something less than the usual experience: the individual is unable to access information or control mental functions that are generally easily accessible or controllable.

Box 5.15: The Case of ANNA O

Anna O. (literary name for Bertha Pappenheim) was a 21-year-old girl with numerous qualities and talents, from a good Austrian family, who found herself having to take care of her seriously ill father neglecting her diet and physical health.

In the winter of 1880, a few months after her father's illness, she began to develop varied and changing symptoms physically, neurologically and psychically, which could not be explained by the medical knowledge of the time. Symptoms included convergent strabismus, paralysis of the right upper limb and then of the two lower limbs, hydrophobia, alterations in the state of consciousness with estrangement from conversations during which she rehearsed events in her imaginative "private theatre", splitting of personality, sudden mood swings, complex hallucinations, mutism.

Admitted at the Salpêtrière hospital, she was treated by Joseph Breuer, who was a luminary in hypnosis and in the treatment of hysteria. Breuer decided to apply hypnosis to "make the patient speak" when she became completely silent: She was able to overcome the speech block, and even the muscular paresis gradually resolved. However, the patient's clinical presentation fluctuated and then worsened after the death of her father. Sigmund Freud also worked at the Salpêtrière hospital and discussed the case with Breuer: They decided to try a new type of therapy, called the "cathartic method", which was making the patient speak freely about everything that went through the mind. When Anna O told a memory surfaced in her mind, the related symptom disappeared, as if she was able to give "free rein" to unpleasant and burdensome emotional contents hidden from consciousness. Thanks to this speech therapy in 1882, Anna was finally free of symptoms.

Dissociative disorders in DSM-5 include the following:

- Dissociative identity disorder
- Dissociative amnesia
- Depersonalization/derealization disorder

5.4.3.2 Dissociative Identity Disorder

We speak of dissociative identity disorder (formerly referred to as "multiple personality disorder") when there are two or more distinct personality states (as in Dr. Jekyll and Mr. Hyde) or an experience of possession: these people do not decide to behave in one way or another, they do not act consciously, because there is a sharp discontinuity in the sense of self and awareness of their actions without the patient exercising their will. Dissociative identity disorder is a failure of identity integration: each of the personality states can be experienced as if it had a personal history, self-image, and distinct identity, including a separate name.

Epidemiology

The disorder has a prevalence of 1.4% in females and 1.6% in males; no prevalence studies are available in individual states.

The onset can take place at any age. In children, there are mainly problems with memory, concentration and attachment, in adolescence, it presents with sudden changes in identity; in adults, they may resemble a late onset of mood disorders, OCD, cognitive disorders due to dissociative amnesia.

Risk Factors

It is often found in association with overwhelming experiences, traumatic events, chronic childhood trauma, physical or sexual abuse in childhood (in general 70–100% of cases, compared to 8–17% in the general population in the USA).

Comorbidity

DDI is often comorbid with other psychiatric pathologies, although the assessment can be sometimes difficult in some studies the main comorbidities are as follows.

- PTSD (79–100%)
- Borderline personality disorder (31–83%)
- Avoidant personality disorder (76%)
- Substance abuse (83–96%)
- Depression (83–96%)
- Somatoform disorder

In such patients, the suicidal risk is very high: over 70% of outpatient patients attempt suicide and multiple attempts are frequent as well as self-harming behaviours.

The functional consequences of the disorder can have varying degrees of impairment (from minimal to profound), occur more frequently in a relational than professional context, and a tendency to minimize the impact of symptoms is common.

Clinical Presentation

There is usually a primary identity that bears the official name of the subject, and which is usually passive, dependent, tending to feelings of guilt and depression; the other identities frequently have different names and characteristics that conflict with the primary identity: for example, they are extremely irritable or hostile, "executive", and aggressive towards others or self-destructive.

There are also important memory alterations: frequent memory gaps about their personal history, both remote and recent. Amnesia is frequently asymmetric through the identities: more passive identities tend to have poorer memories, while more hostile, "executive", or "protective" ones have more complete memories; an identity that does not have control functions may however have access to consciousness through the production of auditory or visual hallucinations (e.g. a voice giving instructions). The demonstration of amnesia can be achieved by witnesses or through the "discoveries" of the individual himself (e.g. the fact of finding at home items of clothing that the subject does not remember having bought).

The symptoms produced by the disintegration of identity are as follows

- Feeling depersonalized observers of their own speeches and actions.
- Perception of voices.
- Strong emotions and egodystonic and disconcerting impulses.
- Sudden changes in attitudes, perspectives, and personal preferences.
- Perception of one's body as different.
- Non-epileptic seizures or other conversive symptoms.
- Dissociative amnesia has the following characteristics.
- Gaps in the remote memory of personal life events.
- Memory errors related to acquired procedures.

- Discovery of evidence of daily actions and tasks that he does not remember having performed.
- Dissociative fugues are common in these patients.

DSM-5 Diagnostic Criteria

A. The patient have two or more distinct personality states. The disruption in identity is characterized by a marked discontinuity in sense of self and sense of autonomy.
B. Recurrent gaps in patient memory for everyday events, important personal information, and/or traumatic events.
C. The symptoms cause clinically significant distress or impairment in social or occupational functioning.
D. The disturbance is not a part of a widely accepted cultural or religious practice.
E. The physiological effects of a substance or another medical condition cannot explain the symptoms.

Differential Diagnosis

- Alcohol or benzodiazepine intoxication (they can both produce amnesia, which is usually anterograde and generalized), cannabis, hallucinogens, ketamine, ecstasy (they can cause depersonalization symptoms).
- General medical conditions, for example, dementia (it is characterized by retrograde amnesia, with progressive loss of memory, usually starting from recent events with preservation of autobiographical information), epilepsy (it causes limited amnesia of the moments during and shortly after a crisis). In particular, in the case of dissociative amnesia, the patient will have a good performance on neuropsychological tests, showing particular deficits in autobiographical information, in contrast to dementia patients.
- PTSD: it is characterized by symptoms of re-experiencing, avoidance, hypervigilance, and arousal.
- Borderline personality disorder, where the fragmentation of identity is only slightly less than in the DID.
- Schizophrenia and other psychotic disorders where hallucinations occur: in the case of DDI, auditory hallucinations and other psychotic symptoms are "customed" on the identity of who is in charge at the moment, and they are more often pseudo-hallucinations (the patient feels them from the inside). The occurrence of auditory hallucinations in patients with DID is already reported from childhood, then it becomes more fluctuating and independent of the other symptoms.
- Bipolar disorder: in DID the fluctuations of emotional states are very rapid, even in a few hours, and abrupt, often in response to environmental stimuli; in contrast, mood swings in bipolar disorder last at least 2 weeks (for major depressive episode) or 4–7 days (for hypo-manic episodes) and their occurrence and recovery is independent of external circumstances.
- Factitious disorder and simulation: the patient deliberately produces physical or mental symptoms to obtain attention and play the role of the sick person, or to

obtain a particular secondary advantage such as a disability pension or avoid certain types of duties. The symptoms are presented "from the textbook" in a very simple way, without complex pictures or the distress and shame that patients with DID feel, insight and judgment are perfectly preserved, unlike DID.

Treatment

Psychotherapeutic approach is fundamental, and involves the combination of various techniques (psychoanalysis, cognitive and behavioural therapy, hypnosis, family therapy) in addition to drug treatment.

Drug therapy aims to reduce secondary depressive symptoms and stabilize mood. Symptoms attributable to the sphere of traumatic disorders (intrusive thoughts, hyperexcitability, and hypervigilance) are partially responsive to drugs.

Eye movement desensitization and reprocessing (EMDR) therapy has been indicated in PTSD in recent years, some case reports suggest that it can be extended to dissociative disorders, but no systematic studies have been conducted.

Among the drugs that can be used are SSRIs, TCAs and MAOIs, β-blockers, clonidine, anticonvulsants, and benzodiazepines. In some individuals with a tendency to irritability and aggression, carbamazepine can be used; atypical antipsychotics (risperidone, quetiapine, ziprasidone, olanzapine) are preferable to first-generation antipsychotics in the control of anxious and intrusive symptoms.

5.4.3.3 Dissociative Amnesia

It is a disorder characterized by the inability to remember autobiographical information, usually limited to a traumatic event, which should be kept in memory and which are usually easily accessible.

A subtype of dissociative amnesia is a dissociative fugue, which manifests as sudden travel or wandering accompanied by anterograde amnesia for this event.

Epidemiology and Course

It has a prevalence of 1% in males and 2.6% in females (in a US sample); the lifetime prevalence is around 6–7% (based on a Canadian and a Turkish sample).

The onset can be sudden in generalized amnesia, less evident in the limited forms and can arise at any age (small children, adolescents, adults).

The natural course of the disease provides that the occurrence of a first episode predisposes to the recurrence of subsequent episodes. The duration of the single episode can vary from minutes to decades and after remission, intense discomfort, suicidal behaviour, PTSD symptoms may remain. Suicidal risk is high: suicidal and self-destructive behaviours are common, particularly after sudden remission with the onset of intolerable memories.

Comorbidity

The disorder can present comorbidities of the dissociative, trauma-related area, depression, substance abuse and personality disorders, significantly worsening the prognosis of patients.

Clinical Presentation

The types of amnesia that can occur are as follows

- Limited amnesia: limited to a limited period.
- Selective amnesia: only about some events of a limited period.
- Generalized amnesia: total memory loss of one's personal history (rare), personal identity, previous knowledge of the world (semantic knowledge) and already known skills (procedural knowledge); has an acute onset with perplexity and disorientation.
- Systematized amnesia: related to a specific category of memories.

The person with dissociative amnesia cannot recall both episodic and semantic memories from the autobiographical memory; on the opposite, the ability to store new information, cognitive functions, procedural memory and language are preserved. The distinction between accessible memories and those that cannot be recalled is typically sharp, unlike the normal functioning of memory (we normally remember recent events more easily and boundaries between different memory domains are softer).

The functional consequences are a varying degree of impairment, from limited to severe.

Although there are few systematized studies on this condition, it seems that the severity of the disorder and amnesia for very early events are correlated with a longer duration and severity of the trauma, typically childhood.

DSM-5 Diagnostic Criteria

A. The patient cannot remember important autobiographical information.
B. The symptoms cause clinically significant distress or impairment in social and/or occupational functioning.
C. The physiological effects of a substance or a neurological or other medical condition cannot explain the symptoms.
D. Other mental disorder cannot explain the symptoms.

Differential Diagnosis

- Normal forgetfulness of autobiographical memory.
- Cognitive impairment.
- PTSD and acute stress disorder.
- Other dissociative disorders.
- Substance abuse.
- Head trauma.
- Fictitious disorder/simulation.

Treatment

The treatment of dissociative amnesia involves primarily a cognitive psychotherapeutic approach. Pharmacological treatment protocols have not been standardized. In cases of acute onset, in a hospital setting, benzodiazepines or barbiturates can be used to "facilitate" the patient's access to information, which can then be consciously processed.

5.4.4 Depersonalization/Derealization Disorder

5.4.4.1 Definition

- Depersonalization is a psychopathological condition in which the individual feels detached from themselves, from aspects of their own self (feelings, thoughts, body or parts of his body, sensations), or divided (out-of-body experience).
- Derealization is defined as a state in which the individual can feel as if they were in the fog, in a dream, in a bubble, as if there was a veil or a glass wall between themselves and the world.

Depersonalization and derealization are frequent symptoms in psychiatric care settings as they might be present almost in all other mental conditions (see "differential diagnosis"), and they are not diagnostic per se. In fact, we might say that the diagnosis of a "pure" depersonalization/derealization disorder is a diagnosis of exclusion, once we have ruled out all other possible mental (and organic) disorders.

5.4.4.2 Epidemiology and Course

A large part of the population experiences these symptoms at least once in their life, without distinction of gender: depending on the studies and the sample, they range from 12% up to about 50%.

The prevalence of depersonalization/derealization disorder, on the other hand, is just 0.8–2.8% in the general population. The onset occurs on average in adolescence; it can be sudden or gradual. The duration of the single episode can vary from hours to years, there can be a persistent course characterized by separate or continuous episodes, the intensity of which can be variable or stable.

Risk Factors

- Acute trauma.
- Other psychiatric disorders, for example, anxiety disorders or depression.
- Substance abuse.
- Childhood trauma.
- The sudden death of a loved one.
- Growing up with a parent with severe psychiatric illness.
- Disturbances or doubts about sexual orientation.

Comorbidity

- Major depressive disorder.
- Anxiety disorders.
- Obsessive-compulsive disorder.
- Avoidant personality disorder.
- Borderline personality disorder.

5.4.4.3 Aetiology

The cause of this disorder is not fully known, but several hypotheses have been formulated:

- Psychodynamic: depersonalization is a defence reaction of the ego in situations of disintegration of the self (e.g. extreme pain, trauma).
- Traumatic stress: up to 60% of people with life-threatening experiences report at least one transient episode of derealization during the traumatic experience or immediately after.
- Neurobiological theory: the association between depersonalization and migraine and the use of cannabis, the positive response to SSRIs and the increase in personalization symptoms in conditions of tryptophan depletion indicate an involvement of the serotoninergic system. Through pharmacological elicitation studies of dissociative states, it seems that the glutamatergic system, through the NDMA receptor, is central to the origin of symptoms.

5.4.4.4 Clinical Presentation

Symptoms of depersonalization are abnormal body experiences, emotional or physical blunting, temporal distortion with abnormal subjective memories.

Symptoms of derealization are subjective visual distortions (blurring, amplified acuity, widening or narrowing of the visual field, macropsia or micropsia) and auditory distortions (amplified or muted voices and sounds). The functional consequences for the subject are represented by the fact that the symptoms can be very distressing and associated with severe morbidity; they can involve relational and professional impairment due to hypo-emotion towards others or a sense of disconnection from everyday life.

DSM-5 Diagnostic Criteria

A. The patient experiences depersonalization, derealization, or both:

1. *Depersonalization*—The patient feels detached with respect to their own thoughts, feelings, sensations, body, or actions.
2. *Derealization*—The patient feels detached with respect to surroundings.

B. Reality testing remains intact also during the depersonalization or derealization experiences.
C. The symptoms cause clinically significant distress or impairment in social and/or occupational functioning.
D. The physiological effects of a substance or another medical condition cannot explain the disturbance.
E. Another mental disorder cannot explain the disturbance.

Differential Diagnosis

Depersonalization/derealization can be a symptom in the context of other mental disorders:

- Schizophrenia.
- Panic disorder.
- Acute stress disorder.
- PTSD.
- Other dissociative disorders.
- Borderline personality disorder.
- Avoidant personality disorder.
- Substance abuse (e.g. cannabis, hallucinogens, ketamine).

It can also be present in other types of disorders of organic origin:

- Temporal lobe epilepsy.
- Vestibular disorders.
- Sleep apnoea.
- Head injuries or infections (e.g. Lyme disease) or autoimmune disease with central involvement.

5.4.4.5 Treatment

In the absence of psychiatric comorbidities, the first-line treatment for DDD is psychotherapy, which can have a cognitive-behavioural approach, but also psychodynamic, hypnotherapeutic, and supportive approach. However, there is a lack of placebo-controlled studies or comparisons between the various options, so the choice will fall on availability in the treatment centre and the patient's desire. Patients can also try stress management techniques, active distraction, reduction of sensory stimulation, relaxation, and physical exercise.

As drug therapy in patients who also have an anxiety or depressive disorder, the first drug choice falls on SSRIs (but also TCAs, mood stabilizers, typical and atypical antipsychotics), while benzodiazepines can be used in the short term to control anxiety symptoms.

Further Reading

Abramowitz JS, McKay D, Storch EA. The Wiley handbook of obsessive-compulsive disorders. New York: Wiley; 2017.

Albert U, Marazziti D, Di Salvo G, Solia F, Rosso G, Maina G. A systematic review of evidence-based treatment strategies for obsessive-compulsive disorder resistant to first-line pharmacotherapy. Curr Med Chem. 2018;25(41):5647–61. https://doi.org/10.2174/0929867325666171222163645.

Association AP. Diagnostic and statistical manual of mental disorders (DSM-5®). New York: American Psychiatric Association; 2013.

Brock H, Hany M. Obsessive-compulsive disorder. Treasure Island: StatPearls; 2020.

Erzegovesi S, Cavallini MC, Cavedini P, Diaferia G, Locatelli M, Bellodi L. Clinical predictors of drug response in obsessive-compulsive disorder. J Clin Psychopharmacol. 2001;21(5):488–92. https://doi.org/10.1097/00004714-200110000-00006.

Frances A. DSM-5 somatic symptom disorder. J Nerv Ment Dis. 2013;201(6):530–1. https://doi.org/10.1097/NMD.0b013e318294827c.

Freud S, Breuer J. Studies on hysteria. Paris: Hachette; 1895.

Grabill K, Merlo L, Duke D, Harford KL, Keeley ML, Geffken GR, Storch EA. Assessment of obsessive-compulsive disorder: a review. J Anxiety Disord. 2008;22(1):1–17. https://doi.org/10.1016/j.janxdis.2007.01.012.

Grant JE, Odlaug BL, Chamberlain SR, Keuthen NJ, Lochner C, Stein DJ. Skin picking disorder. Am J Psychiatry. 2012;169(11):1143–9. https://doi.org/10.1176/appi.ajp.2012.12040508.

Hirschtritt ME, Bloch MH, Mathews CA. Obsessive-compulsive disorder: advances in diagnosis and treatment. JAMA. 2017;317(13):1358–67. https://doi.org/10.1001/jama.2017.2200.

Kroenke K, Swindle R. Cognitive-behavioral therapy for somatization and symptom syndromes: a critical review of controlled clinical trials. Psychother Psychosom. 2000;69(4):205–15. https://doi.org/10.1159/000012395.

Lakhan SE, Schofield KL. Mindfulness-based therapies in the treatment of somatization disorders: a systematic review and meta-analysis. PLoS One. 2013;8(8):e71834. https://doi.org/10.1371/journal.pone.0071834.

Langeland W, Jepsen EKK, Brand BL, Kleven L, Loewenstein RJ, Putnam FW, Schielke HJ, Myrick A, Lanius RA, Heir T. The economic burden of dissociative disorders: a qualitative systematic review of empirical studies. Psychol Trauma. 2020;12(7):730–8. https://doi.org/10.1037/tra0000556.

Lanius RA, Boyd JE, McKinnon MC, Nicholson AA, Frewen P, Vermetten E, Jetly R, Spiegel D. A review of the neurobiological basis of trauma-related dissociation and its relation to cannabinoid- and opioid-mediated stress response: a transdiagnostic, translational approach. Curr Psychiatry Rep. 2018;20(12):118. https://doi.org/10.1007/s11920-018-0983-y.

Lynn SJ, Maxwell R, Merckelbach H, Lilienfeld SO, Kloet DVH, Miskovic V. Dissociation and its disorders: competing models, future directions, and a way forward. Clin Psychol Rev. 2019;73:101755. https://doi.org/10.1016/j.cpr.2019.101755.

Mataix-Cols D. Clinical practice. Hoarding disorder. N Engl J Med. 2014;370(21):2023–30. https://doi.org/10.1056/NEJMcp1313051.

Mataix-Cols D, Rosario-Campos MC, Leckman JF. A multidimensional model of obsessive-compulsive disorder. Am J Psychiatry. 2005;162(2):228–38. https://doi.org/10.1176/appi.ajp.162.2.228.

McKay D, Sookman D, Neziroglu F, Wilhelm S, Stein DJ, Kyrios M, Matthews K, Veale D. Efficacy of cognitive-behavioral therapy for obsessive-compulsive disorder. Psychiatry Res. 2015;225(3):236–46. https://doi.org/10.1016/j.psychres.2014.11.058.

Phillips KA. Understanding body dysmorphic disorder. Oxford: Oxford University Press; 2009.

Poppe C, Müller ST, Greil W, Walder A, Grohmann R, Stübner S. Pharmacotherapy for obsessive compulsive disorder in clinical practice—data of 842 inpatients from the international AMSP project between 1994 and 2012. J Affect Disord. 2016;200:89–96. https://doi.org/10.1016/j.jad.2016.04.035.

Reifter BV. Oxford handbook of psychiatry. J Clin Psychiatry. 2006; https://doi.org/10.4088/jcp.v67n0620b.

Rief W, Martin A. How to use the new DSM-5 somatic symptom disorder diagnosis in research and practice: a critical evaluation and a proposal for modifications. Annu Rev Clin Psychol. 2014;10:339–67. https://doi.org/10.1146/annurev-clinpsy-032813-153745.

Robbins TW, Vaghi MM, Banca P. Obsessive-compulsive disorder: puzzles and prospects. Neuron. 2019;102(1):27–47. https://doi.org/10.1016/j.neuron.2019.01.046.

Ruffini C, Locatelli M, Lucca A, Benedetti F, Insacco C, Smeraldi E. Augmentation effect of repetitive transcranial magnetic stimulation over the orbitofrontal cortex in drug-resistant obsessive-compulsive disorder patients: a controlled investigation. Prim Care Companion J Clin Psychiatry. 2009;11(5):226–30. https://doi.org/10.4088/PCC.08m00663.

Ruscio AM, Stein DJ, Chiu WT, Kessler RC. The epidemiology of obsessive-compulsive disorder in the National Comorbidity Survey Replication. Mol Psychiatry. 2010;15(1):53–63. https://doi.org/10.1038/mp.2008.94.

Sadock BJ, Sadock VA, Ruiz P. Kaplan and Sadock's synopsis of psychiatry: behavioral sciences/ clinical psychiatry. New York: Wolters Kluwer; 2015.

Sadock BJ, Ruiz P, Sadock VA. Kaplan and Sadock's comprehensive textbook of psychiatry. Baltimore: Lippincott Williams & Wilkins; 2017.

Spiegel D, Lewis-Fernàndez R, Lanius R, et al. Dissociative disorders in DSM-5. Annu Rev Clin Psychol. 2013;9(1):299–326.

Stahl SM. Stahl's essential psychopharmacology. Cambridge: Cambridge University Press; 2013.

Stahl SM. Prescriber's guide: Stahl's essential psychopharmacology. Cambridge: Cambridge University Press; 2020.

Stein DJ. Obsessive-compulsive disorder. Lancet. 2002;360(9330):397–405. https://doi.org/10.1016/S0140-6736(02)09620-4.

Steketee G. The Oxford handbook of obsessive compulsive and spectrum disorders. Oxford: Oxford University Press; 2011.

Storch EA, Lewin AB. Clinical handbook of obsessive-compulsive and related disorders: a case-based approach to treating pediatric and adult populations. Cham: Springer International; 2015.

Thorneloe WF. Psychopharmacologic treatment of obsessive-compulsive disorder. J Med Assoc Ga. 1988;77(11):825–6.

Voigt K, Wollburg E, Weinmann N, Herzog A, Meyer B, Langs G, Löwe B. Predictive validity and clinical utility of DSM-5 somatic symptom disorder: prospective 1-year follow-up study. J Psychosom Res. 2013;75(4):358–61. https://doi.org/10.1016/j.jpsychores.2013.08.017.

Walsh KH, McDougle CJ. Trichotillomania. Presentation, etiology, diagnosis and therapy. Am J Clin Dermatol. 2001;2(5):327–33. https://doi.org/10.2165/00128071-200102050-00007.

Personality Disorders

6

A. Somma, L. Fregna, F. Martini, and A. Fossati

6.1 Definitions

Personality disorders (PDs) are very common conditions affecting the interaction between health professionals and patients. Accordingly, they are important to all medical practitioners because of their key role as predictors of treatment outcome, cause of premature mortality, and great cost to society. Therefore, PD should be an important part of every psychiatric assessment, whether done by a qualified expert in PD or a family doctor. However, PD assessment has largely been overlooked in clinical psychiatric practice. For many years (even after the release of the DSM-III in 1980), the PD diagnosis has been used in a pejorative sense, or as a label applied to people who were considered as difficult to treat. Attention to PD in practice has therefore oscillated between attempts to dismiss it as a nondiagnosis, or instead, to regard it as a specialist subject that could be placed outside the realm of "true" psychiatric disorders. The difficulties with the PD diagnosis stem from issues that the scientific and clinical community started to address only in recent years: indeed, nobody doubts the existence of personality, but what constitutes its dysfunctionalities has been difficult to specify. For instance, the DSM-IV-TR and the DSM-5 Section II describe PD general features as "an enduring pattern of inner experience and behavior that deviates markedly from the expectations of an individual's culture. This pattern is manifested in two or more of the following areas: cognition, affectivity, interpersonal functioning, and impulse control. The enduring pattern is inflexible and pervasive across a broad range of personal and social situations, leads

A. Somma · A. Fossati
School of Psychology, Vita-Salute San Raffaele University, Milan, Italy
e-mail: somma.antonella@hsr.it; fossati.andrea@hsr.it

L. Fregna (✉) · F. Martini (✉)
IRCCS San Raffaele Scientific Institute, Milan, Italy
e-mail: fregna.lorenzo@hsr.it; francesca.martini1@hsr.it

R. Cavallaro, C. Colombo (eds.), *Fundamentals of Psychiatry for Health Care Professionals*, https://doi.org/10.1007/978-3-031-07715-9_6

to clinically significant distress or impairment in social, occupational or other important areas of functioning, is stable and of long duration, and its onset can be traced back at least to childhood or early adulthood, and is not better accounted for by other mental disorder or effects of a substance." As it can be easily observed, this statement has three major problems: (a) it does not indicate what represents normal personality and its functions; (b) consequently, which functions should be perturbed to generate PD; consequently, (c) it provides no formal diagnostic criteria for general PD diagnosis. From this poor operationalization of PD diagnosis were likely to originate many of the problems that we will take into considerations in the next sections. As Tyrer and colleagues nicely documented in 2015, the DSM-5 Section II PD diagnoses rely heavily on Schneider's nine pathological personality types. Schneider's nine personality types were based solely on his clinical experience; notwithstanding this fact, they have generally persisted in slightly different forms in all subsequent classifications of personality pathology until DSM-5 Section II. Since the release of DSM-III, operational criteria were used to define ideal or prototypical manifestations that could be considered as exemplars of each PD. Antisocial, borderline, narcissistic, and other qualifying adjectives have proved so enticing to clinicians describing patients that they have often led clinicians to bypass the issue of their scientific foundation. Regrettably, extant research indicates that the DSM-IV/DSM-5 Section II PD categories are neither valid, nor homogeneous types (rather, PDs are better conceptualized as dimensions than as categories), while showing extensive co-occurrence rates with other psychiatric conditions and marked continuities with potentially adaptive personality traits. In an attempt to move the science forward, the DSM-5 Personality and Personality Disorder Work Group proposed the Alternative Model of Personality Disorder (AMPD), which is currently included in the DSM-5 Section III. Consistent with DSM-5 general aim to provide clinicians with transtheoretical operational criteria for mental disorder diagnoses, which are thought to provide clinicians both maximally inter-rater reliable and valid diagnostic criteria, the DSM-5 AMPD strongly relied on scientific evidence in providing a new approach to PD diagnosis that was largely dimensional in nature (although six diagnostic types are still available). Interestingly, the evidence-based dimensional approach to PD diagnosis, leading to abandoning the typological tradition in PD assessment, informed also the development of PD criteria in the ICD-11. Regrettably, the adoption of a typological model, which received few (if any) empirical supports, in PD research is likely to represent a major reason for the (very) limited advancement of knowledge in etiology, pathogenesis, and treatment efficacy in personality pathology.

6.2 Diagnosis

The *DSM-5* Section II PD criteria provide 10 PD categories, which are assumed to represent independent conditions; as we observed in the introduction, the *DSM-5* Section II provides also a general PD description, which is not required to be met for individual PD diagnoses. In other terms, no formal assessment of the general PD criteria is required to carry out one or more specific PD diagnoses. Rather, the *DSM-5* Section II approach assumes that if the criteria for one or more of the individual PD diagnoses are met, then also the general criteria are satisfied. The *DSM-5* Section II proposes that the 10 PD diagnoses may be grouped in three clusters.

1. Cluster A, which includes paranoid, schizoid, and schizotypal PDs and is defined as the "Odd-Suspicious Cluster."
2. Cluster B, which includes antisocial, borderline, histrionic, and narcissistic PDs and is defined as the "Dramatic–Emotional–Erratic Cluster."
3. Cluster C, which includes avoidant, dependent, and obsessive–compulsive PDs and is defined as the "Anxious-Fearful Cluster."

It should be observed that the *DSM-5* Section II 10 PD diagnoses rely on a polythetic format, that is, only a limited number of criteria should be met to receive the PD diagnosis (see Table 6.1). For instance, although a total of 9 criteria are provided in *DSM-5* Section II for schizotypal PD, only five (or more) criteria are needed for schizotypal PD diagnosis. Table 6.1 lists the 10 *DSM-5* Section II PDs, their alleged main presentation, and diagnostic thresholds (i.e., number of criteria that should be met for PD diagnosis). Of course, these conditions should not be the consequence of the physiological effects of a substance (e.g., a drug of abuse, a medication) or another medical condition (e.g., head trauma), and should not be better explained as a manifestation or consequence of another mental disorder. However, the presence of a given PD diagnosis does not exclude the possibility to diagnose one or more other *DSM-5* PDs if diagnostic thresholds are met, as well as of other *DSM-5* Section II psychiatric conditions (e.g., major depression) that may co-occur.

As we stated in the Introduction, the *DSM-5* Section II PD criteria represent the "cut-and-paste" of the *DSM-IV-TR* axis II PD criteria, which in turn were highly similar to those that were reported in the *DSM-III-R* axis II. Thus, more than one generation of clinicians has been trained to PD assessment and treatment based on these criteria. Notwithstanding this traditional appeal, extant research indicates that the *DSM-5* Section II criteria are likely to lack scientific support and to provide PD diagnoses of very limited clinical usefulness. Differential diagnosis among *DSM-5* Section II PDs is often problematic, and high PD covariation rates are the rule rather than the exception. Mostly, a number of studies documented that PD tend to covary rather than to co-occur (i.e., they show systematic patterns of association). Indeed, arbitrarily splitting maladaptive personality dimensions into fuzzy categories based on mixtures of trait-like features and symptom-like features rather than on sound definition of core features of personality functioning and their impairment may represent a pathway leading to PD diagnoses that are provided with few (if any) clinical usefulness. Additionally, the results of the Collaborative Longitudinal Personality Disorders Study suggest that PD criteria are likely to represent an admixture trait-like criteria and symptom-like criteria that are likely to capture behaviors that represent attempts to cope with internal or external demands (e.g., subject's self-harm to reduce affective tension in response to end of a relationship). Even *DSM-5* Section II PD clusters seems to be provided with validity problems, as evidenced by

Table 6.1 The *DSM-5* Section II Personality Disorders: alleged main presentations, diagnostic thresholds, and number of criteria

Personality disorders	Main presentation	No. of criteria	Diagnostic threshold
Paranoid	Lack of confidence in others, suspiciousness and the tendency to interpret others motives malevolent	7	4
Schizoid	Aloofness in social life and limited range of emotional manifestations	7	4
Schizotypal	High discomfort in intimate relationships, distorted cognition and perception, and odd behavior	9	5
Antisocial	Lack of regard for others' rights, with the tendency to violate them	7 (+ 15 for conduct disorder)	3 (+ 3)
Borderline	Instability in relationships, view of oneself, and impulsivity	9	5
Histrionic	Emotional lability and striving for attention	8	5
Narcissistic	Grandiosity, admiration seeking, and empathy deficit	9	5
Avoidant	Inhibition in social relationships, feelings of ineptitude, and high susceptibility to negative evaluations	7	4
Dependent	Submissive and dependent behavior, characterized by and excessive need for caring and a disproportionate need to be taken care of	8	5
Obsessive–compulsive	Preoccupation with perfectionism, being obsessed by order and control	8	4
Residual categories			
Personality change due to another medical condition	An enduring personality disturbance that is directly linked to a medical condition (e.g., cerebral cortex lesion)	–	–
Other specified personality disorder	The pattern of personality fits the general criteria for a personality disorder that is not included in the *DSM-5* classification	Variable	Variable
Unspecified personality disorder	The individual's personality disorder does not fit any specific personality disorder criteria, but criteria for a general personality disorder are met	76 (+ 15 for conduct disorder)	Variable

studies showing poor empirical evidence for the hypothesized three-cluster structure. As a whole, research stemming from clinical perspectives on *DSM-IV* PDs consistently stressed the inadequacy of relying on 10 categories for diagnosing and treating PDs. The unspecified PD diagnosis, which was defined as Mixed PD diagnosis in *DSM-IV/-TR*, represents a major issue in the clinical assessment of PDs. Indeed, it is likely to represent one of the most frequently observed PD diagnoses in clinical populations; however, it lacks any specific clinical meaning and does not allow to tailor treatment to a specific (personality) pathology.

Prominent scholars proposed to dismiss issues of differential diagnosis/PD overlap in favor of identifying core personality functions and assessing the severity of their impairment. Indeed, the severity of impairment in personality functioning, rather than specific PD diagnoses, seems to represent the most relevant factor for clinical decision-making in PD treatment; however, it should be stressed that no specific criterion for PD severity is provided in *DSM-IV*/*DSM-5* Section II. Moreover, consensus emerged among scholars on relying on dysfunctional personality dimensions mapping onto the empirically and cross-culturally validated adaptive personality dimensions to describe the defining characteristics of the subject's personality pathology.

6.2.1 The *DSM-5* AMPD

These considerations led to the development of the *DSM-5* AMPD, which was designed to provide clinicians with PD diagnoses that would be both scientifically valid and clinically useful. In this respect, the *DSM-5* AMPD was thought to provide sound criteria for clinical PD assessment, thus overcoming the massive reliance of the *DSM-5* Section II PD diagnoses on time consuming psychometric measures (on average, semistructured *DSM-5* Section II PD interviews take 1–2 h to be administered), while providing sound measures for assessing both Criterion A and Criterion B features when formal PD assessment is required (certification, forensic assessment, research diagnoses, etc.). One of the major aims of *DSM-5* AMPD was to provide a clear distinction between personality dysfunction (i.e., problems in the core personality functions) and disability (i.e., functional impairments consequent to personality dysfunction). Thus, *DSM-5* AMPD provides a clear definition of core personality functions whose impairment should be identified in PD assessment. In developing the *DSM-5* AMPD, core personality functions were based on personality functioning features that were considered relevant by the majority of personality models.

In summary, the *DSM-5* AMPD asks clinicians to identify impairments in personality functioning and pathological personality traits to make PD diagnoses. Disturbances in *self* and *interpersonal functioning* constitute the core of personality psychopathology and they are evaluated on a continuum ranging from little or no impairment (i.e., healthy, adaptive functioning; Level 0) to extreme (Level 4) impairment. At least moderate (i.e., Level 2) impairment is required for PD diagnosis. Self-functioning involves identity and self-direction; interpersonal functioning involves empathy and intimacy (see Table 6.2). The *DSM-5* AMPD provides a measure (i.e., the Levels of Personality Functioning Scale) for helping clinicians in assessing Criterion A even using a limited amount of clinical work with the client. Table 6.2 lists elements of personality functioning.

The system of the five dysfunctional personality domains and 25 dysfunctional personality traits that were included in the *DSM-5* AMPD Criterion B is listed in Tables 6.3 and 6.4, respectively. This system of dysfunctional personality features has three attractive characteristics: (a) it may be easily observed during routine clinical

Table 6.2 Elements of personality functioning

Self	Interpersonal
Identity. Unified experience of self, with boundaries between self and other people; solidity of self-esteem and self-evaluation; ability to manage a range of emotional experiences	*Empathy*. The ability to understand and appreciate the motivations of others; ability to accept perspectives of views that differ from one's own
Self-direction. The pursuit of consistent and relevant goals; the use of constructive and coherent internal standards of behavior; the ability to self-reflect	*Intimacy*. The ability to make deep and lasting connections with other people; the desire and capacity for intimacy; the mutuality of feelings reflected in relational behavior

Table 6.3 *DSM-5* AMPD dysfunctional personality domains

Negative affectivity	Tendency to frequently experience a wide range of negative emotions (e.g., depression, demoralization, worry, anxiety, anger), often causing behavioral manifestations (e.g., self-injurious acts)
Detachment	Avoidance of social and emotional situations, including a lack of interpersonal interaction and the ability to experience a limited range of affective expressions and in particular a limited ability to experience pleasure
Antagonism	Actions and attitudes that cause frequent conflicts with other people, including an exaggerated sense of superiority and the consequent expectation of deserving special treatment, as well as a lack of consideration for others, a lack of awareness of their needs and feelings, and a tendency to use others for the self's well-being
Disinhibition	Tendency to seek immediate gratification, involving impulsive actions induced by immediate thoughts and external stimuli related to current events, with little consideration for potential future consequences
Psychoticism	Tendency to exhibit a wide range of bizarre or unusual behaviors and thoughts

Table 6.4 *DSM-5* AMPD dysfunctional personality traits

Emotional lability	Tendency for instability of emotional experiences; emotions are easily triggered, intense, and/or disproportionate to events
Anxiousness	Feeling of fear and apprehension about uncertainty; tendency for frequent worry about negative effects of past experiences and negative future scenarios
Separation insecurity	Fear of being left alone because of rejection by significant others, lack of confidence in one's own ability
Submissiveness	Tendency to adapt one's behavior to the desires of others even when it goes against one's own desires
Hostility	Tendency to frequently experience feelings of anger or irritability in response to minor offenses and insults; petty or vindictive behavior
Perseveration	Persistence in tasks or in a particular way of doing things even though that behavior is no longer functional, or has caused repeated failures

Table 6.4 (continued)

Withdrawal	Tendency to be alone rather than with other people; avoidance of social occasions; tendency to avoid opportunities to approach other people
Intimacy avoidance	Tendency to avoid close or romantic relationships, interpersonal attachments, and sexual relationships
Anhedonia	Lack of enjoyment of life's experiences; tendency toward anhedonia and lack of interest in things
Depressivity	Generalized feeling of being down and unhappy; pessimism about the future; shame, guilt, low self-esteem; suicidal thoughts and behavior
Restricted affectivity	Difficulty reacting to emotionally arousing events; limited emotional expression; detachment in situations that normally cause emotional involvement
Suspiciousness	Doubts about loyalty and fidelity of others; feelsings of being mistreated, used, or persecuted by others
Manipulativeness	Tendency to use ruses or subterfuge to control the behavior of other people, including resorting to seduction, charm, or ingratiation
Deceitfulness	Tendency for dishonest and fraudulent behavior; erroneous view of self, with a tendency toward embellishment
Grandiosity	Feelings of self-centeredness, of deserving special treatment; believing to be superior to others and to deserve special treatment
Attention seeking	Tendency to behave in a manner designed to draw attention to oneself and to seek the admiration of others
Callousness	Little consideration for the feelings or concerns of others; lack of remorse for the harmful effects of one's actions on other people
Irresponsibility	Failure to meet obligations, agreements, and promises; negligence with respect to other people's property
Impulsivity	Acting impulsively in response to immediate stimuli, without a plan and with little interest in the outcome; difficulty establishing and following plans; tendency to self-injurious behavior under emotional stress
Distractibility	Difficulty focusing on tasks; with a tendency to divert attention in the presence of external environmental stimuli; difficulty focusing on the goal
Risk taking	Tendency to engage in potentially risky activities for oneself, with little regard for consequences; denial of one's limitations and personal danger
Rigid perfectionism (lack of)	Desire for perfectionism and order, for everything to be flawless and free of errors and imperfections; belief that there is only one right way to do things; difficulty changing points of view
Unusual beliefs and experiences	Belief of having unusual abilities such as telekinesis or mind-reading; unusual hallucination-like experiences may also be present
Eccentricity	Strangeness in behavior, appearance, and thinking that appears unusual, unpredictable, or bizarre
Cognitive and perceptual dysregulation	Strange or unusual thought processes and experiences, such as experiences of depersonalization, derealization, and dissociation

assessment; (b) it does not require sophisticated psychology/psychotherapy training (i.e., it can be easily used also by physicians); (c) it avoids the potentially stigmatizing jargon of the *DSM-5* Section II PD diagnoses. Ideally, observing behaviors suggestive of dysfunctional personality traits and/or capturing indicators of poor self- and/or interpersonal functioning during first examination may be useful in targeting specialized assessment even in nonpsychiatric context, such as general practitioner consultation. Indeed, PD subjects are known to have far higher morbidity and mortality than do those without. In 2015, Tyrer and colleagues reported that in the UK the life expectancy at birth for people suffering from personality dysfunctions is shorter by 19 years for women and 18 years for men than it is in the general population. Increased mortality can be explained partly by increased incidence of suicide and homicide in people with personality disorder; however, increased mortality from cardiovascular and respiratory diseases suggests that other factors are also important, for example, high prevalence of smoking, alcohol, and drug misuse in people with PDs.

To meet PD criteria, impairments in personality functioning and pathological personality traits should be relatively pervasive across a range of personal and social contexts (i.e., Criterion C), relatively stable across time, with onsets that can be traced back to at least adolescence or early adulthood (i.e., Criterion D); moreover, they should not be better explained by another mental disorder (i.e., Criterion E), should not be solely attributable to the physiological effects of a substance or another medical condition (i.e., Criterion F), and should not be better understood as normal for an individual's developmental stage or sociocultural environment (i.e., Criterion G).

The *DSM-5* AMPD provides criteria for PD-Trait specified (PD-TS) diagnosis, which can be diagnosed when specific criteria for the six PD prototypes (i.e., antisocial, avoidant, borderline, narcissistic, obsessive–compulsive, and schizotypal PDs; see Table 6.5) are not met. To diagnose PD-TS, moderate or greater impairment in personality functioning manifested by difficulties in two or more areas (i.e., identity, self-direction, empathy, and intimacy), as well as one or more pathological personality domains (see Table 6.3) or specific trait (see Table 6.4) within domains

Table 6.5 Typical features of the *DSM-5* Alternative Model of Personality Disorders specific personality disorder diagnoses

	Typical features
Antisocial	Incapacity to follow laws and an ethical behavior, and lack of concern for others, associated with irresponsibility, manipulativeness, and risk taking
Avoidant	Avoiding social situations and feelings of ineptitude and incapacity, excessive preoccupation with negative evaluation, and fears of ridicule
Borderline	Lack of stability about self-image, life goals, social relationships, accompanied by impulsivity and risk taking
Narcissistic	Susceptible self-esteem, with attempts at regulation through approval seeking, and either overt or covert grandiosity
Obsessive–compulsive	Difficulties in being able to maintain social relationships, associated with perfectionism and restricted emotional expression
Schizotypal	Impairments in the capacity for close relationships, eccentric beliefs and behavior. Distortions in understanding and perception

are required. Rather, the typical impairments in personality functioning (Criterion A) and characteristic pathological personality traits (Criterion B) for the six *DSM-5* AMPD specific PD are listed in Table 6.5.

6.3 Epidemiology

The epidemiology of PDs is poorly described compared with that of other mental disorders; this is a result of accurate assessments being more difficult to obtain for PDs than other mental disorders in national surveys. Cross-sectional epidemiological studies carried out on community-dwelling participants in North America and Western Europe report a PD point prevalence between 4 and 15%, with a mean prevalence rate of roughly 11%. Differences in prevalence across studies could be attributable to sampling methods, study instruments, and poor diagnostic reliability, especially when based on one interview. Interestingly, epidemiological studies report higher PD prevalence in urban areas than in rural settings. In epidemiological studies based on community samples PD prevalence is usually not affected by participant's gender and ethnicity, although selected PD (e.g., antisocial PD) may be more frequently observed among men.

In clinical practice, PDs are seldom diagnosed and account for less than 5% of all hospital admissions, with borderline PD, antisocial PD, and unspecified PD being the most frequently used PD categories. However, studies involving systematic PD assessment seem to provide a different picture of PD prevalence in clinical populations. Indeed, 25% of patients in primary care and 50% in psychiatric outpatient settings was reported to meet PD criteria. Several reasons may account for these differences, ranging from the cumbersome nature of the *DSM-IV* axis II/*DSM-5* Section II PD diagnosis, resulting in few clinicians assessing all PD components, to clinician's stereotyped thinking (e.g., giving Borderline PD diagnosis to repeatedly self-harming clients, irrespective of the complexity of their issues). In particular, it should be observed that PD subjects rarely seek contact with the healthcare system because of their PD-related problems; rather, they are more likely to ask for treatment because (a) co-occurring conditions (major depressive episode, etc.); (b) acute symptoms which are likely to represent extreme reactions to life events rather than manifestations of a non-PD psychiatric disorder (panic attacks, anger, sleep problems, binge-eating episodes, etc.); (c) self-harming/suicidal behavior and/or aggression; (d) alcohol/drug misuse problems; and (e) general health problems due to problematic lifestyle (obesity, etc.). All these possible presentations require treatment in and by themselves and may mask the underlying personality dysfunctions. Indeed, successful treatment of PD-related problems (even general health problems) definitively benefits from clinician's ability to capture the personality pathology lying behind the client's clinical presentation.

In forensic populations, roughly two-thirds of inmates were reported to meet PD diagnostic criteria. By contrast with community-dwelling samples, PD point prevalence in clinical services has been reported to be higher in women than in men, probably a result of higher rates of help seeking in women than in men.

It should be observed that data on PD epidemiology were based on the *DSM-IV* axis II/*DSM-5* Section II PD criteria, whereas epidemiological data based on the *DSM-5* AMPD criteria are still lacking. However, preliminary findings on adult consecutively admitted psychotherapy participants showed that the point prevalence of *DSM-5* Section II and *DSM-5* AMPD PD diagnoses were pretty similar (76.2% vs. 71.4%) albeit nonredundant (Cohen's $\kappa = 0.69$).

6.4 Clinical Presentation

As we have previously observed, the clinical presentation of PD clients may vary substantially. We will try to give some examples of different clinical presentations of PD clients.

6.4.1 Gordon

6.4.1.1 Referral

Gordon is 52 and is a physics graduate. He has always worked as an executive in the human resources office of top-ranking Italian companies; however, he lost his job several times because of severe conflicts and fights with his senior executives. Gordon's chief complaint is depressed mood and suicidal ideation.

6.4.1.2 Presenting Symptoms

Although Gordon describes himself as depressed, during the interview Gordon appears angry rather than sad; Gordon reports difficulty falling asleep, but he denies any modification in his energy level, pleasure level, or appetite. His speech is fluent, and the tone of voice is appropriate. Indeed, Gordon complains to be the victim of other people's envy. According to Gordon, his innovative ideas could not be understood by "those ignorant clerks (i.e., the company CEO) who were obviously frightened by me and envious of my smart ideas." Despite his wife's recriminations, Gordon does not see himself as an arrogant man; rather he says, "I am a giant in a land of dwarves; should I lower myself to their level?" According to Gordon, his suicidal ideation derives from the lack of consideration that his wife and the people at works had for his "unconventional way of living." Gordon denied any problems with his peers at work, provided that they do not contradict him; when he gets contradicted, Gordon says that he becomes angry and vengeful (e.g., blackmailing or harassing coworkers). Gordon reports a number of extramarital relationships with several employees; according to Gordon, these were not romantic relationships, but "simply a mean to get a preferential line of communication with the company control room." Gordon says that he is facing a marital crisis, since his wife is considering the possibility to divorce. Gordon says that three major factors are making his wife to consider divorce, namely, Gordon's economic instability, his anger outbursts, and inability to share affects with his family. Considering the third point, Gordon said that when he is at home and he is all right he likes to read and listen to music alone "without the annoying presence of my wife and my son."

6.4.1.3 Additional Background Information

Gordon asked for a psychiatric consultation 5 years ago when he became severely depressed after having being fired for the fourth time. At that time, he was sad, had no energy, lost more than 10 kg in less than a month, and attempted suicide by poison.

DSM-5 Section II PD Diagnosis: Narcissistic PD

DSM-5 AMPD Profile

Level of Personality Functioning Scale: Moderate (2). Personality Disorder Domains: Negative Affectivity (+); Detachment (++); Antagonism (+++); Disinhibition (−); Psychoticism (−)

DSM-5 AMPD PD Diagnosis: Narcissistic PD

6.4.2 Elisabeth

6.4.2.1 Referral

Elisabeth is 22 years old woman who was attending a residential program for opiate addiction treatment. Elisabeth has been referred because of her problems with rules; Elisabeth induced another young woman attending the rehabilitation program to have a sexual intercourse with her while a third patient was taking pictures, "just to rock the boat."

6.4.2.2 Presenting Symptoms

During the interview, Elisabeth complains that rules have always been a problem for her; she started to be frequently truant during junior high school; truancy was so frequent that she failed her first year. Elisabeth says that her "love for freedom" started soon afterward; although Elisabeth comes from a wealthy family, she is currently homeless. Elisabeth says that she ran away from home for the first time when she was 12; Elisabeth says that she definitively left home when she was 16 living in homeless shelters or on trains. Elisabeth denies suicidal ideation; rather, she says, "I do not think about killing myself; I simply do it!" Indeed, Elisabeth tried to kill herself four times by injecting heroin overdose; all four times intensive care treatment was necessary to save Elisabeth's life. Elisabeth says that her mood changes abruptly from depression to anger or anxiety during a typical day since she was a teenager; Elisabeth says, "When I get mad burning my harms or my breasts with cigarettes usually works, it calms me down." According to Elisabeth drug abuse is closely related to her mood swings. She started to drink alcohol when she was 13; since then, she tried a number of different psychotropic drugs, including MDMA, LSD, and heroin. Elisabeth says that she developed an opiate (heroin) addiction when she was 17. Elisabeth reports to be highly irritable and aggressive; Elisabeth says that she has been involved in a number of fights, and that she has been arrested two times for having tried to stab with a knife "disrespectful people." Elisabeth complains to feel frequently bored or empty, "but I have my way to deal with these feelings. I borrow (i.e., steal) a car and a drive like a mad driver, the faster the better". Elisabeth had four car accidents in the last year, two of which required hospitalization.

6.4.2.3 Additional Background Information

Elisabeth intelligent quotient (IQ) was 110, but she was unable to complete high school; she occasionally resorted to prostitution both to obtain money and to "feel powerful, strong, and desired."

DSM-5 Section II PD Diagnoses: Antisocial PD, Borderline PD, Histrionic PD

DSM-5 AMPD Profile

Level of Personality Functioning Scale: Severe (3). Personality Disorder Domains: Negative Affectivity (+++); Detachment (++); Antagonism (+++); Disinhibition (+++); Psychoticism (−)

DSM-5 AMPD Diagnosis: Antisocial PD

6.4.3 Gregory

6.4.3.1 Referral

Gregory is a tall, markedly overweight 35-year-old unemployed man who lives in a small town. He asked for psychiatric treatment on a voluntary basis 10 years before the psychological consultation. At that time, Gregory complained of being unable to work, as well as to being unable to engage in any leisure activities since he spent all his time controlling repeatedly the disposition of objects within his room, washing himself repeatedly because of fears of contamination and trying to drive intrusive sexual images away. Recently, he also started to be preoccupied with doubts concerning almost everything (for instance, he was constantly preoccupied with the doubt of not telling all he meant to say when he talked to someone), had to look persistently at people in order to be sure to maintain their images in his memory. He was diagnosed with obsessive–compulsive disorder (OCD), but several interpersonal difficulties that could not be ascribed to the OCD symptoms emerged and that led to Gregory's referral to the psychotherapy unit.

6.4.3.2 Presenting Symptoms

During the consultation, Gregory looked grim and aloof; however, his affect was not blunted. Although he rarely smiled during the interview, he seemed anxious or manifested anger at times. His mood was neutral. He did not display any sign of mental confusion and did not display any indications of delusions or hallucinations. His speech was circumstantial and included a lot of irrelevant details. Sometimes, the meaning of his words was somewhat vague and obscure; for instance, he said that he was frightened by a supernatural being that he called "the entity." Indeed, Gregory was afraid that the "entity" could reach him also in the hospital where none of his family could protect him. When the interviewer asked him if this "entity" was something like a ghost or a devil, he said, "You are completely wrong. The entity is neither a ghost nor a devil, and this is the reason why it frightens me. You can exorcise evil spirits, but what can be done in the case of the entity?" Gregory said that that he never saw or heard the "entity"; rather, he felt it as an impending, threatening presence that was moving toward him, "The entity is something that I cannot see; I can only feel it when it comes near to me… I can only ask my mother to stay near the door of my room to protect me." This "entity" was frightening Gregory from his early childhood; since then, Gregory's mother had to stand guard—particularly when Gregory took his daily shower—to avoid having the "entity" come too close to him. Although they caused marked anxiety, these episodes did not represent persistent ideas, thoughts or images that are perceived as intrusive (i.e., obsessions); rather, they seemed to represent frequent unusual perceptual experiences (i.e., illusions).

Gregory said that he was not superstitious at all; rather, he has been deeply involved since late childhood in what he called "the study of ancient sciences," that is, magic and paranormal activities.

Gregory describes himself as a loner, "... someone who prefers to stay on his own, on guard ... You know, doctor, all you say can be used against you .... The less people know of you the less they can damage you. I do prefer that other people mind their own business and not my own business." Gregory complained also to be unable to confide his feelings and thoughts to anybody—with the partial exception of his mother and his older brother—because of the "fear of being cheated and betrayed." Gregory said that "I can trust only my family ... all other people at a first glance may look nice and polite, but they are there only to take advantage of you or to cheat you."

6.4.3.3 Additional Background Information

In a sense, Gregory lived with his family. However, he always lived alone in a separate room, spending the majority of his time there. At best, he stayed with his parents and his brother only for lunch and dinner. When Gregory was asked if there is something wrong with his family he said, "No, I love them! They have always done their utmost for me. Simply, after a while I am uncomfortable to have them near me. ... I told you, I have always preferred to stay on my own. I prefer calling them when I need something; you know my mother lives downstairs, and my older brother lives within reach ...."

DSM-5 Section II PD Diagnoses: Avoidant PD, Paranoid PD, Schizotypal PD

DSM-5 AMPD Profile

Level of Personality Functioning Scale: Extreme (4). Personality Disorder Domains: Negative Affectivity (++); Detachment (+++); Antagonism (+); Disinhibition (−); Psychoticism (+++)

DSM-5 AMPD PD Diagnosis: Schizotypal PD

Far from giving an exhaustive overview of all possible PD presentations, these clinical vignettes aim at helping the reader to appreciate the importance of appropriate assessment of dysfunctional personality features that may lay behind the acute clinical symptoms, which frequently trigger clinical consultation.

6.5 Etiopathogenesis

Notwithstanding the impressive number of studies that were carried out on PDs since 1980, no established etiological factor has been reported in the literature for any PD. Moreover, it should be observed that the large majority of studies were carried out on schizotypal PD, antisocial PD, and borderline PD, with few (if any) studies investigating other PD manifestations, with the possible exception of psychopathy. Psychopathy should not be considered synonymous of antisocial PD and is not included in the *DSM-5* Section II, although a psychopathic specifier was provided for *DSM-5* AMPD Antisocial PD profile.

The most consistent etiological research findings are related to the genetic connection of schizotypal PD to schizophrenia. Indeed, a large body of literature suggested that schizotypal PD is moderately heritable, and genetically associated with schizophrenia. For example, adoption data have demonstrated that schizotypal PD is overrepresented in the first-degree biological relatives, but not first-degree adoptive relatives, of probands with schizophrenia. Moreover, individuals with schizophrenia and individuals with schizotypal PD manifest deficits in working memory and executive functioning, high rates of smooth-pursuit eye movement dysfunction, and diminished frontal lobe grey matter volume, although these abnormalities are less pronounced in schizotypal PD than in schizophrenia.

Twin and adoption studies have demonstrated that antisocial personality disorder, and chronic antisocial behavior more generally, are moderately heritable by a marked shared environmental component, meaning that it is influenced by environmental factors shared within families. Although data are still controversial, studies of monozygotic twins discordant for a history of early maltreatment have pointed to higher rates of antisocial behavior in abused twins, lending credibility to the possibility that such maltreatment is directly causal.

Although a biosocial model of borderline PD has been proposed (which postulates a complex set of developmental transactions between genetic vulnerabilities to emotional dysregulation and psychosocial factors, particularly an invalidating environment provided by parents and others), it received inconsistent support from empirical literature and the etiology of borderline PD remains unknown. Indeed, twin studies have indicated that borderline PD is at least moderately genetically influenced; however, the magnitude of heritability varies substantially across studies, perhaps reflecting the heterogeneity of this condition. Molecular genetic studies suggested that genetic factors may contribute to the development of borderline PD; however, no specific genes have yet been clearly identified as causative.

Early brain imaging studies suggested that people with borderline PD exhibit amygdala overactivity when judging others' emotions. Although reduced volume in the amygdala has been reported in some studies with structural magnetic resonance imaging, evidence about the specificity of reductions in amygdala volume in patients with borderline PD seems to be inconsistent. Recent positron emission tomography and functional magnetic resonance studies suggested that borderline PD may be characterized by a dysfunctional frontolimbic network; however, studies on the specificity of these findings for borderline PD are still lacking; thus, further studies are needed before considering these preliminary reports, particularly studies including healthy controls, participants with other psychiatric disorders, or participants with other PDs.

Research data suggest that borderline PD subjects report elevated rates of childhood sexual and physical abuse. However, data from studies of monozygotic identical twins discordant for borderline PD, a design that allows investigators to control for genetic influences, found little or no evidence for a direct causal effect of trauma, including early emotional, sexual, or physical abuse, on later borderline PD traits. Indeed, participants with borderline PD report many negative psychosocial factors during childhood and substantially more adverse events than do subjects with other PDs; however, no close association between these experiences and the development of psychopathological changes in adulthood has been documented.

In summary, findings from twin studies, molecular genetics, and epidemiological research suggest that joint consideration of multiple genetic and environmental factors has greater explanatory power than separate studies of genetic or environmental causation. Thus, multifactorial gene–environment interactions are likely to be a generic mechanism involved in the majority of cases of mental illness, which is only partially tapped by existing gene–environment studies.

6.6 Treatment

Up to now, the evidence base for the effective treatment of PDs is insufficient with the large majority of the available literature on PDs focusing on the treatment of borderline PD. Psychotherapy treatment is considered the treatment of choice for PDs, although firm evidence for its efficacy is still missing. Indeed, the average duration of treatment was short, follow-up were scares, and poor control of coexisting psychopathology was performed; rather, the number of outcome measures was very large, particularly in relation to the small number of participants. The psychobiological model of PD remains untested despite researchers reasonably suppose that behavioral traits associated with PD could respond to drugs. Accordingly, pharmacotherapy should only be used when integrated into psychotherapy treatments, should be time limited to manage specific symptoms, and withdrawn when these are resolved. Moreover, most clinical trials investigating the effect of drugs on PD were poorly designed, and focused almost exclusively on borderline PD, with most of trials being underpowered in terms of sample size.

6.6.1 Psychotherapy Treatment of PDs

No data are currently available as to the efficacy of psychotherapy treatment for subjects suffering from *DSM-5* Section II Cluster A disorders, with the partial exception of schizotypal PD which has been suggested to benefit from cognitive therapy. No randomized controlled trials on the efficacy of drugs for subjects with schizoid or paranoid PD are currently available; individuals with schizotypal PD have been studied in a few small, usually open-label studies using typical and atypical antipsychotics. Although schizotypal PD subjects showed some improvement in overall symptom severity, the risk to benefit ratio is still unclear.

Notably, relatively more studies have been conducted on *DSM-5* Section II Cluster B PDs than on other PDs, although findings are still controversial. There is very limited evidence available on psychological interventions for adults with antisocial PD; specifically, only three studies showed some evidence that contingency management, schema therapy, and dialectical behavior therapy may be more effective than the control condition in addressing aggression, reconviction, global state/functioning, social functioning, and adverse events as main target variables; however, no intervention reported compelling evidence of change in antisocial behavior.

Over the last decades, a variety of psychological interventions for borderline PD have been developed. Although dialectical behavior therapy and mentalization-based treatment were the most studied psychotherapies, other treatment are available, including schema-focused therapy, transference-focused psychotherapy, and systems training emotional predictability problem-solving. Recent meta-analytic data showed that psychotherapy reduced the severity of borderline PD symptoms and suicidality and may reduce self-harm and depression while improving psychological functioning compared to usual treatment. However, it should be observed that all available findings were based on low-quality evidence; moreover, most trials did not report adverse effects. No controlled psychological or pharmacological intervention studies on histrionic PD and narcissistic PD are currently available.

Meta-analytic data suggested that cognitive and psychodynamic treatment resulted in medium to large positive effect size for Cluster C disorders, although it was unclear which of the personality disorders benefited most from treatment. Preliminary controlled studies have suggested that cognitive–behavioral treatments may be efficacious for Avoidant PD; moreover, group treatments seemed to be beneficial. Currently, there are no controlled psychological or pharmacological intervention studies on dependent personality disorder, whereas one controlled study suggested superior outcomes for interpersonal therapy as opposed to cognitive therapy among depressed patients meeting criteria for obsessive–compulsive PD.

6.6.2 Pharmacological Treatment of PDs

Few small sample studies, usually based on open label design were carried out to evaluate the efficacy of typical and atypical antipsychotics on schizotypal PD; although schizotypal PD participants showed some improvement in overall symptom severity, the risk-to-benefit ratio remains unclear. No randomized controlled trials have been carried out for pharmacological treatment of schizoid or paranoid PDs; thus, no robust evidence for drug efficacy in these PDs is available at present.

There is a dearth of studies of drug treatment of histrionic PD and narcissistic PD, with most of the evidence focusing primarily on borderline PD and to a lesser extent on antisocial PD. Cochrane review gave no evidence for the efficacy of SSRIs, while showing that mood stabilizers (in particular, topiramate, lamotrigine, and valproate) could diminish affective dysregulation and impulsive–aggressive symptoms in borderline PD. Moreover, antipsychotic drugs (in particular, aripiprazole and olanzapine) showed some efficacy in improving cognitive–perceptual symptoms and affective dysregulation. However, concerns were raised as to the fact that the trials showing positive outcome provided unreliable data. Mostly, risk-to-benefit ratio should be taken into account in drug treatment of borderline PD.

Based on these considerations, the following recommendations for the use of drugs in the treatment of borderline PD clients were proposed: (a) Drugs should not be used as the primary treatment choice for borderline PD; (b) the time-limited use of drugs that showed evidence for efficacy in randomized clinical trials can be considered as an adjunct to psychosocial treatment, to manage specific symptoms; (c) clinicians should be extremely cautious in prescribing drugs that could be lethal in overdose or associated with substance misuse; (d) the use of drugs can be considered in acute crisis situations but should be withdrawn once the crisis is resolved; (e) drug treatment should be considered when a client with borderline PD has active comorbid psychiatric disorders; (f) at the opposite, if borderline PD clients have no comorbid illness, efforts should be made to reduce or stop the drug.

A recent meta-analysis on pharmacological interventions for antisocial PD, based on 11 studies involving 416 participants indicated that many participants who received an antisocial PD diagnosis in the original studies presented primarily with substance abuse problems. Although 11 different drugs were compared with placebo, data for antisocial PD participants were available only for phenytoin, desipramine, nortriptyline, bromocriptine, and amantadine. Thus, available evidence is insufficient to draw any conclusion about the use of pharmacological interventions in the treatment of antisocial PD; moreover, data on pharmacological treatment of antisocial PD came from single, nonreplicated studies, which suffered from severe methodological issues. In other terms, available evidence indicates that pharmacological interventions should not be routinely used in treating of antisocial PD or its associated behaviors.

Finally, no data from randomized controlled trials of pharmacological treatment of participants satisfying the full criteria of any cluster C PD have been reported in the scientific literature. However, studies in patients with social phobia suggested a therapeutic effect of antidepressants as compared to placebo, pointing at a possible effectiveness of these drugs in participants with avoidant PD. It should be observed that although social phobia and avoidant PD may share a common genetic liability, some studies documented that avoidant PD captured a broader constellation of symptoms and personality features pointing toward more severe personality dysfunction when compared to social phobia. These considerations suggest caution in extending social phobia data to pharmacological treatment of avoidant PD.

Hopefully, new, evidence-based approaches to PD diagnosis (e.g., *DSM-5* AMPD), as well as better understanding about the underlying biological and psychosocial developmental processes that lead to the manifestation of dysfunctional personality will result in developing specific psychotherapies and drugs in the future for specific dimensions of personality dysfunction.

Further Reading

Bateman AW, Gunderson J, Mulder R. Treatment of personality disorder. Lancet. 2015;385(9969):735–43. https://doi.org/10.1016/S0140-6736(14)61394-5.

Clark LA. Assessment and diagnosis of personality disorder: perennial issues and an emerging reconceptualization. Annu Rev Psychol. 2007;58:227–57. https://doi.org/10.1146/annurev.psych.57.102904.190200.

Gibbon S, Khalifa NR, Cheung NH, Völlm BA, McCarthy L. Psychological interventions for antisocial personality disorder. Cochrane Database Syst Rev. 2020;9(9):CD007668. https://doi.org/10.1002/14651858.CD007668.pub3.

Haslam N. Unicorns, snarks, and personality types: a review of the first 102 taxometric studies of personality. Aust J Psychol. 2019;71(1):39–49. https://doi.org/10.1111/ajpy.12228.

Kendall T, Burbeck R, Bateman A. Pharmacotherapy for borderline personality disorder: NICE guideline. Br J Psychiatry. 2010;196(2):158–9. https://doi.org/10.1192/bjp.196.2.158.

Khalifa NR, Gibbon S, Völlm BA, Cheung NH, McCarthy L. Pharmacological interventions for antisocial personality disorder. Cochrane Database Syst Rev. 2020;9(9):CD007667. https://doi.org/10.1002/14651858.CD007667.pub3.

Krueger RF, Markon KE. The role of the DSM-5 personality trait model in moving toward a quantitative and empirically based approach to classifying personality and psychopathology. Annu Rev Clin Psychol. 2014;10:477–501. https://doi.org/10.1146/annurev-clinpsy-032813-153732.

Leichsenring F, Leibing E, Kruse J, New AS, Leweke F. Borderline personality disorder. Lancet. 2011;377(9759):74–84. https://doi.org/10.1016/S0140-6736(10)61422-5.

Lenzenweger MF. Epidemiology of personality disorders. Psychiatr Clin N Am. 2008;31(3):395–403. https://doi.org/10.1016/j.psc.2008.03.003.

Lilienfeld SO, Latzman RD. Personality disorders: current scientific status and ongoing controversies. In APA handbook of psychopathology: psychopathology: understanding, assessing, and treating adult mental disorders. 1st ed. Washington, DC: American Psychological Association; 2018. p. 557–606.

Stoffers JM, Völlm BA, Rücker G, Timmer A, Huband N, Lieb K. Psychological therapies for people with borderline personality disorder. Cochrane Database Syst Rev. 2012;2012(8):CD005652. https://doi.org/10.1002/14651858.CD005652.pub2.

Torgersen S. Epidemiology. In: Widiger TA, editor. The Oxford handbook of personality disorders. Oxford: Oxford University Press; 2012. p. 186–205. https://doi.org/10.1093/oxfordhb/9780199735013.013.0009.

Trull TJ, Widiger TA. More thoughts on a shift to a dimensional model of personality disorders. Soc Pers Psychol Compass. 2008;2(2):949–67.

Tyrer P, Reed GM, Crawford MJ. Classification, assessment, prevalence, and effect of personality disorder. Lancet. 2015;385(9969):717–26. https://doi.org/10.1016/S0140-6736(14)61995-4.

Widiger TA, Trull TJ. Plate tectonics in the classification of personality disorder: shifting to a dimensional model. Am Psychol. 2007;62(2):71–83. https://doi.org/10.1037/0003-066X.62.2.71.

Adult Consequences of Neurodevelopmental Disorders

7

M. Bosia, F. Seghi, G. Bigai, F. Martini, L. Fregna, V. Fazio, and R. Cavallaro

7.1 Autism Spectrum Disorders (ASD)

7.1.1 Definition

The term autism derives from the Greek "*autos*" (meaning "self"), and it was first used in psychiatric nomenclature by Eugen Bleuler in 1910 to describe a condition of detachment and withdrawal into one's self that he observed in some patients with schizophrenia. Paving the way to the modern definition of autism, in 1943, Leo Kanner labeled as "inborn autistic disturbances of affective contact" a syndrome detected in children, mainly characterized by the lack of engagement toward their external environment. Currently, autism spectrum disorder is considered a neurodevelopmental disorder, characterized by a core impairment in social and communicative abilities, and thus in social interactions, accompanied by repetitive and restricted behavioral patterns. While the onset of symptoms occurs during childhood, typically before the age of 3, and the diagnosis is often made at that time, the manifestations of the disorder frequently persist during adulthood and may affect different domains of functioning through the entire life span.

M. Bosia · R. Cavallaro
IRCCS San Raffaele Scientific Institute, Milan, Italy

Vita-Salute San Raffaele University, Milan, Italy
e-mail: bosia.marta@hsr.it; cavallaro.roberto@hsr.it

F. Seghi · F. Martini (✉) · L. Fregna (✉)
IRCCS San Raffaele Scientific Institute, Milan, Italy
e-mail: seghi.federico@hsr.it; francesca.martini1@hsr.it; fregna.lorenzo@hsr.it

G. Bigai · V. Fazio
Vita-Salute San Raffaele University, Milan, Italy
e-mail: bigai.giorgia@hsr.it; fazio.valentina@hsr.it

R. Cavallaro, C. Colombo (eds.), *Fundamentals of Psychiatry for Health Care Professionals*, https://doi.org/10.1007/978-3-031-07715-9_7

7.1.2 Diagnostic Criteria

The view of autism has changed many times through the course of history, and its diagnostic criteria reflected this convoluted pathway. If autism was first used to define features of schizophrenia, the second edition of the Diagnostic and Statistical Manual of Mental Disorders (DSM-II) identified autism as a form of childhood schizophrenia in 1952. Based on empirical research on neuropsychology, genetics, and brain imaging, the theoretical background for autism has significantly improved throughout the following DSM revisions, redefining it as a "pervasive developmental disorder" and classifying it within a severity spectrum of five distinct disorders. According to the fifth and latest edition of DSM (DSM-5), all the disorders lying under the autistic definition have been reunited within a single dimensional condition, namely "autism spectrum disorder." This change aimed to improve the sensitivity and specificity of the diagnosis and put the focus on the specific impairments varying in severity across the spectrum, as the main treatment targets.

As reported in Box 7.1, DSM-5 identifies two core domains of symptoms: (1) deficits in social communication and interaction; (2) restricted repetitive behaviors, interests, or activities. Moreover, DSM-5 specifiers allow to define the individual characteristics, such as the presence or absence of specific impairments and/or conditions, and to rate the severity level, based on the amount of support needed.

The criteria for autism proposed by the latest International Classification of Diseases (ICD-11) of the World Health Organization (WHO) follow a similar approach to DSM. Accordingly, autism is included in the category of neurodevelopmental disorders, its symptoms are grouped into two main domains, and, although with some differences, a subdivision according to the severity is also included. In contrast to DSM, the hyper- or hyporeactivity to sensory input is not listed among diagnostic symptoms.

Box 7.1: DSM-5 Diagnostic Criteria for Autism Spectrum Disorder

- Enduring deficits in **social communication and interaction** that are constant and involve multiple contexts
- **Restricted repetitive behaviors, interests, or activities**, at the present time or in the past.
- The onset of the symptoms must occur at the beginning of development (however they may only become evident when social demands outweigh limited skills, or strategies may be learned to hide deficits).
- Symptoms have to impact important areas of the personological functioning, such as social and occupational areas.
- Intellectual disability or development delay cannot explain this condition. Codiagnosis of autism spectrum disorder and intellectual disability is frequent and can be done if social communication is at a lower level than expected for general development.

Note:

Individuals who have symptoms that are not in line with the criteria for autistic disorder, Asperger's disorder, or developmental disorder, but have difficulties in social communication should be evaluated for social communication disorder.

Currently, there are no specific criteria to diagnose autism spectrum disorder in adults. Given that symptoms may change through the life span and may be masked by compensatory strategies, the diagnostic criteria may be met based on historical information, but the current presentation must cause significant impairment.

7.1.3 Epidemiology

Currently, the prevalence of autism is estimated to be around 1–2% in children and adolescents, while in adults is about 1% without variation across age ranges. These figures show a certain variability between studies due to methodological heterogeneity and, most importantly, to challenges in diagnostic procedures. Despite this, it is clear that the prevalence significantly increased in the past decades (from 0.04% in the 1970s) and even a conservative estimate would suggest that autism spectrum disorder currently affects four and a half million people in the European Union. This rise in prevalence may rely on several factors, including biological and environmental ones, but the main reasons are increased awareness not only in clinical practice but also among parents and teachers, and the changes in diagnostic classification that broadens the criteria and allows for codiagnosis of attention-deficit/hyperactivity disorder.

Concerning the gender differences in prevalence, while historically autism was considered a male condition, now the estimated male-to-female ratio ranges from 4 to 3:1. Moreover, it is possible that this proportion does not reflect a real difference in the disorder, but rather a diagnostic gender bias, as criteria are more centered on male features and even girls who meet criteria are at risk of not receiving a clinical diagnosis.

The onset of first symptoms is typically between 18 and 24 months; however, the mean age of the diagnosis varies from 38 to 120 months and may also be in adulthood (based on history).

7.1.4 Etiology

Autism is a multifaceted disorder characterized by high heterogeneity in manifestations and severity throughout the life span. The etiology is not yet fully unraveled, being very complex and involving several genetic and environmental factors as well as their interplay.

The link between *genetics* and autism has been extensively investigated. Twin studies clearly showed heritability estimates ranging from 40 to 90%, with a polygenic basis. Genome-wide association studies (GWAS) have identified potential genes associated with ASD, most of which are involved in the development and function of the nervous system. Moreover, a significant proportion of ASD cases display numerical or structural chromosomal alterations.

Several *environmental factors* have also been suggested to contribute to the risk of autism, such as prenatal and perinatal complications and older paternal age. Other factors have been investigated, including drugs and toxic exposure, but with a less clear association with ASD.

Epigenetic mechanisms (i.e., processes through which gene expression is altered without modification in the DNA sequence) mediate the interaction between environmental and genetic factors, as the role of epigenetic aberrations (i.e., DNA methylation, microRNAs) is supported by the identification of genetic mutations in imprinted regions and genes that control epigenetic processes. Environmental factors may also cause oxidative stress in brain cells, leading to epigenetic changes affecting neural function.

Finally, the gut *microbiome* has recently gained attention for its role in postnatal development and maturation of brain function and behavior, mainly through interaction with the immune system, as suggested by the observation of a higher prevalence of gastrointestinal illnesses and an altered intestinal microbial composition in people with ASD.

7.1.5 Clinical Manifestations, Comorbidities, and Functional Outcome in Adult ASD

Autism spectrum disorder is a highly heterogeneous condition presenting with a variety of signs and symptoms that differ in severity not only between individuals but also across the life span. Evidence indicates that symptoms may change from childhood to adulthood; however, the trajectory is not so clear. Some studies report, in a proportion of people, a decrease in the severity of ASD symptoms with aging, others even suggest that a minority of children diagnosed with ASD, typically high-functioning, do not meet diagnostic criteria as adults. On the contrary, for children with a codiagnosis of autism and intellectual disability, ASD symptoms seem to worsen over time. Finally, it is also to notice that, although autism is most often diagnosed in toddlers, some people with less severe symptoms and higher levels of functioning, especially females, may receive the first diagnosis during adulthood.

Despite the abovementioned variability, in general, autism is a lifelong condition, and adults show persisting symptoms in the three main areas affected by ASD, that is, social interactions, verbal and nonverbal communication, and repetitive behaviors, with an impact on global functioning.

Concerning *social and communicative domains*, adults with ASD may present difficulties in interpreting social cues, understanding others' thoughts or feelings, reading facial expressions, looking people in the eyes when talking to them, comprehending figurative language expressions, such as metaphors, proverbs, and idioms. Moreover, they may create their own descriptive words and phrases, talk mainly about one or two favorite topics and use flat and monotone speaking patterns that do not express feeling nor adjust to the context. Because of these features, people with ASD often have difficulties in participating in conversations, as well as in building and maintaining close relationships. Adults with ASD often present also difficulties in *dealing with emotions*, with problems in regulating and responding to them, which may manifest as emotional outbursts or meltdowns when something unexpected happens, objects are moved or rearranged or any changes to the routine occur. Concerning *behavior*, typically, repetitive behaviors and rituals persist into adulthood, as well as apparent inappropriate behaviors, such as making noises in places where silence is expected. Other *less specific signs* that may be present in adults with ASD include altered sensitivity to visual, auditory, olfactory, gustatory, or tactile input; clumsiness and difficulties in coordination; deep knowledge in specific areas of interest or field of study; brilliant achievement in one or two academic subjects with poor results in others. Moreover, impairment in both nonsocial and social *cognitive domains* is associated with ASD. In adults, while the IQ is usually intact, variable degrees of deficit are often observed in theory of mind, emotion perception and processing, processing speed, and verbal learning and memory.

It is to note that a relative proportion of adults, despite continuing to experience some ASD-related difficulties, do not display symptoms in certain contexts, showing appropriate social interaction and no obvious restricted interest nor repetitive behaviors. This presentation of ASD is due to *compensatory strategies*, that is, schemes that people consciously employ during everyday social reciprocity applying learned rules, such as copying gestures of others. Compensation appears to be more frequent in people with better general cognitive abilities and in females, and it has been linked to a late first diagnosis.

Only recently, studies focused on *functional outcome* of ASD in adulthood and it is difficult to draw a precise picture with reliable percentages, also because of differences in the methodologies applied across studies as well as in the definition of criteria to classify outcomes. Overall, the outcome for adults with ASD in terms of global functioning and quality of life is poorer than for their same-age peers, but it shows a very variable pattern. While a minority of people are able to attain good employment, reach full independence and engage in friendships and romantic relationships, the majority has variable degrees of impairment across these domains, and some may not be able to live independently. The outcome depends on several factors, and studies suggest higher intellectual and verbal functioning in childhood and lower symptom severity are associated with better outcomes in adult life, while comorbidities compromise the functional outcome.

Children, adolescents, and adults often experience a number of comorbid medical and psychiatric conditions that need to be identified and appropriately treated in order to improve global quality of life and to better recognize possible subtypes groups with a stronger genetic basis.

The main *medical comorbidities* include the following:

- *Seizure disorders*
  The prevalence of seizures in adult subjects with ASD is estimated between 20 and 30%. Complex partial seizures are the most frequent, but all types of seizures may present and usually become apparent by adolescence. Seizures may be associated with low cognitive performance, dysmorphic features, and motor impairment.
- *Sleep disorders*
  Sleep disturbances are very frequent in ASD, with prevalence rates ranging from 40 to 80%. The most commonly reported problems involve sleep onset, sleep maintenance, and sleep duration. However, other types of disturbances, including parasomnias and sleep-related movement disorders, are also observed.
- *Gastrointestinal disorders*
- Currently, the prevalence of gastrointestinal disorders in ASD is largely unknown, with estimates varying from 9 to 70%, partially due to the difficulties in the clinical evaluation as these disorders may have an atypical presentation in ASD, such as tapping or putting pressure on the abdomen. The gastrointestinal disorders most frequently associated with ASD include gastroesophageal reflux disease, gastritis, esophagitis, inflammatory bowel disease, celiac disease, Crohn's disease, and colitis.

Main *psychiatric comorbidities* include:

- *Attention deficit/hyperactivity disorder*
- As detailed in this chapter, ASD and ADHD are both neurodevelopmental disorders that typically manifest in childhood but persist in adulthood. While previous versions of the DSM excluded the possibility of the dual diagnosis, the DSM-5 amended this exclusionary criterion, leading to growing research on this topic. The still scarce literature estimates the presence of a codiagnosis of ADHD in 10–43% of patients with ASD and suggests that the co-occurrence of the two is associated with a general increased psychiatric morbidity.
- *Anxiety disorders*
  Epidemiological studies indicate that up to 40% of patients with ASD may have a comorbid anxiety disorder. This increased prevalence may depend on risk factors for anxiety that are related to ASD, such as social skill deficits, cognitive rigidity, and heightened physiological arousal. The most commonly reported anxiety disorders in ASD are generalized anxiety disorder, panic disorder, and social phobia.
- *Depressive disorders*
  Prevalence rates of depression in ASD vary from 10 to 37%. Depressive disorders may be challenging to recognize in patients with ASD due to their difficulties in verbalizing feelings and may manifest more overtly at the behavioral level with a reduction of interest in special interests, decreased self-care, and regression of skills.

- *Bipolar disorder*

  ASD, especially the high-functioning type, is also associated with an increased risk of bipolar disorder, with prevalence estimates ranging from 5 to 9%, thus suggesting that the two conditions may share common vulnerability genes.
- *Obsessive-compulsive disorder*

  The rates of lifetime co-occurring obsessive-compulsive disorder in ASD are estimated from 9 to 22%. Making a codiagnosis of obsessive-compulsive disorder is particularly challenging as many features share similarities. For example, rituals and worries in ASD may resemble compulsions and obsessions, and they may either overshadow each other or be erroneously interpreted as symptoms of different disorders.
- *Psychotic disorders*

  Recent reviews and meta-analyses indicate an increased prevalence of psychotic disorders among patients with ASD, with very variable rates ranging from 4 to 60%. However, their true co-occurrence is debated, as autism and psychosis spectrum present a significant overlap in symptoms. Both disorders are associated with deficits in core social cognitive domains and communicative abilities, symptoms like self-talking, magical thinking, and disorganized speech under stress may present in ASD, as well as the restricted affect, which may mimic the negative symptomatology of schizophrenia.

As a global comment to psychiatric comorbidity rates suggested by the literature, it is crucial to distinguish concomitant clinical manifestation in light of the underlying psychopathological features, as they are more often syndromic conditions secondary to the intrinsic difficulty in coping with the environment observed in people with ASD. Even if these symptoms, when taken together, are sufficient to satisfy diagnostic criteria for other disorders, they usually mimic different clinical phenotypes and should be reinterpreted as reactive behaviors to environmental stimuli to avoid an overestimation of genuine comorbidities. Other factors should be considered in assessing a proper distinction, including global functioning and intelligence quotient (IQ). ASD subjects with low functioning and IQ are more likely to show syndromic manifestations of different dimensions of psychopathology as a maladaptive pattern of response to external stimuli, while in those with higher levels of functioning and IQ full comorbidity can be diagnosed more frequently. This distinction has great importance in defining proper therapeutic strategies.

7.1.6 Diagnostic Assessment and Evaluation Scales

Currently, there are no standardized criteria to diagnose autism in adulthood, and the diagnostic criteria for children (reported in Box 7.1) are used as a reference for adults. A relevant number of people with ASD, especially those with fluent language and normal-range intellectual level, may reach adulthood without a prior diagnosis. Undiagnosed adults often experience functional and emotional problems related both to the autistic symptoms per se and to the fact that their difficulties are misunderstood and poorly supported. Conversely, adults who receive a diagnosis often report different benefits associated with it, such as greater self-understanding and self-acceptance, more support from others, as well as access to appropriate services, and the opportunity to join a community of adults with ASD. It is therefore essential to make an appropriate diagnosis in adulthood; however, different reasons make it also particularly challenging. Issues for diagnosis include the *need for accurate historical information* that may be hard to attain, the presence of *compensating strategies* that may camouflage the symptoms, the *difficulties* that some adults with ASD may experience *in recognizing their symptoms*. Moreover, in addition to the previous considerations about comorbidity, *comorbidity with other mental health disorders* may present with symptoms may either overshadow those related to ASD or be mistaken for features of autism.

Based on these premises, the diagnostic process for adults with autism should involve a multiprofessional team, including not only trained psychiatrists but also psychologists, speech therapists, occupational therapists, and other specialized clinicians depending on the individual presentation (for example a neurologist for comorbid seizures). The diagnostic evaluation mainly consists of interviews with the patients and their families, partners, or carers and clinical examinations. Standardized assessment tools are also available and may be used to complete and deepen the evaluation, although they are not strictly required for diagnosis.

Key elements to assess ASD include the following.

7.1.6.1 Clinical History

It represents the first step of diagnostic assessment. It should be detailed and should include the following.

- Information on *family history* for neurodevelopmental disorders; *pre/perinatal conditions*, such as complications during pregnancy or delivery and screening for metabolic and congenital disorders;
- *Developmental history* evaluating psychomotor, linguistic, and social milestones, the presence of unusual behaviors, and history of a neurodevelopmental condition such as learning disabilities;
- *Medical history*, paying special attention to auditory, visual, and sensory difficulties, neurological disorders and genetic conditions,
- *Family and social functioning*.

7.1.6.2 Autism Symptomatology

It should be identified through direct observations, self-report, and information from carers. As already mentioned, symptoms may be less easy to detect in adulthood; however, the following signs should be assessed as they may be suggestive of autism in adulthood.

- Persistent difficulties in social interaction and social communication;
- Repetitive behaviors, rigid routines, and resistance to change or restricted interests, problems in obtaining or maintaining employment or education;
- Difficulties in initiating or sustaining social relationships.

Symptoms of ASD may be even harder to detect in women because they are more often able to use compensatory strategies. Females tend to try to socialize, may have a few friends, use imaginative play and emotional language, and also their special interests may have social content. Making the first diagnosis of ASD in an adult woman is thus particularly challenging and misdiagnosis with different psychiatric disorders, especially borderline personality disorder, is still frequent.

The assessment of autism symptoms may be completed also through standardized questionnaires. One of the most frequently used is the Autism Spectrum Quotient—10 items (AQ-10), a short comprehensive, self-administered assessment for autism especially indicated for adults without a moderate or severe learning disability. A score of 6 or higher is indicative of ASD. Moreover standardized questionnaires are available also for families and caregivers. One example is the Developmental, Dimensional and Diagnostic Interview—Adult Version (3Di-Adult), a brief structured, informant-report interview assessing the person's past and current autism-relevant characteristics, providing quantitative scores for each element of the ASD dimensions identified by DSM-5.

7.1.6.3 Clinical Examination

It is needed to detect the possible presence of accompanying features and/or comorbid conditions. In particular, the assessment should include a thorough *physical examination* along with *visual and auditory testing* and a specialistic neurological exam, after which EEG and additional neuroimaging controls may be required. Because autism may be associated with specific genetic syndromes (e.g., fragile X syndrome and tuberous sclerosis complex), as well as identified genetic variations, the assessment should also include genetic counseling and, if indicated, specific genetic investigations.

7.1.6.4 Assessment of Emotional, Behavioral, and Cognitive Difficulties

A mental state examination and a psychiatric interview taking into account also personal and family history should be conducted and, if required, completed with specific questionnaires and rating scales, to identify possible comorbid psychiatric conditions, especially anxiety and mood disorders. Since difficulties in expressing and comprehending emotions, disturbing behaviors, and deficits in cognitive and sociocognitive domains, especially theory of mind, are also typical manifestations of autism, they must be properly evaluated to define individualized rehabilitative treatment targeting them. Finally, *socioenvironmental information* should be gathered to detect environmental factors that may hamper quality of life and well-being and to reduce environmental stress.

7.1.7 Management of Adult ASD

As reported through the chapter, ASD is a very heterogeneous condition, with a highly variable outcome in adulthood. Global health and functional outcome are strongly influenced by the possible psychiatric and medical co-occurring disorders previously mentioned, which need to be properly treated in order to reduce their burden. Net of comorbidities, the strongest prognostic predictor, is the global intellectual level. Individuals with moderate to severe cognitive impairments often need support with employment and daily living through the life span, while individuals with cognitive abilities within the normal range usually reach a fair level of functioning and independent living, even if subtle difficulties in social interaction and communication and emotional control persist, and a subgroup no longer meets diagnostic criteria in adulthood.

Despite occasional claims in various media, currently, there is no "miracle" treatment for ASD. Actually, there is no medication, nor "alternative treatments" with evidence of efficacy to treat core autism features. The interventions indicated for ASD vary from subject to subject and over time within the same subject. As for the diagnostic process, the treatment approach should be integrated. It should include a multidisciplinary team involving psychiatrists, clinical psychologists, nurses, occupational therapists, speech and language therapists, social workers, and support staff, for example, to follow the patient in access to housing, educational, and employment services. As a general principle, the entire team should dedicate time to build a trusting and supportive relationship with the adult suffering from ASD, and, where appropriate, also with family and partners with the aim to promote personal autonomy, active participation in care, and self-management. The setting is also particularly important, as some environmental factors may trigger behavioral problems, such as bright colors, loud noise, and restricted personal space. A wide range of interventions may be considered, and the following factors should be taken into account in order to choose the best individualized strategy.

- Response to previous interventions if available
- Type and severity of symptoms
- Strengths
- Degree of functional impairment deriving from the autism and possible comorbid mental and/or physical disorders
- The presence, nature, severity, and duration of such coexisting disorders
- The identification of social or personal factors that may further contribute to functional impairment

As already highlighted, there is no pharmacological, physical, or dietary intervention recommended for the treatment of core features of autism, while evidence is growing on the possible benefits of *psychosocial interventions*. For adults with ASD without a severe learning disability who have identified difficulties in social interaction, social learning programs should be considered. These programs can be delivered either in group settings or individually and include modeling, feedback, discussion and decision-making, and strategies for dealing with socially difficult situations. For adults who require support with daily activities, psychosocial interventions focused on life skills are indicated. This type of training specifically addresses different issues and aspects of daily living, including structured leisure activity programs, anger management, anti-victimization interventions, and supported employment programs.

Behaviors that challenge (i.e., behaviors that negatively affect the person's or other people's quality of life and/or put safety at risk) such as stereotypic rocking or hand flapping, anger, aggression, self-injury, and disruptive or destructive behavior, must be comprehensively assessed and specifically addressed for treatment. The first step is to identify factors that may trigger or maintain the behavior, such as physical disorders, comorbid mental conditions, social and physical environment, communicative difficulties, and changes to routines or personal circumstances. If present, such factors need to be properly targeted, by treating the co-occurring physical and/or mental conditions, making changes or accommodations to the physical environment, and providing interventions to advise and support family and carers. Once considered these factors, psychosocial interventions, systematically targeting the challenging behavior(s) and focusing on quality of life-related outcomes, should be offered. Off-label use of antipsychotic medication can also be considered, either alone (when other interventions cannot be delivered due to the severity of the behavior that challenges) or in conjunction with a psychosocial intervention (when response to the intervention is limited). Antipsychotic treatment should be frequently monitored for efficacy and side effects and discontinued if there is no indication of a clinically significant response at 6 weeks.

As discussed through the chapter, co-occurring conditions, including seizures, gastrointestinal disorders, sensory impairments, anxiety, and depressive disorders, are frequent in people with ASD and need to be treated in order to promote global health. Comorbid mental disorders, as described in the previous paragraph, are especially challenging because of the symptoms overlap and the possible impact of ASD core features on the treatment of the coexisting condition. When indicated, delivery of cognitive and behavioral interventions for a specific mental disorder should be adapted for adults with ASD. Pharmacological treatment for coexisting mental conditions should be offered according to the guidelines for the specific comorbid disorder.

Box 7.2: Pragmatic Language Disorder

Pragmatics refers to the ability to use language in the context of communication, including the capacity to understand the speaker's intended meaning, beyond the meaning of single words and sentences, and to conduct appropriate discourse. A disruption of this ability, hampering social communication, represents a core feature of ASD and may also be present in different neurological and psychiatric disorders. Although ASD is the primary diagnostic hypothesis for people presenting pragmatic deficits, a distinct diagnosis of Social (Pragmatic) Communication Disorder has been included in the fifth version of DSM. This neurodevelopmental disorder is characterized by persistent difficulties in the social use of verbal and nonverbal communication causing significant limitations in social functioning. Specific symptoms include deficits in the person's ability to use appropriate greetings, adapt communicative style based on the setting, enter a conversation and respect the turns, understand figurative language, humor, and not explicitly stated (inferential) meanings, and interpret the verbal and nonverbal signals of others during an interaction.

Pragmatic Language Disorder should be considered in the differential diagnosis process of ASD and carefully evaluated in the absence of the restricted/repetitive patterns of behavior, interests, or activities that characterize ASD.

Box 7.3: Autism in Older Adults

Autism is a lifelong condition. However, information on older adults with ASD and on aging-relevant topics is still lacking. The limited data available suggest that people with ASD experience both autism-related aging effects and nonspecific aging effects as the general population with a negative impact on health and functioning. While the course of autism-specific symptoms is not clarified (some studies report a reduced severity but others the contrary), there is consistent evidence showing increased medical morbidity and poorer health outcome, with a higher prevalence of epilepsy, Parkinson's disease, diabetes, and heart conditions compared to age-matched controls. Similarly, mental health issues seem to increase through aging, with a higher prevalence of comorbid psychiatric disorders in older adults with ASD compared to younger ones, as well as an increased risk of self-injury and suicidal ideation, further linked to the social isolation often faced by this population. Concerning cognitive deterioration, the very few studies examining this topic suggest different patterns of decline for different cognitive processes, for instance, with more preserved visual memory and immediate recall but poorer executive functioning compared to the general population. Further research is needed to improve outcome in older people with ASD, defining specific assessment and intervention strategies, also addressing access to health-care services and social support.

7.2 Attention-Deficit/Hyperactivity Disorder (ADHD)

7.2.1 Definition

ADHD is a neurodevelopmental disorder typically occurring during childhood, with symptoms persisting, at least partially, during adolescence and adulthood in a significant number of subjects. The disorder relies on two core features: inattention and hyperactivity/impulsivity, whose main manifestations usually co-exist, causing marked impairment in different areas of functioning. The etiopathogenic construct of ADHD has been debated for decades. If descriptive traces of "hyperactive syndromes" date back to the early 1900s, the first comprehensive appearance of attention deficit disorder in the modern diagnostic systems is documented in DSM-III (1980). Ever since, a growing body of neurobiological evidence has accumulated in support of childhood and adult ADHD, involving genetic and heritability studies, neurotransmitters, and neural networks, reshaping the paradigm of this disorder from a longitudinal perspective. If original formulations of the diagnostic criteria were centered on childhood, the latest revision of DSM-5 ADHD criteria addressed the possibility to combine criteria specifically for subjects aged 17 years and older, aiming at individuating clinical manifestations unique to this age.

7.2.2 Adult ADHD Diagnostic Criteria

Ever since the acknowledgment of ADHD as a unique diagnostic entity, the focus was put on its neurodevelopmental etiology. However, clinical manifestations of ADHD in adolescents and adults highlighted the need to include criteria that could specifically target these individuals. The Wender Utah Criteria were first developed to fill the gap between previous DSM criteria (specifically designed for children) and the post-pubertal population. According to this diagnostic set, three conditions should be considered when assessing the presence of ADHD in adults, namely a retrospective diagnosis of ADHD in childhood, typical clinical features, and the exclusion of other confounding disorders. However, their reliability diverged from the recent revisions of the disorder conceptualization. Main issues included the use of retrospective analysis to assess childhood, the need for both core domains to be met in childhood and adulthood, and the identification of several concomitant symptoms in adult age that are poorly specific to ADHD.

The evolution in the DSM classification of ADHD has overcome most of these limitations. The latest version presented in DSM-5 (Box 7.4) represents a consistent effort to integrate childhood and adulthood diagnoses into a single, versatile diagnostic entity. For the first time, subjects aged 17 or older were addressed with specific clinical criteria (Criterion A), requiring a cutoff of five symptoms to assess the presence of each core domains. Moreover, the maximum limit for the age of onset was raised up to 12 years (Criterion B), reducing the risk for recollection bias at adult ages, especially by subjective self-reports. Criterion B also plays a critical role in all those subjects presenting with clinical manifestations of ADHD at any age but without a clear-cut presentation in childhood. By stating "Several inattentive or hyperactive-impulsive symptoms," no definitive cutoff is indicated, with subthreshold symptoms before the age of 12 still being considered suggestive of a full-blown ADHD later in life. Some patients, especially with a high intelligence quotient (IQ) and supportive familiar and scholastic settings, might overcome the interference derived from the disorder (Criterion D) with compensatory behaviors earlier in life. However, a consistent change in the environmental conditions (i.e., moving to a new home, leaving school) can result in the rise of a sufficient number of symptomatic manifestations and impairments in different areas of functioning.

Box 7.4: DSM-5 Diagnostic Criteria for ADHD in Children and Adults

A. A pattern characterized by inattention and/or hyperactivity-impulsivity which are persistent and lead to a decrease in functioning or development.
   - **Inattention**: at least six inattentive symptoms which have continued for 6 months or more and that are not compatible with development level, leading to a deficit in social, academic, and working functioning.
   - **Hyperactivity and impulsivity**: at least six hyperactive-impulsive symptoms have persisted for 6 months or more and that are not compatible with development level, leading to a deficit in social, academic, and working functioning.

     Note: For adolescents older than 17 years old are necessary at least five symptoms.

B. Certain number of symptoms must occur before the age of 12.

C. Certain number of symptoms must occur at least in two contexts, such as home, school, and work.

D. These symptoms impaired the quality of life.

E. The symptoms are not only present in case of schizophrenia or another psychotic disorder and other mental disorder cannot explain this condition.

7.2.3 Epidemiology

The increased awareness of adult ADHD determined a shift from a childhood-centered to a lifelong disorder. The prevalence of ADHD in children is estimated between 5 and 7% in the general population, without significant variations globally; adult ADHD, instead, presents a 2.5–4% prevalence, with local epidemiological studies showing significant variability across countries. The lack of internationally standardized diagnostic criteria before DSM-5 and the underestimation of the disorder, are the main reasons for the inconsistencies among studies. However, not all children diagnosed with ADHD persist in this condition through adolescence and adulthood. Recent evidence from longitudinal studies suggests a persistence rate of about 40–50%, in line with the differences in prevalence between age ranges.

The risk of developing ADHD seems higher among relatives of affected subjects, with heritability estimated to be approximately 75% in children. It is unclear whether ADHD persistence in adult age is suggestive of stronger genetic determinants.

No gender differences have been observed in adult ADHD, while in the childhood population a 3–4:1 male/female ratio is described. Referral bias might explain the gender variation in children, at least partially, considering that young girls are less likely to present the hyperactivity subdomain of symptoms.

7.2.4 Etiology

ADHD is a complex disorder whose etiology relies on both genetic and environmental factors. According to the neurodevelopmental hypothesis, no distinguishing etiological features have emerged between childhood and adulthood. Even if extensive knowledge on ADHD etiology is currently lacking, evidence points toward the disruption of multiple neural circuitries. Structural and functional abnormalities in the prefrontal cortex, the caudate, the cerebellum, and other structures in the cortico-limbic pathways sustain deficits in various cognitive, motor, and behavioral domains. Hypoactivity in these regions reflects slower maturation processes and reduced functional connectivity within neurocircuitries, leading to inattentiveness and emotional dysregulation. However, concomitant hyperactivation in the default mode and visual networks has been observed, highlighting the imbalance in cerebral connectivity.

At a molecular level, evidence of neurotransmitters impairment might explain the aforementioned neural disruptions. In particular, reduced dopaminergic and noradrenergic activities, through reduced receptor density, genetic polymorphisms, and receptor functional disruptions, have been proposed to justify ADHD core symptoms. However, not all agents with direct stimulation properties improved symptoms, adding complexity to the original hypothesis. Indeed, single-nucleotide mutations in the dopamine transporter causing increased dopamine release were linked to ADHD, suggesting the need for a neurotransmitters modulation rather than a simple stimulation or inhibition. Accumulating evidence on ADHD etiology is crucial to identify new, effective agents that can rebalance cerebral dysregulation, ameliorate clinical symptoms, and eventually improve quality of life.

7.2.5 Clinical Manifestations of Adult ADHD

As a childhood condition, clinical manifestations of ADHD are mostly based on behavioral features observed by parents, teachers, or other caregivers. Shifting the paradigm from a childhood-centered to a persistent disorder through the life span required a recontextualization of the symptoms at different areas of functioning. If children usually manifest their impairment in the scholastic and domestic settings, adolescents and adults might show difficulties in other areas such as academic and working functioning, family care, and social interactions (friends and relatives). However, the severity of symptoms and their functional consequences appear strongly interconnected with the environmental context. Different conditions might be protective against the significant impairment caused by ADHD symptoms, such as being closely monitored, frequent working travels and changes in the workplace, and brief, one-to-one social interactions versus long group meetings.

The assessment of ADHD in adults demands a careful revision of all clinical manifestations across different areas of functioning using both external observers and self-reports, being tailored to every single patient.

Comparable to children one, adult ADHD revolves around the same core domains (inattentiveness, and hyperactivity/impulsivity), even though they can develop independently. According to DSM-5, adult ADHD may have a predominantly inattentive or hyperactive/impulsive presentation, as well as a combined one. *Inattention* refers to the difficulties in maintaining the focus for long timespans, usually leading to inability to concentrate, distractibility for external stimuli, mind wandering for unrelated thoughts, and failure in managing sequential tasks. Adult patients might complain about difficulties in organizing work with frequent inaccuracies and missing details, inability to sustain attention during long group meetings, or easy forgetfulness about deadlines (i.e., calls, appointments, paying bills). These symptoms should occur at least in two different settings, usually causing poor time management and frequent loss of items for daily activities, and should not be better explained by hostile, defiant behaviors or lack of comprehension. Instead, the *hyperactivity/impulsivity* domain is expressed through excessive motor activity in inappropriate situations on the one hand and being involved in thoughtless, risky actions that may lead to harmful or unwanted consequences without a careful balance of the risk–benefit ratio on the other hand. Hyperactive/impulsive presentations in adults are usually characterized by extreme restlessness and impatience, with difficulties in remaining seated or still in different situations, such as the workplace, meetings, restaurants, movie theatres, churches, lectures, or even in queue. These patients might appear increasingly talkative and incapable of respecting interpersonal boundaries, including waiting for their turn during conversations and intruding in other's actions and activities. Decision-making is usually impaired, possibly leading to hasty behaviors like excessive spending, acting without thinking, taking or leaving jobs, or abruptly interrupting relationships, lacking an adequate evaluation of possible consequences.

Inattentive and hyperactive/impulsive behaviors can lead to a wide range of *functional consequences,* especially in those adults with a history of full-blown ADHD during childhood. Many personal and relational life goals can be affected. Incapability to sustain the focus and forgetfulness about deadlines can often result in academic underachievement, instability in the working functioning, and frequent job changes or unemployment. Hyperactivity and novelty-seeking can determine easy irritability, mood lability, and emotional dysregulation, usually impacting long-term relationships. Family duties are often neglected, with consequent poor, dysfunctional parent–child relationships. Distractibility usually results in driving or job-related accidents potentially harmful to self or others. Restlessness and impulsivity might lead to impaired stress tolerance, with frequent emotional outbursts, interpersonal conflicts, and violent behaviors up to aggressiveness and criminal acts.

Unsurprisingly, different trajectories of evolution have been described in children with ADHD. The *co-existence of multiple psychiatric disorders* with ADHD is quite frequent, with estimates up to 65% of children and 75% of adults. Substance use disorder, conduct disorder in adolescents, and personality disorders in adults, especially antisocial, are common, as well as autism, learning, and other neurodevelopmental disorders. Moreover, mood fluctuations sometimes evolve into full-blown mood and anxiety disorders in adults, even though clinical manifestations of

ADHD persist beyond mood episodes. Instead, those subjects with prominent impulsive features and low tolerance to frustration can show disruptive and impulse-control disorders as comorbidities in adult age. In turn, these conditions create a vicious circle with ADHD, sustaining symptoms and impairment in global functioning and possibly contributing to increased risk of suicide in early adulthood.

In conclusion, ADHD is a polymorphic disorder. A large variety of clinical manifestations can occur at different settings, expressed at sub- or supra-threshold levels according to many factors, including academic, working, social, and familiar context, and environmental triggers. As a mostly behavioral disorder, large interindividual differences can be described, as various combinations of symptoms can be integrated into a high number of clinical phenotypes showing different functional consequences and prognoses. A careful revision of all symptoms across the life span is crucial to evaluate presentations of the disorder that are unique to every single patient.

7.2.6 Diagnostic Assessment and Evaluation Scales

ADHD is a highly heterogeneous clinical condition with considerable variability in functional impairment, according to the severity of symptoms, protective environmental factors, personal resources, and other comorbidities. Despite its lifelong duration, fluctuations have been described in behavioral and clinical manifestations across the life span, leading to frequent changes in the expression and pervasiveness of symptoms. As a result, personal insight into the disorder is usually variable, and external reports (e.g., family, friends) play a crucial role. Despite advances in the research field in both children and adults, no specific neurobiological marker has emerged as distinctive of ADHD. To date, the assessment of ADHD relies only on clinical information. Therefore, the diagnostic algorithm is a complex, multistep process requiring a multimodal and multidisciplinary approach. Based on DSM-5 diagnostic criteria, the initial assessment of ADHD in adults is composed of three steps:

7.2.6.1 Step 1: Screening for Childhood-Onset and Current Symptoms

A systematic revision of the familiar and the personal psychiatric history before the age of 12 is necessary, addressing evaluation scales together with previous neuropsychiatric documentation and the caregivers' (e.g., parents, relatives, teachers) point of view whenever available. Notes, school records, or other relevant reports should be carefully reviewed in search of signs of inattentiveness and hyperactivity/impulsivity. If a childhood-onset is confirmed, the clinician should perform extensive assessments for current symptoms of ADHD, based on both the patient's and external observers' (e.g., partner, relatives, colleagues, close friends) perspective, with the aid of age-related evaluation scales and relevant documents. As most of the behavioral manifestations are not specific to ADHD, this step is essential in ruling out other relevant differential diagnoses with overlapping presentations, such as mood, anxiety, personality, neurodevelopmental, and cognitive disorders

7.2.6.2 Step 2: Clinical Characterization of ADHD

A global characterization of the subtype is required, assessing the predominant presentation (inattentive, hyperactive/impulsive, or combined), the severity of symptoms, full-blown vs remission status, and evaluation of global functioning and disability. A description of concomitant symptoms typical of adult age is also crucial, including mood lability, sleep disorders, impaired stress tolerance, and executive functions disorders.

7.2.6.3 Step 3: Evaluation of Comorbid Psychiatric and Medical Conditions

This step covers all the possible trajectories of evolution of ADHD by assessing psychiatric and medical comorbidities, with a special focus on those conditions with a higher prevalence in ADHD subjects. Mood and anxiety disorders, substance abuse and addiction, metabolic and cardiovascular conditions should all be reviewed to provide the most suitable treatment plan.

The diagnostic process of ADHD in adults is a composite procedure involving broad-spectrum assessments and a wide range of concomitant conditions. A multidisciplinary approach is usually required, with psychiatrists working alongside neurologists, neuropsychologists, speech therapists, and other internal medicine specialists. Given the low specificity of symptoms and the lack of validated biomarkers, all relevant information should be collected through age-specific evaluation scales, including self-reports and semistructured interviews. Different assessment tools have been approved in the adult population, among which the most frequently used are the Adult ADHD Self-Report Scale (ASRS-v1.1), the Wender Utah Rating Scale (WURS), and the Diagnostisch interview voor ADHD bij volwassenen (DIVA). ASRS-v1.1 is a 6-item self-report screening for the general population, with 12 additional optional items covering the core features of ADHD, including inattention, behavioral disorders, and emotional problems. DIVA is an 18-item semistructured interview covering the main symptoms of the core domains of ADHD, based on DSM-5 criteria. When choosing the most suitable assessment scales across countries, it is important to consider tools validated in the local language, especially for self-reports.

7.2.7 Management of Adult ADHD

The mainstay of adult ADHD management is the combination of pharmacological and nonpharmacological treatments, including psychoeducation, psychological therapies, and cognitive rehabilitation. Both approaches are strictly interrelated, providing complementary effects: if medications target ADHD symptoms, the other interventions allow for functional recovery in the impaired settings. However, the choice of the most appropriate treatment plan should be individualized according to the specific presentation of each patient and requires direct involvement of the partner, the family, close friends, or other significant acquaintances. Despite the lack of internationally accepted guidelines, an updated European Consensus

Statement on the management of adult ADHD has been recently published. The treatment plan of ADHD in adults relies on three main aims, involving multidisciplinary support:

7.2.7.1 Management of Comorbid Disorders

A proper evaluation and treatment of mental and medical comorbidities is essential, considering the consistent impact on the symptomatic dimensions of ADHD and the level of global impairment. Psychotic, mood, anxiety, and substance disorders demand immediate treatment, as improvements are usually associated with a concomitant reduction of ADHD burden. However, milder conditions that do not reach the full criteria for a comorbid disorder, such as mood swings and impulsive behaviors, might ameliorate after targeting ADHD symptoms directly.

7.2.7.2 Improvement in Symptoms of Inattention or Hyperactivity/Impulsivity, According to the Main Presentation

The clinician should carefully review the medical history whenever available in search of previous treatments considering the usual childhood onset. Those subjects diagnosed with ADHD earlier in life are recommended to continue pharmacological agents that showed a beneficial effect in reducing symptoms and improving quality of life.

Patients diagnosed with ADHD for the first time in late adolescence or adult life should follow different steps. After a comprehensive evaluation and an appropriate treatment of comorbid conditions, they should begin with a *psychoeducational program* to increase awareness of ADHD, reduce the social stigma, and improve the quality of interpersonal relationships. If no significant control over symptoms and functional repercussions is achieved, most guidelines recommend stimulant and nonstimulant *pharmacological agents* unless the patient favors psychological approaches. *Stimulants are the first choice* of medication; among them, methylphenidate (MPH) and amphetamines have the approval for ADHD. MPH is a norepinephrine and dopamine reuptake inhibitor with a short half-life (approximately 2 h), showing the highest improvement in symptoms, especially on mortality rates by accidents, illicit acts, suicidal behaviors, and substance misuse. Studies of efficacy on MPH showed both short- and long-term significant effects. Extended-release formulations recently made compliance more feasible by stabilizing pharmacokinetics and reducing the frequency of daily administrations. Even if effective, several possible adverse reactions should orient the choice of MPH. Neuropsychiatric (motor and verbal tics, sleep troubles, mood swings, headache, anxiety, and psychotic symptoms), cardiovascular (hypertension, tachycardia, palpitations, arrhythmias, myocardial infarction), cerebrovascular conditions (stroke), and appetite changes are among the most frequent side effects, undermining adequate compliance. Besides, the dopaminergic activity can lead to misuse of the drug beyond prescriptions, making it less advisable in addictive disorders. The psychiatrist should take into account all indications and contraindications when prescribing MPH for ADHD, according to the risk–benefit ratio, and carefully review potential side effects arising from the treatment monthly. Considering the high burden of comorbid medical disorders among adults, a recent meta-analysis suggested amphetamines (dexamfetamine and lisdexamfetamine) as first-line treatment given the comparable efficacy but favorable tolerability profile. If stimulant agents do not show adequate efficacy or are contraindicated, nonstimulants are second-choice medications. Atomoxetine (ATX) is a nonstimulant agent approved for ADHD, acting as a presynaptic norepinephrine reuptake inhibitor with no influence on dopaminergic transmission. Despite lower efficacy than MPH, ATX shows a primary indication in the case of substance abuse due to a lower risk of addiction and in the case of co-occurring anxious manifestations thanks to its selective effect on social anxiety. However, close monitoring of liver dysfunction and suicidal ideation is necessary. Other medications have been investigated as monotherapies and combination treatments, including clonidine, guanfacine, bupropion, SSRIs, and reboxetine. However, evidence supporting their use is not strong enough, with no approval for ADHD in Europe. When prescribing pharmacological agents for adult ADHD, it is mandatory to consider indications of use under local approval. Most of these drugs, especially MPH, are licensed for childhood ADHD only (while off-label in the adult population), showing high variability across European countries.

7.2.7.3 Reduction of Functional Impairment Arising in Different Settings

Together with treatments aimed at reducing the severity of clinical symptoms of ADHD, functional approaches should be offered to compensate for impairment in daily life and disabilities from academic and working deficits. However, evidence supporting their role is limited to the use as adjunctive therapies with ongoing medications. Given that no single intervention showed superiority over the others, the choice of the appropriate therapy depends on the predominant manifestations of each patient. *Cognitive-behavioral therapies* (CBT) represent the primary approach addressed at adults with ADHD, delivered through structured procedures in individual and group sessions. If behavioral training targets time management, planning skills, and organizational strategies, cognitive restructuring aims at developing problem-solving skills and correcting automatic negative thinking and self-confidence. As a result, CBT deals with dysfunctional coping schemes and cognitive distortions such as avoidance conducts, social anxiety, low self-esteem, and procrastination. In case of predominant hyperactive/impulsive behaviors, derivatives of CBT, including *dialectical-behavioral therapy* (*DBT*) and *mindfulness-based therapy*, should be integrated. If DBT has been first developed to treat borderline personality disorder (BPD), evidence points to an active role in ADHD when comorbid with BPD or primarily characterized by irritability, anger outbursts, and impulsivity. Instead, mindfulness-based therapy targets emotional regulation and compensative coping strategies. When subjects present with predominant inattentive features, attentional remediation interventions aim at improving neurophysiological and neuropsychological processes. *Cognitive remediation* (CR) is a training program delivered via multiple computerized sessions acting on cognitive deficits, especially working memory. Even though different studies showed a short- and long-term improvement in objective measures of working memory after CR, most of them failed to capture a concomitant effect on inattentive symptoms and daily functioning. Instead, neurofeedback (NFB) is a neurophysiological technique based on electroencephalographic (EEG) monitoring. The device provides instant positive and negative feedback through visual or auditory stimuli to induce better control over EEG signals and brain functioning. Although evidence of efficacy in childhood ADHD traces back to the end of the last century, high-quality studies assessing NFB on adults are currently lacking, causing limited application.

7.3 Intellectual Disability

7.3.1 Definition

Intellectual disability (ID), also known as developmental disability (DD) or mental retardation (MR), indicates a group of individuals with cognitive impairments, with onset at birth or in infancy (before 18 years), leading to consequences in later ages. This condition is due to a combination of environmental and genetic factors that result in functional life-skill deficits and significant limitations in adaptive behavior (conceptual, social, and practical skills).

In order to make a diagnosis of ID, the first necessary step is an assessment of the intelligence quotient (IQ), followed by an evaluation of the environmental adaptation. Evidence shows that patients with the same intellectual level do not all have the same levels of adaptive and social function, thus indicating that the global functioning of an individual is determined by the interaction between cognitive capabilities and external stimuli from the surrounding.

Box 7.5: DSM-5 Diagnostic Criteria for Intellectual Disability

The diagnosis of intellectual disability must include the three following criteria

A. Both clinical evaluation and standardized intelligence tests detect deficiency of intellectual functioning, such as reasoning, planning, abstract thinking, and learning.
B. Deficits in adaptive abilities lead to limited daily functioning and determine a failure to achieve the sociocultural standards of autonomy and responsibility.
C. Symptoms must occur during the development period.

7.3.2 Diagnostic Criteria

Once Intellectual Disability is recognized, the degree of severity is determined by the level of adaptive functional impairment, which, in turn, associates with the level of support that the patients require in everyday life. According to DSM-5, there are four different severity levels of ID.

- *Mild Intellectual Disability* (85% of patients with ID): it usually relates to IQ range from 50 to 70. It is often identified during the first or second grade when academic demands increase. People can develop normal social and communication skills so that many adults with mild intellectual disability live independently with minimal support. The required support may increase when in social or external stress.
- *Moderate Intellectual Disability* (10% of patients with ID): in this case, IQ level usually ranges from 35 to 50. Individuals can acquire language and adequate communication abilities during early childhood, but they early show academic difficulties and do not achieve above a second to third-grade level. They usually need social and vocational support and may be able to perform semiskilled work under appropriate supervision.
- *Severe Intellectual Disability* (4% of patients with ID): it associates with IQ levels from 20 to 35. Patients may manifest poor motor development, little or no communication skills, and usually acquire minimal self-care abilities in a controlled environment. In adulthood, patients may be able to perform work-related tasks, but only under supervision.
- *Profound Intellectual Disability* (1–2% of patients with ID): it associates with an IQ level less than 20. They present cognitive and sensorimotor impairments in association with gross global functioning disability, which often relate to identifiable underlying organic causes. Patients need constant aid and supervision, including health care and nursing assistance. As adults, they may have acquired some base self-care and communication skills, but only with appropriate training earlier in life.
- *Unspecified Intellectual Disability* is a term, included in DSM-5, reserved for individuals over the age of 5 who have sensory/physical impairments (e.g., blindness, deafness, severe motor disturbances) and concurrent mental disorders, which lead to adaptive functional disabilities. Though these patients are difficult to evaluate, they are strongly suspected of having intellectual disability.

7.3.3 Epidemiology

Approximately 1–3% of the general population suffers from ID, and prevalence increases in developing countries, where 10–15 per 1000 born meet the DSM-5 diagnostic criteria for ID. The reasons may be found in the higher number of births, lack of adequate health-care services (which leads to poor delivery assistance with a higher risk of perinatal complications), and poor specific scholastic support. However, even in more developed societies, the prevalence of ID may be underestimated, above all in its lower severity degrees (mild and moderate). This may be ascribed to the fact that patients show significant impairments when academic and social skill demands increase, with the peak of diagnosis occurring between the ages of 10 and 14 years. In general, ID is about 1.5 times more frequent among males than females.

7.3.4 Etiology

Although the etiology of ID remains largely unclear in most cases (idiopathic), it is possible to determine a biological cause in approximately 25% of affected individuals.

Usually, ID has a multifactorial etiology, resulting from the interaction of different determining factors, which can be subdivided as follows.

- Genetic causes: they include genetic and chromosomal inherited alterations, which often determine both physical and intellectual disabilities. The most common are trisomy 21 (Down Syndrome), fragile X syndrome, Prader–Willi syndrome, cat's cry (cri-du-chat) syndrome, phenylketonuria, Rett syndrome, neurofibromatosis, and enzyme deficiency disorders such as maple syrup disease.
- Developmental causes: they relate to prenatal or perinatal insults, including infections (rubella, syphilis, toxoplasmosis, herpes simplex virus, HIV), fetal distress (alcohol or drugs fetal exposure, hypo/anoxia, maternal uncontrolled diabetes, placenta damages), premature births, high fever, traumatic brain injury.
- Environmental causes: they include long-term exposure to environmental and sociocultural factors, such as malnutrition, severe neglect or abuse, and poor scholastic access.

7.3.5 Psychiatric Syndromes in Adults with Intellectual Disabilities

Epidemiological surveys indicate that up to 40–70% of adults with ID show patterns of concomitant psychiatric symptoms, 3–5 times more frequently than the general population. The prevalence of psychopathologic conditions increases with the severity of Intellectual Disability. The symptomatologic pictures presented are similar to those of the primary psychiatric disorders, however, they are usually secondary to the neural and functional alterations responsible for the Intellectual Disability itself. The most common psychiatric syndromes associated with ID are described below:

- *Behavioral and Conduct Disorders* (40% of patients)

  Many patients who suffer from ID, in particular in its mildest degree, show dysfunctional psychomotor activity early in life, characterized by hyperactivity and attention disorders, motor repetitive stereotypical behaviors (e.g., toe-walking), poor frustration tolerance, rigid problem-solving ability, which may lead to self-injurious behaviors (e.g., head-banging and self-biting).

  The poor social skills that these patients can acquire during growth may lead, later in adulthood, to more structured conduct disorders, in particular Disruptive

Conduct Disorder. In mild ID patients, the progressive acknowledgment of global impairment compared to the rest of the population, the repeated failures, and the inability to satisfy social expectations often lead to low self-esteem, inadequacy, anxiety, and feelings of anger, which translate into social isolation or aggressive behaviors.

- *Affective and Mood Alterations* (40% of patients)

  Adults with milder forms of intellectual disability often show unspecific depressive tracts such as negative self-image and low self-esteem, but, in general, they manifest depressive symptoms that are similar to those of the unimpaired population (e.g., depressed mood, sleep disorders, lower food intake, increased self-injurious acts, poor attention, and concentration). On the contrary, in more severely impaired patients, depressive conditions are usually characterized by atypical features (e.g., aggressive behavior, agitation, and anxiety). In general, subsyndromic depressive symptoms, manic-like symptoms with or without psychotic features have all been reported with higher frequency in patients with ID. A full-blown depressive or manic episode may rarely occur, and it usually results more difficult to diagnose due to the underlying behavioral manifestations.
- *Autism Spectrum Disorders* (5% of patients)

  Evidence underlines the presence of a positive correlation between ID and the incidence of Autism Spectrum Disorders (ASD). It can be very difficult to distinguish Autism from ID, especially in its most severe form. Patients who suffer from both ID and ASD typically exhibit poor social interaction, lack of communication, low response to external stimuli in favor of self-stimulation, and repetitive behaviors.
- *Psychotic Disorders* (10% of patients)

  Individuals with ID may present schizophrenia-like symptoms, in particular, positive symptoms (e.g., delusions, hallucinations), which occur more frequently than in patients without ID. In addition, many individuals who suffer from ID show less severe psychotic-like behaviors that impact thought and language (e.g., audible self-talk, repetitive overvalued ideas, imaginary friends, echolalia) and motor activity (e.g., mannerisms and stereotyped movements). Negative symptoms may also occur, especially a lower psychomotor initiative and autism-like interactions.
- *Anxiety* (30% of patients)

  Patients with ID often manifest anxiety symptoms, presenting with both psychological and physical features. The onset and/or exacerbation of anxiety typically occur related to stress factors, such as changes in the environment and lack of external support, or associated with depressive symptoms. The most frequently observed subtypes of anxiety disorders are generalized anxiety, panic attacks, and specific phobias (e.g., agoraphobia, social phobia), and obsessive–compulsive traits.

7.3.6 Diagnostic Assessment of Intellectual Disability and Concomitant Psychiatric Comorbidities

As mentioned earlier, Intellectual Disability is usually diagnosed during childhood. However, in its mildest degrees, it may be undiagnosed since late adolescence and young adulthood, when the individuals show a certain inability to reach a proper level of social and global functional skills. In order to achieve a valid diagnosis, it is mandatory to conduct a thorough investigation including the following.

- *Clinical History Gathering*. It may evidence the presence of both organic and environmental determining risk factors and alterations in behavior or cognitive impairment that can support the hypothesis of ID. In this regard, it is extremely useful to investigate the results obtained in the academic career.
- *Physical Examination*. An accurate evaluation of external appearance may help to recognize the presence of organic conditions that may have affected the patient's ID. In fact, some physical features are typically associated with genetic and chromosomal diseases (e.g., Down syndrome, fragile X syndrome), prenatal and perinatal events (e.g., alcohol fetus exposure), and long-term exposure to environmental factors (e.g., malnutrition).
- *Neurological Examination*. Neurological disturbances increase in incidence and severity in direct proportion to the degree of ID. Sensory impairments frequently occur in severe degrees of ID, with a rate of presentation that is about four times that of the general population. Sensory alterations include hearing impairment, visual defects (deficits in spatial and body image concept, design recognition, till blindness in most severe cases). Epilepsy disorders occur in one-third of those with severe intellectual disability. Typically, ID patients manifest motor disturbances ranging from poor coordination and clumsiness to more severe alterations of muscle tone (e.g., rigidity, spasticity, hypotonia, and lack of muscles strength), reflexes (hypo/hyperreflexia), and involuntary movements (e.g., choreoathetosis, tremors, and tics).
- *Mental Status Examination*. It is important to conduct an accurate psychiatric interview, evaluating: the level of consciousness, cognitive functions (e.g., attention, concentration, memory, orientation); external aspect (e.g., lack of self-care and hygiene); psychomotor activity (e.g., reactions to external stimuli, social interaction, stereotyped movements); mood and affect (e.g., a tendency to depression, anger, feelings of inadequacy, anxiety, aggressive behavior); thought and language form and content (e.g., stereotyped thoughts and wording, echolalia, lack of spontaneity in communication); willfulness and impulse control issue (lack of impulse control, intolerance to changes and easy frustration, poor problem-solving capacity); neurovegetative functions (e.g., appetite, sleep–wake cycle); ability of judgment and awareness of illness.

- *Intelligence*. ID always relates to a varied range of subthreshold intelligence levels. Intellectual impairment in adulthood may be diagnosed with the help of specific clinical tools such as the Kaufman Adolescent and Adult Intelligence Test and the Wechsler Adult Intelligence Scale—Fourth Edition (WAIS-IV). Most severe degrees of mental retardation might not be able to perform a reliable WAIS testing, leading to the development of alternative scales. The Beta-4 assessment is a nonverbal measure of general intelligence consisting of five clinician-administered subtests that can be performed in low-functioning or low-skilled individuals.
- *Laboratory*. They result useful to evaluate the presence of organic conditions and include complete urine and blood analysis and chromosome studies, which are important for after-birth diagnosis (e.g., fluorescent in situ hybridization, karyotype analysis).
- *Instrumental Investigations*. These include neuroimaging studies such as computerized tomography (CT) and magnetic resonance imaging (MRI), which may evidence alterations in brain structure, in association with functional MRI (fMRI) and diffusion tensor imaging (DTI) which evaluate the presence of alterations in neuronal connectivity. Electroencephalography (EEG) is indicated in case of concomitant seizure disturbances or its latent condition to consider the risk of treatment-induced seizures.

The diagnosis of psychiatric comorbidities in adults who suffer from ID may be insidious and challenging for many reasons. For instance, patients with ID often show difficulties in recognizing and communicating self-emotions and internal feelings. This difficulty, associated with language deficits, builds a barrier that cannot be easily overpassed during a psychiatric interview. In addition, there is a widespread tendency to attribute any behavioral and psychopathological alteration to impairment due to ID, underestimating the presence of a primary comorbid psychiatric disorder. Specific clinical rating scales may help the clinicians to evaluate the presence of psychiatric comorbidity. For example, the Aberrant Behavior Checklist (ABC), the Developmental Behavior Checklist (DBC), and the Behavior Problem Inventory (BPI) are useful tools to rate behavioral abnormalities such as self-injurious conducts and aggressive and stereotyped behaviors. More specific scales, such as the Psychopathology Inventory for Mentally Retarded Adults (PIMRA), represent precious tools to identify the presence of comorbid psychiatric disorders. It is a 56-item scale that refers to DSM-III classification and includes 8 subscales corresponding to different mental disorders (Schizophrenia, Affective Disorder, Psychosexual Disorder, Adjustment Disorder, Anxiety Disorder, Somatoform Disorder, and Personality Disorder, Inappropriate Mental Adjustment). Other assessment scales for mental disorders are the Brief Symptom Inventory (BSI), the Psychiatric Assessment Schedule for Adults with a Developmental Disability (PAS-ADD), and the mini-PAS-ADD. The evaluation of emotional disturbances in subjects with severe disabilities is possible using the Diagnostic Assessment for the Severely Handicapped II (DASH-II).

7.3.7 Management of Psychiatric Comorbidities in Adult Patients with ID

Since the etiology of mental disturbances in ID adult patients is often multifactorial, a complex approach combining psychological, behavioral, and pharmacologic treatments is recommended.

7.3.8 Pharmacological Treatments

Frequently, a straightforward psychiatric diagnosis is not possible. No pharmacological agent is indicated in the treatment of ID, but empirical pharmacotherapy should be considered in case of concomitant medical and psychiatric symptomatology. Evidence underlines that a single-drug treatment is better to avoid possible pharmacological interactions and reduce the occurrence of adverse drug reactions. It is recommended to start the treatment with a single medication, at the lowest effective dose, followed by slow uptitration of dosage. Only if monotherapy fails, dosage increase or combination with more medications should be considered. The benefits achieved with pharmacotherapies may not last for reasons that remain largely unclear. In case of loss of efficacy, it is mandatory to discontinue the treatment and substitute it with same class medications only if necessary (i.e., in case of persistence of symptoms). Then, if there is no treatment response, it is recommended to consider a different drug category. In order to prevent the occurrence of adverse drug reactions and intolerance to the treatment, it is mandatory to undertake physical, laboratory, and instrumental (e.g., ECG and EEG) examinations. These investigations must be repeated at regular intervals in time to monitor the patient and conduct appropriate pharmacovigilance. Although pharmacological treatment in monotherapy is preferred, severe mental disorders often require a combination of medication. Given below is a description of pharmacological treatments related to the most frequent psychiatric syndromes is provided.

- *Aggressive behavior, psychomotor hyperactivity, manic symptoms*: first choices are mood stabilizers and anticonvulsants such as valproate, carbamazepine, oxcarbazepine, lamotrigine, and gabapentin. These treatments are useful also in case of concomitant epilepsy disorders. In case of persistence of symptoms or association with psychotic features, low-dose atypical antipsychotics, and beta-blockers may be considered.
- *Depressive-like states:* it is indicated to use SSRIs (e.g., fluoxetine, fluvoxamine) and tricyclic antidepressants. In case of poor response, it is possible to associate them with lithium or another mood stabilizer. In some cases, stimulants such as methylphenidate, atomoxetine, and bromocriptine also represent treatment options.
- *Psychotic disturbances:* it is recommended to consider atypical antipsychotics as first-line treatment (e.g., risperidone, olanzapine, quetiapine, aripiprazole). In case of prominent positive symptoms, typical antipsychotic agents, such as halo-

peridol, have the highest efficacy, even though they are more likely to induce extrapyramidal side effects, compared to non-ID population. In the case of refractory illness, treatment with clozapine must be considered, given the lower incidence of motor side effects and the evidence of a reduction in psychiatric admissions.

7.3.9 Psychological and Behavioral Treatments

Since patients with ID manifest communication deficits and a tendency to social retirement, recent evidence shows that these types of interventions may be useful to shape dysfunctional behavior and enhance social interaction. These may be useful also in the treatment of depressive disorders and anxiety disorders, as primary treatment or, more often, in association with appropriate pharmacological treatment.

Box 7.6: Benzodiazepines Paradoxical Effect

The use of benzodiazepines as anxiolytics or in states of psychomotor agitation should be avoided in the case of ID. Their use in this type of patient has in fact proved not only ineffective but deleterious. The paradoxical effect of benzodiazepines in patients with ID is well documented and leads to real crises of psychomotor agitation, aggression, outbursts of anger, and increased anxiety.

This effect seems to derive from the persistence of GABA-A subunits with a different conductance (cationic instead of anionic) resulting in excitatory activity when bound by benzodiazepines.

Box 7.7: Tips for Mental Health Nursing Management in Adult with Intellectual Disability

Individuals with ID and additional mental illnesses represent a heterogeneous group of patients, whose characteristics largely vary depending on the severity degree of ID and the association with psychiatric disorders.

The interventions required for mental health care in adults who suffer from ID are not so different from those applied to the general population. What is challenging is the identification of primary psychiatric symptoms, not simply attributing them to altered behaviors connected with ID. The difficulty is due to the communication deficiency presented by patients, especially those with severe degrees of ID, who often use nonverbal communication to express themselves (e.g., gestures, body movements, unspecific sounds, or facial expressions).

Usually, the health care of people with ID is directed to the management of challenging and possibly aggressive behavior. Focusing on emotional contact and direct relationship may help to ameliorate mental illness symptoms, for example, by decreasing anxiety symptoms and increasing mood level, helping to prevent and promptly decrease feelings of anger and frustration, which often lead to aggressive behavior in patients with ID.

Further Reading

American Psychiatric Association. Diagnostic and statistical manual of mental disorders. 5th ed. Arlington: American Psychiatric Association; 2013.

Antonacci DJ, Attiah N. Diagnosis and treatment of mood disorders in adults with developmental disabilities. Psychiatry Q. 2008;79:171–92.

Asherson P, Buitelaar J, Faraone SV, Rohde LA. Adult attention-deficit hyperactivity disorder: key conceptual issues. Lancet Psychiatry. 2016;3(6):568–78. https://doi.org/10.1016/S2215-0366(16)30032-3.

Bakken TL, Sageng H. Mental health nursing of adults with intellectual disabilities and mental illness: a review of empirical studies 1994–2013. Arch Psychiatr Nurs. 2016;30:286–91.

Barahona Corrêa B, van der Gaag RJ. Autism spectrum disorders in adults. Cham: Springer; 2017.

Castrén E, Elgersma Y, Maffei L, Hagerman R. Treatment of neurodevelopmental disorders in adulthood. J Neurosci. 2012;32:14074–9.

Ferrell RB, Wolinsky EJ, Kauffman CI, Flashman LA, McAllister TW. Neuropsychiatric syndromes in adults with intellectual disability: issues in assessment and treatment. Curr Psychiatry Rep. 2004;6:380–90.

Fuentes J, Hervás A, Howlin P; (ESCAP ASD Working Party) (2021) ESCAP practice guidance for autism: a summary of evidence-based recommendations for diagnosis and treatment. Eur Child Adolesc Psychiatry 30:961–984.

Häβler F, Thome J, Reis O. Polypharmacy in the treatment of subjects with intellectual disability. J Neural Transm. 2015;122:93–10.

Kendall K, Owen MJ. Intellectual disability and psychiatric comorbidity: challenges and clinical issues. Psychiatr Times. 2015;32:60–3.

Sadock BJ, Sadock VA, Ruiz P. Kaplan & Sadock's synopsis of psychiatry: behavioral sciences/clinical psychiatry. New York: Wolters Kluwer; 2015.

Solanki J, Khetan J, Gupta S, Tomar D, Singh M. Oral rehabilitation and management of mentally retarded. J Clin Diagn Res. 2015;9:ZE01.

Sonido M, Arnold S, Higgins J, Hwang YIJ. Autism in later life: what is known and what is needed? Curr Dev Disord Rep. 2020;7:69–77.

Weibel S, Menard O, Ionita A, Boumendjel M, Cabelguen C, Kraemer C, Micoulaud-Franchi JA, Bioulac S, Perroud N, Sauvaget A, Carton L, Gachet M, Lopez R. Practical considerations for the evaluation and management of Attention Deficit Hyperactivity Disorder (ADHD) in adults. Encéphale. 2020;46(1):30–40. https://doi.org/10.1016/j.encep.2019.06.005.

Eating Disorders

8

S. Erzegovesi, D. Pratesi, F. Martini, L. Fregna, M. Carminati, and M. C. Cavallini

8.1 Definition and Epidemiology

Eating disorders are among the most common health problems in adolescents and young adults in Western countries. The incidence of anorexia nervosa (AN) is at least 8–9 new cases among women, while for men it is between 0.02 and 1.4 new cases per 100,000 people a year. Even more relevant is bulimia nervosa (BN), which every year records 12 new cases per 100,000 people among women and about 0.8 new cases per 100,000 people in 1 year among men (Italian Ministry of Health, 2020). DSM-5 defines feeding and eating disorders as follows: "Feeding and eating disorders are characterized by a persistent eating disorder or diet-related behaviors that result in impaired consumption or absorption of food and which significantly damage physical health or psychosocial functioning." However, the DSM-IV did not provide a precise definition of eating disorder, thus causing numerous issues in settling the diagnostic boundary of Eating Disorders Not Otherwise Specified (EDNOS). Thanks to the integrated approach (dimensional approach and categorical approach) of the fifth edition of the DSM, new diagnostic categories have been included, and some diagnostic criteria have been modified. This allows patients to receive a diagnosis that accurately define their symptomatology, to establish an adequate treatment plan. They, therefore, include, according to the new DSM-5: pica, rumination disorder, avoidant/restrictive food intake disorder (ARFID),

S. Erzegovesi · F. Martini (✉) · L. Fregna (✉) · M. C. Cavallini
IRCCS San Raffaele Scientific Institute, Milan, Italy
e-mail: erzegovesi.stefano@hsr.it; francesca.martini1@hsr.it; fregna.lorenzo@hsr.it; cavallini.cristina@hsr.it

D. Pratesi · M. Carminati
Vita-Salute San Raffaele University, Milan, Italy
e-mail: pratesi.donata@hsr.it; carminati.matteo@hsr.it

R. Cavallaro, C. Colombo (eds.), *Fundamentals of Psychiatry for Health Care Professionals*, https://doi.org/10.1007/978-3-031-07715-9_8

anorexia nervosa, bulimia nervosa, binge eating disorder (BED), other specified feeding or eating disorders (OSFED).

8.2 Etiopathogenesis

Eating disorders are multifactorial disorders both in terms of the risk factors that determine them and in terms of treatment. This implies that taking in charge of the subjects who are affected must be carried out, from the beginning of the therapeutic path, on several fronts: psychiatric, medical, psychological, alimentary, familiar, and relational, in order to reduce the probability of the chronicization of the disorder and likewise decrease the likelihood of bad outcomes. Risk factors for EDs include social, family, psychological, developmental, and biological factors.

Among the *biological factors*, we find some obstetric complications such as maternal anemia, diabetes mellitus, preeclampsia, placental infarction, neonatal heart problems. It has been observed that the greater is the number of adverse neonatal events, the higher is the risk of developing an eating disorder. The importance of genetics in determining a vulnerability to DCA was then also emphasized by several studies. Several papers report, for example, that the concordance rate for monozygotic twins is significantly higher (about 50–60%) than that for heterozygous ones, underlining the weight of genetic factors. Genetic factors could influence at the time of puberty the production of ovarian hormones, in particular estradiol, involved in the genetic transcription of neurotransmitters, such as serotonin, which regulate mood and appetite. Empirical data show that the *onset of puberty* must be considered as a relevant risk factor, and these data emphasize the association between anticipation of pubertal development and increased risk of developing an eating disorder. During early pubertal development, there is a state of greater anxiety related to physical change leading to a doubling of the fat mass in the body and the appearance of feminine forms. This change does not coincide with the maturation of the patient's identity and makes the subject vulnerable to the comments of parents and peers. The new body structure determines a conflict with the dominant aesthetic ideals and, in a condition of increased anxiety and impulsivity, can generate hyper-control behaviors that lead to a condition of fragility in which an eating disorder can easily develop. *Tolerance to distress* represents another element linked to a complex interaction between genetic and environmental factors. According to some authors, tolerance to distress is "the ability to resist and accept a negative emotion, to be able to develop a problem-solving process." A risk factor is represented precisely by a lack in the ability to choose cognitive and behavioral strategies for the management of experiences linked to both positive and negative emotions.

Other predisposing risk factors include the presence of familiarity for eating disorders (ED), premorbid temperamental predisposition to perfectionism (which indicates a greater risk for the restrictive subtype of AN), tendency to be excellent, fear of others' judgment related to fear to disappoint, overeating in the family (indicating a greater risk for the binge-purge subtype). Overeating in the family is a risk factor shared by both BN and BED.

The *sociocultural and psychological aspects* are factors of primary importance in the development of the disorder as well. Among these, we can consider the female models present in our historical-cultural period, accentuated by the use of the female image in social networks, low nuclear self-esteem, marked interpersonal difficulties, and intolerance to emotions.

Among the *triggering causes* we can commonly observe: relational problems that have arisen within the family (separations, quarrels, high conflict), sentimental or school or dynamic delusions established within social networks, competition in some sports, comments between peers, and, very frequently, the starting an "occasional restrictive diet" or even choosing "healthier foods" in one's diet. Another relevant trigger is social exclusion, which can influence interpersonal relationships and interrupt the normal development of identity, potentially increasing the salience of competition, the pursuit of perfection, and a greater internalization of the ideals of beauty as a standard.

Among the *maintenance factors* of an organic type, the main place is occupied by the effects of fasting. Weight loss accentuates the mental focus on food, causes a worsening in the distorted evaluation of body image, and alters the perception of internal hunger and satiety regulation signals. It negatively affects the mood and accentuates obsessive ruminations and social isolation. Finally, it can trigger hyperphagic crises that increase anxiety and the need for subsequent control.

According to the transdiagnostic cognitive-behavioral theory of eating disorders, developed by the Center for Research on Eating Disorders of the University of Oxford, today we tend to consider all eating disorders within a single diagnostic continuum, rather than separate disorders, since anorexia, bulimia, and binge eating disorder have predominantly common clinical characteristics and sometimes the patients who suffer from them tend to alternate various symptoms (from restrictive behaviors to bulimic crises and compensatory behaviors) migrating thus from one diagnostic category to another.

8.3 Clinical Presentation

The psychopathological core of eating disorders consists of an overestimation of the body image and shape (an *excessive preoccupation with the body image*), responsible for a profound alteration in the way in which the subject experiences the relationship with his own body, with one's weight and with food. The overestimation of the weight and shape of the body affects the evaluation of one's self-esteem and the performance perceived by the subject in other domains of life. This symptom seems to have central importance in all three eating disorders (AN/BN/BED) and seems to affect the presence of the other symptoms and the global correlation between them, regardless of the specific diagnosis of the eating disorder. The overestimation of body weight and shape is speculated to be the direct causative factor of many, if not most, of the eating disorder symptoms. It thus becomes a key goal to consider in transdiagnostic treatment and a potentially useful severity specifier for the binge eating disorder.

The patient often overestimates the size of his body and adopts dysfunctional behaviors due to dissatisfaction and fear of gaining weight. Among these behaviors, according to the trans-diagnostic model of development and maintenance of eating disorders, we find "body checking" (BC) and "avoidance of body image" (BIA). Both are due to excessive worry and the need to control the shape and weight of the body, as well as the food to be consumed. Body checking and repeated behaviors aimed at evaluating one's shape, size or weight, perpetuate the pathology both by increasing affective dysregulation and reinforcing the belief that the continuous check of the various parts of the body facilitates the numerical control of weight. "Body checking" refers to the following aspects: the frequent measurement of the dimensions of the various parts of the body with a tape measure; weight control several times a day; the continuous confrontation with the mirror and with the body of other people. Although some report feeling reassured after confirmation that their body shape has not changed, this does not stop their worry about the body shape and the frequent repetition of this behavior. On the contrary behaviors aimed at avoiding body image (BIA) include avoiding looking in the mirror or exposing the body, refusal to be weighed, or a tendency to camouflage one's shape with baggy clothes. The other characteristic symptoms are related to the pervasive and obsessive thought of food linked to the "drive for thinness". Most of the patients collect recipes, count calories, have diaries in which they write the kcal consumed during each meal, spend hours eating, and take care of feeding family members, cooking for them. Some food rituals are also common, such as cutting food into small pieces, hiding the food, filtering the oil, and eating very slowly.

As symptoms persist, patients become more irritable, depressed, and socially isolated, and obsessive–compulsive symptoms worsen. In most cases, obsessive symptoms are subsequent and probably consequent to the reduction in caloric intake and conditioned by weight changes; however, in a minority of cases, psychiatric symptoms arise earlier. Affective symptoms and impaired interpersonal functioning are found in all changes in eating behavior. Depressive and anxious symptoms appear to manifest themselves homogeneously in the various disorders and play a central role in maintaining the symptomatic network. The transversal presence of these general psychiatric and interpersonal domains, in addition to the core symptoms described above, confirms the validity of the expanded transdiagnostic theory of ED. Furthermore, interpersonal distrust and a sense of personal ineffectiveness, together with impaired proprioceptive awareness and the urge to thinness, have been shown to be at the center of the network of symptoms in anorexia, bulimia, and BED, and they seem to have a role in determining the prognosis and the outcome at a 5–10 years follow-up. The family context must also be assessed in the evaluation of the maintenance factors of the disorders. Often, during the illness, the relationship with the parents becomes tense and hostile. Investigating the various models of family classes within which the disorder has developed and the impact that the disorder has had on the environment is thus crucial to set up a treatment aimed at changing family interactions and supporting the relatives with psychoeducational interventions about how to take charge even at home during meals and about the prevention of any critical comments that could negatively influence the course of the disorder. Most patients do not have insight into the disorder and/or experience it in an egosyntonic way but do not recognize how weight loss/gain can be a threat to their survival. Therefore, most of the time, they manifest disinterest or open resistance to undertaking a therapeutic path.

8.3.1 Prognosis

After 21 years

- Anorexia
  - 51% of patients in complete remission
  - 21% partial remission
  - 10% chronic
  - 16% death
- Bulimia: 70% remission in 11–12 years
- Binge eating: 60% in remission after 6 years.

However, we frequently encounter the phenomenon of crossover that is the passage from one clinical picture to another, especially between subtypes of anorexia (from restrictive to binge).

Up to 30% of people with a history of eating disorder have a crossover phenomenon.

8.4 Anorexia Nervosa

8.4.1 Definition

In the most accredited international classification for eating disorders, the International Classification of Diseases (ICD 10), and in the Diagnostic and Statistical Manual of Mental Disorders (DSM-5), two main psychopathological characteristics are recognized for anorexia nervosa (AN): "fear fatness/fear of gaining weight" and "distortion of body image/disturbance in the way one's body weight is perceived." Fear of gaining weight is generally recognized as a central feature of anorexia, but its dependence on the patient's stage of development (e.g., less readily detectable in underage patients), culture, and stage of the disease question its clinical utility as a diagnostic criterion. For these reasons, DSM-5 extends the "fear" criterion to include "fearless reported" anorexia: even when the patient does not verbalize an "intense fear of gaining weight or becoming fat", the clinician can make a diagnosis of AN if "persistent behavior that interferes with weight gain is detected." On the other hand, ICD-10 provides a more psychopathological definition of fears: it defines fear of fatness as a pervasive idea, intrusive and overrated, suggesting a strong link (corroborated by the literature) between AN and obsessive-compulsive spectrum disorders. As far as the distortion of the body image is concerned, both ICD-10 and DSM-5 define it as a core aspect of the psychopathology of anorexia nervosa and require it as a necessary criterion for the diagnosis of AN.

Both classifications, however, do not investigate the usefulness of insight, or awareness of the disease, in AN. Scales such as The Yale–Brown–Cornell Eating Disorders Scale (YBC-EDS) and Brown Assessment of Beliefs Scale (BABS) may be useful to evaluate this construct.

In anorexia, a frequently present physical symptom is amenorrhea, traditionally considered a milestone in the diagnosis of AN and described by the ICD-10 within the category of "diffuse endocrine disorders involving the hypothalamus–pituitary–gonadal axis". However, in the DSM-5, the requirement of amenorrhea was eliminated as a diagnostic criterion for various factors, including the increasingly common use of contraceptives among women, the presence of women with AN in menopause or prepubertal age, the small but relevant percentage of anorexic males, and, above all, the minority of women who continue to menstruate but meet the other criteria for AN.

8.4.2 Epidemiology

The disease affects about 0.3% of adolescents and young adults in Western countries (although the rates of a subclinical disorder would be higher), while it is rare in developing countries where there is less social pressure toward thinness. The disease is more common in women than in men with a woman-to-man ratio of 10:1. The psychopathological onset takes place between the ages of 13–14 and 25, with peaks between 15 and 18 years. Currently, the disease rate is homogeneous in the different social classes and there is no evidence, as in the past research, of a higher prevalence in the middle-upper class.

Box 8.1: DSM-5 Diagnostic Criteria for Anorexia Nervosa

1. Restriction of dietary intake leading to weight loss and underweight relative to age, sex, and health status.
2. Behavior focused on avoiding weight gain accompanied by constant worry that this event may occur.
3. Distortion of body image with difficulty in recognising the seriousness of the current state of health, giving weight and physical appearance too much importance in the assessment of the self.

Subtypes

Restricting type: In the past 3 months, there were no recurrent episodes of binge eating or purging behaviors (such as voluntarily provoked vomiting, misuse of medications such as laxatives, diuretics, or enemas), but weight loss is achieved primarily through a severely negative caloric intake by diet, fasting, and exercise.

Binge-eating/purging type: In the past 3 months, there were recurrent episodes of binge eating or purging behaviors (such as voluntarily provoked vomiting, misuse of medications such as laxatives, diuretics, or enemas).

Severity

Mild: BMI more than 17
Moderate: BMI 16–16.99
Severe: BMI 15–15.99
Extreme: BMI less than 15

8.4.3 Etiopathogenesis

Anorexia nervosa has a multifactorial and multidimensional etiopathogenetic model typically characterized by the interaction of three classes of risk factors: *sociocultural*, *individual*, and *familiar*.

Concerning sociocultural factors, anorexia has been defined by the French Devereux as an "ethnic disorder," that is, as a phenomenon capable of expressing the anxieties and contradictions of society such as those typical of industrialized countries. However, although teenagers of Western civilizations are subjected to the psychosocial pressure of thinness, only a small part develop EDs.

We can, therefore, deduce that there are other factors, such as *individual* ones, that must coexist with environmental and social ones for a disorder such as anorexia nervosa to develop. It has been found that patients suffering from restrictive anorexia typically have an obsessive-compulsive personality, with social inhibition and emotional restraint, and a strong drive for perfectionism. On the contrary, patients with bulimic–purgative anorexia have an outgoing and impulsive personality. Among the predisposing factors to EDs, we also find reduced self-esteem. Regarding the *familiarity* in the etiopathogenesis of eating disorders, it is still not clear which genes are involved in the genesis of the disorder, but studies on twins highlight a higher incidence of the disease in homozygous twins compared to heterozygotes. As for the nongenetic family aspects, the family of the anorexic patient has traditionally been described, since the early 1980s, as an "anorexogenic family," characterized by the presence of an anxious, sometimes "oppressive" mother and an absent father. More recent scientific studies consider the cause-effect relationship between family characteristics and anorexia nervosa simplistic, emphasizing the point of view, universally valid for scientific studies, for which "correlation is not causation."

8.4.4 Clinical Presentation

The onset of anorexia nervosa is often gradual and insidious, with a progressive reduction in food intake. In most cases, the disorder occurs because of a low-calorie diet that started to change the body weight and shape, excluding carbohydrates and fats from the daily diet. Sometimes even the selection of "healthy foods" can be related to the onset of anorexia nervosa. In other cases, the symptoms may appear after digestive difficulties, diseases (including depression), surgery, or traumas. Before the presentation of the clinical picture, stressful events, or life changes (e.g., losses, separations, job failures) are often observed. Patients decrease caloric intake by reducing food servings or by skipping snacks between meals and then, gradually, also the main meals, resulting in a significant weight loss.

"In the first months, there is a phase of subjective well-being, due to the weight loss, the associated improvement of one's image, and the feeling of omnipotence produced by the ability to control hunger, while awareness of the problem is low and there is no request for help (the so-called honeymoon phase with the disorder). Later, concerns about body shape and weight become pronounced, and the fear of gaining weight does not diminish with weight loss".

To reduce weight, some people use self-induced vomiting or abuse of laxatives, diuretics, or, more rarely, anorectic drugs. This subgroup (bulimic/purgative subtype anorexia) has a worse prognosis, more frequent medical complications, and, from a psychopathological point of view, greater comorbidity, with a greater frequency of impulsive behaviors, self-harm, suicide attempts, and alcohol, or other substances', abuse.

Family relationships can often deteriorate, becoming substantially blackmailing. Vicious circles are built between parents and patients: parents try to stimulate their daughter to eat more, spy on her, control her, limit her, so exasperating the behavior of the girl, who hides the food or acquires a particular skill in leaving the plate to give them the impression that the food has been consumed (chop, crumble, drain oil, etc.).

Despite the starved condition, the anorexic patient maintains physical hyperactivity, dedicating herself to long walks (up to 20–30 km per day) and exhausting workouts. She does not seem to feel fatigued and cold until the most advanced stages of the disease. In fact, in these patients, there is a difficulty in recognizing, perceiving, and responding to somatic needs: the patient does not feel hunger and cold or perceives them in an altered way. Another characteristic typically found in anorexia is the distortion of the body image, with a persistent tendency to see oneself fat, or with a physical appearance that does not satisfy the "ideal" one.

Anorexia nervosa has significant psychiatric comorbidity. The most frequently associated disorders are major depression, obsessive–compulsive disorder, and other anxiety disorders, alcohol or substance addiction. Among the personality disorders, the most frequent are borderline, narcissistic, and avoidant disorders. Psychiatric comorbidity tends to decrease with weight recovery and normalization of nutrition.

8.4.4.1 Physical Complications and Laboratory Tests

- *Liver:* liver function is impaired, with high transaminases and very low proteins. If, on the other hand, we were faced with a patient in a phase of anorexia in which the muscle tissue is still minimally intact, we would find high values of proteins. Cortisol is the hormone that carries this proteolytic response and, in fact, subjects with anorexia in initial phases are always hypercortisolemic.
- *Blood:* mixed anemia is noted in the blood count, with low iron due to decreased intake (in a condition of anorexia, however, amenorrhea is present in the female sex, and therefore the amount of iron lost will be less). White blood cells are only increased in severe infections but are usually low due to prolonged fasting.
- *Kidney:* renal function is impaired, with low urea and creatinine, due to the reduced protein metabolism. Only in case of dehydration or intense physical activity, the values can be high.
- *Muscle:* in some cases, the CK increases much due to intense muscular exercise carried out on weakened muscles.
- *Gonads:* sex hormones are significantly reduced. Amenorrhea is frequently present in the female sex.
- *Hydro-electrolyte balance:* sodium is low in patients who drink a lot of water, while it is increased in patients who use many laxatives and are dehydrated. In the first case, the water intake will have to be restricted, because, if the patient continued to drink a lot, she would risk reaching a condition of cerebral edema due to hyponatremia.
- *Thyroid:* a state of hypothyroidism is almost always present since the body tends to save energy by slowing down metabolic functions (first, it eliminates the menstrual cycle, useless for survival in a starved condition). TSH and FT4 tend to be normal, while FT3 is low, being the hormone most closely related to the activation of peripheral metabolism. Therefore, since the activities that waste energy are reduced or eliminated, the active transport systems through cell membranes will also be automatically reduced. Therefore, in the case of administration of drugs, even of a psychotropic type, poor efficacy or paradoxical effects related to reduced membrane activity could occur.

8.4.4.2 Course and Prognosis

The course of AN is variable: in about 50% of cases there is complete recovery, in about 30% of cases a partial but satisfactory remission, and in about 20% of cases there is a chronic course, without recovery of a sufficient quality of life. In the case of serious and chronic patients (20%), the risk of death from malnutrition and suicide is estimated at around 10%. Several prognostic factors have been identified:

- *Negative prognostic factors*
  - Delay in treatment
  - Onset in older age
  - Bulimic-purgative subtype
  - Poor family support

- *Positive prognostic factors*
  - Short duration of illness
  - Good premorbid socio-working adaptation

8.4.4.3 Differential Diagnosis

In the differential diagnosis, it is necessary to consider all organic causes of starvation, such as neoplastic diseases, endocrine disorders (diabetes mellitus, hyperthyroidism, Addison's disease), and gastrointestinal disorders (peptic ulcer, ulcerative colitis). Anorexia must also be distinguished from weight losses associated with other psychiatric conditions, for example, depressive disorders, somatoform disorders, social phobia, obsessive-compulsive disorder, panic disorder, body dysmorphism disorder.

8.4.5 Treatment

The anorexic patient rarely asks spontaneously to be treated. This refusal to treatment is part of an extreme desire for control over herself, over her own body, and is dictated by the egosyntony of psychopathological symptoms.

The treatment involves a multidisciplinary approach: dietary, psychological, and medical/pharmacological. From a nutritional point of view, with a BMI <10, enteral or parenteral nutrition must be set, given the too high risk for the patient. In cases with less severe BMI, an oral nutritional rehabilitation is set up in hospital, with the setting of customized meals: in this regime, the patient cannot choose the food to eat but can define three foods to be completely excluded from the diet. The nutritionist can define, in line with the medical and psychological staff, a weight recovery goal considered "acceptable" by the patient from a psychological point of view, and sufficient by the carers from a metabolic point of view. For this reason, in the first stages of treatment, meals must not aim at "fattening" but must offer a caloric intake slightly above the basal metabolic rate. Carbohydrate loads must be avoided, which otherwise could cause a sudden increase in the minimum amount of insulin present in the body, increasing the risk of refeeding syndrome. In countries where there is famine, UNICEF offers bars called PlumpyNut, composed of 50% milk powder, 25% white sugar, and 25% peanut butter: this suggests how, in conditions of emaciation, the intake of fibers (e.g., from vegetables, whole grains, or legumes) should be minimized to prevent bloating and reduced absorption of other nutrients, and refined carbohydrates, proteins, and fats should be preferred.

From a medical point of view, in the presence of underweight subjects, the general patient conditions must always be assessed, namely vital parameters, ECG, blood sugar, electrolytes, and kidney function. The medical condition of a patient with an eating disorder is very precarious and could worsen at any time, so it is essential to make the basic clinical, instrumental, and laboratory tests.

Another issue is disease insight: it is often difficult to establish a good working alliance with patients with poor insight. In these patients, the sense of imminent death and the perception of danger are absent. Therefore, it makes no sense to motivate the subjects with phrases such as "if you don't eat, you'll die," but rather we need to focus on issues such as school efficiency, aesthetics, and, more generally, quality of life, which are more taken into consideration by subjects with anorexia nervosa.

A further problem is the patient's manipulation of the environment ("if you forbid me to do this, I won't eat"), which must be managed by providing specific support to the caregivers and the patient's family members. With a seemingly "peaceful" position, like that of fasting, the patient puts those around her in a condition of total powerlessness. In these cases, the clinical staff must try to make patients understand how this behavior is not a real challenge to others but is instead a symptom of the disease itself that must therefore be managed and dealt with by the patient. The working alliance is very important and, to be able to achieve it optimally, it is necessary to create a neutral environment, without the judgments and reinforcements usually present in the family context. In addition, it must be remembered that anorexia is a multifactorial disease, so it is incorrect to look for a single "responsible" in the family or society (although they may be some of the factors to be taken into consideration).

Finally, it is necessary to find common goals with the patient, such as

- The improvement of attention and concentration.
- The reduction of the obsessive ideation about food and body image.
- The return to acceptable levels of global functioning, given that the patient often feels in a condition of absolute solitude.

Concerning psychopharmacological treatments, the role of Serotonin Reuptake Inhibitors (SSRIs) in some phases of treatment and for the prevention of relapses has been highlighted. Antipsychotic drugs are used to exploit their sedative and weight gain effects related to their intake. The use of these drugs is also supported by the hypothesis that there is an accentuated dopaminergic tone in the anorexic pathology. In general, there is currently no etiological psychopharmacological therapy for eating disorders. However, drugs can be used for the management of psychopathological symptoms concomitant with anorexia (anxiety, depression, obsessive–compulsive symptoms, and poor insight into illness).

8.4.5.1 The "Marsipan Study"

The English MARSIPAN (Management of Really Sick Patients with Anorexia Nervosa) Study defines the clinical guidelines for managing a patient with anorexia nervosa of considerable severity when this comes to the attention of the nonspecialist doctor (e.g., in the case of access to the emergency room or medical departments not specialized in EDs).

The study stems from the observation that the anorexic patient runs, in extreme opposites, two types of risks

1. In institutions experienced in treating frail and underweight subjects (e.g., medical departments that follow cancer patients or elderly patients), the risk of dying from malnutrition is much greater than for other sick people, because the clinical

unit, accustomed to treating critically ill patients suffering from various diseases, tends to refeed the anorexic patient excessively slowly, so slowly that there is a risk that the patient will die from emaciation.
2. On the contrary, the anorexic patient who goes to a psychiatric unit, a unit accustomed to treating physically younger and healthier patients, runs the exact opposite risk, namely that of dying based on an excess of nutrition.

From these two opposed criticalities, from the UK Royal College of Physicians and the Royal College of Psychiatrists arose the need to establish guidelines for the diagnostic orientation and treatment of a severely underweight person who is at the same time attempting to resist medical treatment. Since these are life-threatening patients, treatment must be insisted upon, and it is impossible to give up in the face of the patient's refusal of treatment.

The MARSIPAN guidelines, therefore, evaluate six key points:

1. *Risk Assessment:* evaluates the risk, or how much the patient is at risk of life, through clinical, instrumental, and laboratory investigations.
2. *Avoid Refeeding Syndrome:* provides all the recommendations aimed at minimizing the risk of the re-nourishment syndrome, more easily found in psychiatric units, less accustomed to managing organ criticalities that can also affect young patients suffering from a chronic eating disorder.
3. *Avoid Underfeeding Syndrome:* provides all the recommendations aimed at minimizing the risk of undernutrition, which is often reached in the fear of overloading the patient from a nutritional point of view.
4. *Manage Behavioral* Problems (e.g., Sabotaging Nutrition): addresses behavioral problems. These are subjects who often "sabotage" the treatment, not following medical instructions and, for example, disconnecting the drip or throwing away the food supplement provided to them.
5. *Treat Under Compulsion* (in Italy, Trattamento Sanitario Obbligatorio or TSO): evaluates the mandatory treatment.
6. *Manage Family Concerns:* evaluate family aspects. In eating disorders, more than in any other medical or psychiatric pathology, it is essential to provide care and information also to family members. Family members often add important anamnestic information, left out by patients due to their lack of awareness of the disease.

8.4.5.2 Risk Assessment

In the risk assessment, it is important to assess the BMI, weighing the patient in each case, even when the assessment of weight is considered a stressful event. When the BMI is below 13, the subject is in critical condition and may be metabolically unstable, even if the blood tests reveal nothing of significance. Other elements of risk are as follows.

- Heart rate < 40 BPM.
- Body temperature <35°; it must be considered that almost all anorexic patients are hypothermic.

- Muscle strength: assessable with the Sit-Up-Squat-Stand test (SUSS Test), which evaluates the ability to flex in bed and to do a squat. When the score obtained is <2, or the patient is unable to do these exercises, it can be considered as a risk factor.
- A QTc on ECG >450 ms. We must therefore be careful when administering certain drugs.
- Blood pressure variation: it is necessary to consider that anorexic patients have rather low blood pressure. On the other hand, pressure variation (e.g., orthostatic hypotension) is worrying and constituting a risk factor rather than hypotension as such.
- Low sodium (<130 mmol/L): it should suggest an occult infection or an excess of water assumption.
- Low potassium (<3.0 mmol/L): it should make us think about purging conducts, such as the abuse of laxatives or induction of vomiting.
- Hypoglycemia (glucose <54 mg/dL): when associated with low albumin and/or high CRP, it should lead to suspicion of an occult pulmonary infection.
- Urea and creatinine: they must be particularly monitored as these patients, being undernourished, often show low plasma values of urea and creatinine and, for this reason, finding them normal can already indicate a deficit of renal function.
- Leukopenia or high transaminases: these are two findings that often alarm general practitioners, who hypothesize hepatic or hematological problems and prescribe further tests, with the risk of delaying proper renutrition. In reality, these findings are directly related to fasting and/or physical hyperactivity.

8.4.5.3 Refeeding Syndrome and Underfeeding Syndrome

Nutrition must be reintroduced gradually, and the gastric tube is useful in these cases because it makes it less easy for the patient to sabotage the "therapy". Calories are introduced very slowly (5–10 kcal/kg/day). The guidelines recommend checking even twice a day those blood chemistry indices that most easily suggest a refeeding syndrome. Refeeding syndrome arises when an organism accustomed to surviving in a condition of chronic food deficiency receives an excess of nourishment, especially sugars: this leads to cellular reorganization, following the arrival of insulin, which causes electrolytes such as magnesium and phosphorus to enter the cell. The electrolytes then enter the cell, causing hypomagnesemia, hypophosphatemia, and consequent risk of cardiac arrest. It is, therefore, necessary to check the plasma values of glucose, phosphorus, potassium, magnesium, calcium and increase nutrition, up to 40 kcal/kg/day, only if the electrolytes remain stable. If the wasting conditions are not extreme, a higher renourishment dosage, for example, 20 kcal/kg/day, can be started, and calories can be increased more rapidly in the following days. As far as phosphorus is concerned, high-concentration supplements (e.g., potassium phosphate supplements) or foods, such as Parmigiano Reggiano or other aged cheeses, which contain high and readily bioavailable calcium and phosphorus, can be very useful. Since the intracellular migration of electrolytes is caused by the increase in insulin resulting from the intake of carbohydrates with a high glycemic load, it is necessary to implement a relatively low-sugar and high-fat renutrition to limit the problem. Finally, hyponatremia can be aggravated by hyperhydration, and very low sodium (<120 mmol/L) can put the patient at risk of cerebral edema and epileptic seizures. For this reason, in the first days of renutrition, it is often necessary to limit the patient's water supply.

8.4.5.4 Behavioral Problems, Treatment Under Compulsion, Family Concerns

Regarding the tendency of patients to control the feeding, it is often necessary for a nurse to be specifically dedicated to a single patient, in order to minimize the risk of sabotage of renourishment. It is equally important to implement complete and transparent communication, both among the treating staff members and between staff and family members. As regards compulsory treatment, differently, for example, from the UK, in Italy there are no specific rules for the treatment of the most severe anorexic patients against their will. The current Italian law on Compulsory Health Treatment (TSO) does not consider the specific needs of anorexic patients. It would be, therefore, desirable, for the future, to revise TSO legislation to allow a more effective and timely intervention in the case of patients suffering from severe anorexia and with little or no awareness of the disease.

8.5 Bulimia Nervosa

8.5.1 Definition

Compared to the classification of bulimia nervosa (BN) made in DSM IV, in DSM-5 there is:

- a reduction in the minimum required average frequency of binge eating and inappropriate compensatory behaviors (from 2 to 1 times per week for at least 3 months);
- the disappearance of subtypes (purging and nonpurging).

Box 8.2: DSM-5 Diagnostic Criteria for Bulimia Nervosa

A. Repeated episodes of binge eating. A binge eating episode is defined by both of the follows:
   - Ingesting an excessive amount of food considering the time taken and the adequate portions of a meal.
   - Feeling of loss of control over the act of eating or the amount or nature of the food ingested.

B. Preventing weight gain by practicing inappropriate compensatory behaviors (such as voluntarily provoked vomiting, misuse of medications such as laxatives, diuretics, or enemas, but also fasting, or excessive exercise).

C. The frequency of inappropriate eating behaviors occurs, on average, at least once a week for 3 months.
D. Body shape and weight improperly influence self-assessment.
E. Dysfunctional eating behaviors do not occur only during episodes of anorexia nervosa.

Severity

Mild: An average of 1–3 episodes of inappropriate compensatory behaviors per week.

Moderate: An average of 4–7 episodes of inappropriate compensatory behaviors per week.

Severe: An average of 8–13 episodes of inappropriate compensatory behaviors per week.

Extreme: An average of 14 or more episodes of inappropriate compensatory behaviors per week.

BN can therefore be characterized by the presence of binge eating and purging behaviors.

In common language, the term binge is used to refer to an abundant and rich diet; in clinical practice, the term binge does not define only an excessive amount of food that the subject ingests but also refers to how the food is taken. The term binge means eating in a defined period (e.g., 2 h) a significantly greater amount of food than most people in the same conditions would eat in that same time. The subject consumes the food quickly and voraciously, swallowing it without tasting it, and, sometimes, choosing foods that, under normal conditions, he/she does not even consider appetizing. During the episode, the subject has the feeling of losing control (e.g., feeling unable to stop eating or to control what or how much you eat). As for the eliminatory behaviors, these can occur through the use of laxatives, diuretics, enemas, self-induced vomiting, compensatory fasting, or motor hyperactivity.

8.5.2 Epidemiology

BN is an eating disorder that affects 1–3% of the general population, predominantly female subjects, with a male to female ratio of 1:10. Males represent only 10–15% of patients. Onset is around 18 and after 25 years of age with an age range of 12–35 years. The average duration of the disease at the time of referral is about 5 years. Some years pass before the patients come to the attention of a specialist because, unlike patients suffering from Anorexia Nervosa, whose physical impairment is more pronounced, patients suffering from bulimia nervosa are frequently normal-weighted.

8.5.3 Etiopathogenesis

As for AN, the etiopathogenesis of BN is multifactorial, determined by the interaction between sociocultural, familial, and individual factors. However, in the pathogenesis of this disorder, some personal aspects acquire extreme relevance. Patients with BN are prone to experience frustrations as threats to their self-esteem, resulting in a feeling of discomfort toward their bodies. Compared to anorexic patients, bulimic patients are more impulsive, extroverted, and choleric. Narcissism, intended as the demand for attention from others, is considered among the risk factors for BN. As far as family structure is concerned, we frequently find conflicts within the family, and bulimic patients often feel neglected and rejected, abandoned by parental figures.

8.5.4 Clinical Presentation

The age of onset of BN is slightly higher than AN and the clinical presentation at onset is sometimes overlapping: in about 25% of cases, the diagnostic criteria for anorexia are met for the first period after onset, then binge eating episodes begin to interrupt dietary restriction, creating a cycle which tends to perpetuate over time. The onset of BN may occur after a strict restrictive diet for weight reduction or as a result of personal and emotional difficulties in managing situations of loss. In the early stages, the patient maintains absolute secrecy about her behavior, and, sometimes, years may pass before a family member notices the problem.

Binge episodes can be characterized by "objective binge eating" or "subjective binge eating," in which the feeling of loss of control is not associated with objectively high food intake. In both cases, however, the food ingested during the hyperphagic binge is high in sugars and fats, avoided during the dietary restriction, high in energy density, easy to ingest, and often lower in cost (Ministry of Health, 2017).

Binge eating is triggered by so-called emotional hunger linked to states of dysphoric mood, conflicts, stressful events, feelings of emptiness and loneliness, but also feelings of boredom and low tolerance to frustrations. It is also defined as "comfort eating" or "stress-induced eating." According to this theory, deficits in emotion regulation should translate directly into eating alterations. Overeating has been associated with unhealthy or extreme weight control conducts, dieting, nonsuicidal self-harm, body dissatisfaction, low self-esteem, and depressive symptomatology.

The relationship between negative affect and binge eating in subjects with BN is well recognized, and current treatment strategies help normalize disordered eating conducts by promoting adaptive emotion regulation skills. Interestingly, studies have shown that dysfunctional eating behaviors in patients with BN not only include negative affect-induced overeating but also positive affect-induced underfeeding risk. This finding highlights the importance of maintaining a regular eating schedule as a primary goal, since individuals with BN may tend to underfeed when their mood improves during treatment. The idea that food may represent a tool used to regulate emotions (but those emotions can also regulate food intake) holds a prominent place among current theories on emotional eating and is the basis of several psychotherapeutic approaches to eating disorders.

Following the binge, most subjects, to compensate for feelings of guilt, shame, self-evaluation, inadequacy, and fear of gaining weight, use compensatory behaviors such as self-induced vomiting, improper use of laxatives or diuretics, excessive physical activity, and fasting, spitting out food after chewing it but before swallowing it, intake of drugs that control appetite. In a subgroup of people, self-injurious behaviors are present, sometimes aimed at pushing away the discomfort following the binge or punishing themselves for it. NSSIs (nonsuicidal self-injurious behaviors) are often used as maladaptive coping strategies of emotional dysregulation. Such acts are seen more frequently in bulimic patients with greater body image distortion and body dissatisfaction. They are related to negative body-related feelings, reduced sensitivity to pain, and reduced ability to cope with distress.

Box 8.3: Clinical Case

C. is a 27-year-old girl who at the first visit reports vague physical problems such as weakness and muscle cramps. She immediately admits that this is due to her "daily habits": 1–2 episodes of self-induced vomiting and heavy abuse of laxatives (an average of 40 tablets of senna derivatives every 2 days).

About 5 years before, after a "bad sentimental disappointment", C. discovers the immediate but temporary effect of a binge: she opens the fridge and quickly eats all the sweets she finds. After a momentary relief, she begins to experience bloating, weight in the stomach, and acid regurgitation; shortly after, she feels "empty inside" and guilty for what she did.

It is possible to observe typical aspects of bulimic disorder: the tendency to normalize the pathological behavior, the lack of awareness of the disease, the abuse of laxatives, the onset that goes back several years before the patient turns to the clinician, emotional and traumatic life events that trigger symptoms, the resort to impulsive actions to cope with the emotional tension arising from the trigger event.

In the following days, C. says to herself that she is well aware of the heavy discomfort resulting from the binge; nevertheless, after a phone call to a friend, with whom she remembers the disappointment experienced, she finds herself acting on impulse a new binge, this time more "chaotic": she first ingests sweets, then cold cuts, then sauces with bread, then more ice cream, then vegetables seasoned with lots of oil and vinegar... to finish two glasses of bitter "to digest".

In the space of 2 months, C. gained 15 kg (from 50 kg for 160 cm height and BMI: 19.5–65 kg AND BMI: 23.4).

At this point, she appears unable to control her binges and seeks alternative methods to "counterbalance" her food intake: she vomits immediately after a binge, takes laxatives every night and every few days fasts for a whole day. Moreover, she started drinking alcohol.

Her weight drops a little (from 65 to 60 kg) but she starts complaining about physical problems, such as acid reflux, muscle problems, increasing asthenia, caused by repeated vomiting.

Recourse to alcohol is not uncommon since it represents facilitation to emptying through emesis and on the other hand, it is a risk factor for further complications related to BN.

Among the compensatory behaviors, some patients choose complete abstention from food. However, this behavior only serves to reinforce the vicious circle because the subject, in a condition of fasting, becomes even more vulnerable to the urge to eat in an uncontrolled way (fasting → uncontrolled eating → guilt → eliminatory behaviors → fasting).

Table 8.1 Medical complications of bulimia nervosa

Medical consequences of bulimia nervosa
• ECG alterations (due to potassium depletion)
• Salivary glands hypertrophy
• Erosion of dental enamel and dental caries (dentists are sometimes the first to notice the presence of emesis)
• Fatigue and weakness
• Dehydration, hypotension, and dizziness
• Abdominal swelling and meteorism, intestinal motility alteration (laxative abuse can lead eventually to paralytic ileus)
• Ulcers and lacerations of the oral and pharyngeal mucosa
• Gastric cramps, gastroesophageal reflux, gastritis
• Numbness and tingling limbs
• Russel's sign: calluses caused by the activity of gastric juices on the skin of the subject's hands

In most patients the presence of compensatory behaviors has medical consequences, detailed in Table 8.1.

8.5.4.1 Physical Complications and Psychiatric Comorbidities

- *Electrolyte imbalance:* One of the most dangerous complications of BN is the development of electrolyte imbalance, could result in cardiac arrhythmias. In particular, self-induced vomiting with the loss of gastric acid can lead to hypochloremic alkalosis and resultant hypokalemia. Laxative abuse may result in hypokalemia as well.

- *Subconjunctival hemorrhage and/or epistaxis:* They are caused by the rupture of small blood vessels induced by the increase in pressure that occurs during vomiting.
- *Dental complications:* Chronic exposure to stomach acid can cause dental damage (perimyolysis), in particular to the buccal and lingual surfaces. Other oral/dental consequences of self-induced vomiting include tooth discoloration, increased tooth sensitivity, oral mucositis, and cheilitis.
- *Larynx and vocal cords damage:* These organs may also be inflamed by acid exposure, leading to hoarse voice, chronic sore throat and cough, and difficulty swallowing.
- *Salivary glands:* Parotid and other salivary gland enlargement due to vomiting occurs in about 10–25% of patients.
- *Gastrointestinal tract disturbances:* Vomiting can lead to acid reflux, dyspepsia and dysphagia. Less frequently, it can cause small tears in the esophagus, and, rarely, esophageal rupture. Laxative abuse can instead lead to local effects such as chronic constipation or diarrhea, rectal prolapse, hemorrhoids, and hematochezia.

Finally, individuals with bulimia nervosa tend to be more prone to psychiatric complications. Comorbid psychiatric disorders can be detected at the time of diagnosis of BN or can develop later during the course of the disorder. The most common comorbidities are anxiety disorders (53%) and mood disorders (43%). Literature reports that approximately 90% of subjects with BN reported at least one episode of a mood disorder in their lifetime, usually a depressive one. The most frequent personality disorder associated with BN is Borderline personality disorder, particularly common in patients with history of childhood emotional trauma. At least 30% of patients with BN have consumed alcohol or stimulants in an attempt to control their appetite and weight. Besides, bulimia can increase the risk of suicide in patients.

In general, individuals with BN commonly have comorbid conditions that include acting-out behaviors, compared to the other subjects with eating disorders.

8.5.4.2 Course and Prognosis

The course is usually chronic or remitting: often the disorder persists for several years, either continuously (chronicity occurs in 20% of subjects) or with alternating phases of remission and flareups. 5–10 years after the onset, 30–50% of patients have some eating disorder of clinical relevance, although, in many cases, it is an atypical form (forms of binge eating, eating disorder NOS, sometimes anorexia). Concerning severity and risk of hospitalization, we can say that BN is associated to a greater extent with self-harming behaviors and, as mentioned, with specific personality disorders, especially borderline. Therefore, in general, these are subjects with high levels of impulsivity that can manifest through self-injurious and sometimes anticonservative behaviors.

No unequivocal predictors of favorable prognosis have been identified; however, indicators of worse prognosis include childhood obesity, low self-esteem, and personality disorders.

Comorbidity with other disorders worsen the prognosis of these patients:

- Substance abuse that occurs in 55%
- Alcohol abuse 46%
- Agoraphobia 27–34%
- Social phobia 15–55%
- Depression 31–90%
- Generalized anxiety disorder 23–70%

8.5.5 Treatment

The effect of pharmacological treatments in eating disorders is recognized as limited by the main international guidelines:

- APA (2006): drugs are not the first-choice treatment for BN; however, FDA recognizes the utility of fluoxetine's use for the treatment of bulimic disorders at dosages generally higher than those used for the treatment of depression (up to 60 mg). Similar utility, although with weaker evidence, is described for sertraline. However, the presence of purging behaviors sometimes does not guarantee the stabilization of plasma levels of pharmacotherapies. When concomitant anxiety or depressive symptoms are present, the use of pharmacotherapy helps to mitigate their effects.
- NICE (May 2017): medications are not the first choice in the treatment of EDs, but they may have a role as adjuvants to psychological therapies. It is recognized, however, that they can reduce the frequency of binge and purging in both BN and BED.

How then to manage the treatment of these patients?

1. First of all, it is necessary to implement a motivational intervention to the treatment, consisting of helping patients recognizing that they need specialist care and maintaining their motivation to get well over time. This goal is primary given the frequency of their reluctance to treatment.
2. Restoration of dietary regularity usually leads to a substantial improvement in the patient's general state.
3. The third aspect of the intervention consists of dealing with the poor evaluation that the patient has of his body shape and weight and helping her recognizing and managing dysfunctional eating habits.
4. Specialist treatment passes through the integration of different approaches (pharmacotherapy with antidepressants or antipsychotics, psychotherapy). Family therapy seems to be the most useful therapy for younger patients and is, therefore, the approach mainly used with adolescents.

An essential point in the psychiatric management of the patient with bulimia nervosa is a multidisciplinary approach, which aims to do the following:

- Offer and coordinate different types of care to the patient.
- Involve different professional figures: psychologists, psychiatrists, dieticians, internists, dentists, and school staff in the management of the patient.
- Promote team discussion and supervision.

This integrated intervention shows several advantages over a single intervention, such as the possibility of increasing the continuity, consistency, and effectiveness of the diagnostic process; the possibility of shortening the time of the subsequent therapeutic intervention; not to waste resources and energies of both therapists and patients; not to promote the maintenance of the state of disease.

Concerning the psychological interventions, the approach changes in the case of an adult or a minor patient.

In the case of subjects *more than 18 years old*, at first, self-help groups focused on bulimia can be proposed. This approach consists of providing material with cognitive-behavioral indications of self-help. This program can be supplemented with short support sessions (4–9 sessions of 20 min each for 16 weeks, starting at the beginning of the week).

If the treatment is ineffective after 4 weeks, or poorly accepted, or contraindicated, another possible approach is cognitive behavioral therapy focused on bulimia (CBT-ED) (20 sessions in 20 weeks). In the first phase, CBT consists of motivational and educational interventions aimed at establishing a regular eating pattern, encouraging, counseling, and supporting the patient as he/she engages in this process. Then, CBT should focus on addressing the psychopathology of the eating disorder, such as extreme food restriction, concerns about body shape and weight, and the tendency to overeat in response to difficult thoughts and feelings.

Toward the end of treatment, meetings become less frequent and focus on reinforcing and maintaining positive change and reducing the risk of relapse. If therapists find it helpful, significant figures for the patient are involved for one-on-one support.

There are other possible psychotherapeutic approaches: interpersonal psychotherapy, dialectical behavior therapy, acceptance and commitment therapy, integrative cognitive-affective therapy, psychodynamic therapies, and family therapies.

Dialectical behavior therapy (DBT), is one of the most effective treatments in subjects with borderline personality disorder and has been extended, with appropriate modifications, to all disorders characterized by emotional dysregulation as a nuclear symptom, including eating disorders.

The term "dialectical" refers to the possibility of being in two different positions at the same time. For example, it is important for the patient to accept himself as he/she is but also to be motivated to change.

The treatment includes individual and group meetings, during which the patient learns skills to manage symptoms (preventing binge eating and refraining from eliminatory behaviors). It is a manualized treatment with versions available for both adolescent and adult clinical populations.

In the case of *minor patients*, the starting approach is represented by Family therapy, a family and patient support course that is developed in 18–20 sessions over 6 months.

It aims to establish a good therapeutic relationship with the patient, the family members, and caregivers and support and encourage the family to help the patient, maintaining a nonjudgmental approach.

It also aims to: give information about body weight regulation, dieting, negative effects of self-induced vomiting, laxative abuse, or other compensatory behaviors; foster a collaborative approach between the patient and his/her parents to establish regular eating habits and minimize compensatory conducts. It includes regular one-on-one meetings with the patient during treatment, interventions to encourage self-control of bulimic behaviors, and discussion sessions with family members.

In the later stages of treatment, when the eating behavior has stabilized, the focus of care is directed toward supporting the person to establish a level of independence appropriate to the developmental stage of the subject. In the final phase of treatment, the focus shifts to dealing with any problems the patient and his/her family may have and relapse prevention.

8.6 Binge Eating Disorder

8.6.1 Definition

Binge-eating disorder (BED) was first described in 1959 by psychiatrist Albert Stunkard to describe the characteristics of a subgroup of subjects with obesity and recurrent episodes of excessive and uncontrolled eating: a behavior he called binge eating. However, its existence as a distinct diagnostic entity was ignored until the second half of the 1980s, when research on the prevalence of BN in the population highlighted a conspicuous subgroup of patients who did not adopt compensatory behaviors after binge eating episodes. At the same time, it was observed that about one-fourth of individuals seeking obesity treatment reported recurrent episodes of binge eating but did not meet the diagnostic criteria for bulimia nervosa. In 1994, the American Psychiatric Association included BED among eating disorders not otherwise specified and, in Appendix B of the Diagnostic and Statistical Manual of Mental Diseases (DSM-IV), provided a list of diagnostic criteria for further study. Subsequent studies confirmed that BED has distinctive features compared to bulimia nervosa and obesity, and supported the validity and clinical utility of the diagnosis of BED. However, it was not until 2013 that the disorder was recognized by DSM-5 as a distinct diagnostic category within nutrition and eating disorders.

Binge eating disorder (BED) is a psychopathological condition that affects 2.5% of adults and 1.6% of adolescents. This disorder is characterized by recurrent episodes of uncontrolled eating.

Box 8.4: DSM-5 Criteria

A. Repeated episodes of food dyscontrol called binge eating, characterized by a huge amount of food ingested, in a limited amount of time and the feeling of loss of control over the act of eating or the amount or nature of the food ingested.

B. The eating behaviour characteristic of the binge eating episode is characterized by the following:
   - Fast eating
   - Continuation until the sensation of discomfort from excessive fullness
   - Dissociation of the episode from the feeling of physical hunger
   - Isolation, due to the embarrassment in respect of the episode
   - Feelings of guilt, disgust or depression for having succumbed to the impulse

C. Binge eating elicits distress in the subject

D. The frequency of episodes of binge eating is, on average, at least 1 day a week for 3 months

E. Binge eating is not regularly followed by practicing inappropriate compensatory behaviours (such as voluntarily provoked vomiting, misuse of medications such as laxatives, diuretics, or enemas, but also fasting, or excessive exercise).

8.6.2 Epidemiology

BED has a prevalence of 0.7–4.6% in the general population; 20–30% of obese patients, in general, have BED, while up to 70% of obese patients who come to hospitalization have BED. Symmetrically, 75% of BED patients are overweight or frankly obese. The male–female ratio is 1:1.5. The onset occurs between 30 and 40 years, regardless of diets or other specific reasons.

8.6.3 Etiopathogenesis

Binge-eating disorder is more common in female sex. Although people can develop a binge-eating disorder at any age, the onset often occurs in the late teens or early 1920s.

Factors that can increase the risk of developing a binge-eating disorder include the following:

- *Family history.* An individual is much more at risk of developing an eating disorder if parents or brothers and sisters have an eating disorder. Family history can indicate a genetic predisposition but it can also represent an environmental risk factor (parent modelling).

- *Dieting.* Many people with a binge eating disorder have a history of unsuccessful diets. Diet or limitation of calories intake during the day can cause a need for a binge to eat.
- *Psychological issues.* Many people suffering from a binge eating disorder feel negative about themselves and undervalue their capacities and achievements. Triggers for a binge eating episode can include stress, poor self-image, and the availability of favorite foods.

Factors that have been linked to the maintenance of the disorder are emotional, social, and cognitive dysfunctions. Binge-eating episodes are more likely to occur within in individuals with reduced emotional awareness, difficulties in emotion regulation, and frequent interpersonal problems. Moreover, binge eating is associated with neurocognitive dysfunction, including difficulties in inhibitory control and reward processing.

8.6.4 Clinical Presentation

Binge eating disorder frequently co-occurs with obesity and metabolic syndrome. 40% of patients with binge eating disorder have a BMI > 30, and, of this percentage, 15% have a BMI > 40. More than 40% of patients with BED have metabolic syndrome.

Because of this comorbidity, individuals with BED are at increased risk for obesity-related medical consequences, such as type 2 diabetes mellitus, hypertension, and dyslipidemia, as well as premature mortality. Moreover, individuals with BED show a higher prevalence of other health conditions, including asthma, gastrointestinal symptoms, sleep disorders, musculoskeletal problems, neurologic problems, and gynecologic conditions.

However, BED is distinct from obesity for several reasons. Individuals with BED, compared to those with obesity without BED.

- Consume more calories in food intake studies and are generally more sedentary.
- More frequently report overestimation of weight and body shape, concern for their appearance and body weight, dissatisfaction with their physical appearance.
- Show greater cognitive deficits on neuropsychological tests.
- Have greater functional impairment and psychological distress, worse quality of life, and a higher prevalence of mental health disorders.

Indeed, BED is frequently associated with psychiatric comorbidities. 50% of BED patients have at least one depressive episode in their lifetime and, in general, increased recurrence of dysthymia, Panic and other anxiety disorders, borderline personality disorder, and substance and alcohol abuse.

8.6.4.1 Course and Prognosis

The average age of onset of BED is about 21 years. However, BED has a wide distribution of age of onset, from 14 to 30 years. In typical cases, BED begins with episodes of binge eating, often associated with stressful events. Episodes of binge eating determine, in most cases, an increase in weight, which leads some individuals to undergo diets to try to lose weight, but generally without obtaining lasting results. This process is the opposite of what happens in bulimia nervosa, where usually the diet precedes the appearance of episodes of binge eating. Sometimes, however, BED can also begin after a period of strict dieting. The disorder generally has a chronic course. In most cases, individuals alternate prolonged periods characterized by recurrent episodes of binge eating with others characterized by a good control over eating. Although little information is yet available on the course of BED, it appears that migration of the disorder to bulimia nervosa and anorexia nervosa or other eating disorders is rare. In contrast, a longitudinal study of adolescent females observed that BED increases the risk of developing overweight, obesity, and depression by approximately twofold. Finally, it has been estimated that the average persistence of BED is approximately 16 years. Overall, 40–50% of patients experience an improvement, but with an 80% chance of relapse. Integrated treatment, consisting of pharmacological treatment with SSRIs, psychological and nutritional intervention, is associated with a better prognosis.

8.6.5 Treatment

As in other eating disorders, to ensure that the patient has a good chance of achieving an effective and lasting recovery, it is essential to provide a multidisciplinary approach. This should be based on the coordinated involvement of internists (to manage the organic disorders and define a dietary plan appropriate to the weight loss required) and psychiatrists (to correct the mental and behavioral patterns typical of the disorder).

The treatment starts from "assumptions" different from those for the treatment of bulimia nervosa.

The patient must work on the following

- Motivation to reduce weight and not binge.
- Willingness to carry out diets without eliminatory behaviors, although a dietary restriction is a rare event in these patients.
- Concerns about overweight.

The *psychotherapeutic approach* that seems to give the best long-term results is cognitive-behavioral therapy (CBT), aimed at redefining the relationship with food and providing the patient with the tools to react favorably to negative stimuli that can be commonly encountered in everyday life and represent the main trigger for binge eating.

NICE guidelines (2017) have recommended CBT as the treatment of choice for this disorder. In particular, CBT-based Guided Self Help (GSH) should be considered as the intervention of the first choice. It consists of working through a book about binge eating, and having short sessions with a therapist (between 4 and 9 sessions, each lasting 20 min) to monitor developments. If this treatment is not accepted, is contraindicated, or ineffective, group CBT-ED (behavioral therapy for eating disorders) may be considered. If group CBT-ED is not available or the person refuses it, individual CBT-ED should be considered. The same indications are provided for binge-eating disorder in children and young adults as for adults.

Some research supports other specialized psychological treatments for BED, most notably interpersonal psychotherapy (IPT). In general, psychological treatments result in remission of binge eating episodes in approximately 50–55% of individuals affected and improvement in associated eating disorder and depressive symptoms. These beneficial effects are maintained at 2 and 4 years. Unfortunately, the main disadvantage of these approaches is that they generally do not produce significant weight loss.

Depending on the severity of the disorder, in the first phase, it may be necessary to provide a hospitalization of a few weeks or a period of Day hospital, followed by periodic psychotherapeutic sessions for several months.

A model of treatment aimed at producing a permanent weight loss is represented by the following, which provides two types of intervention that can also be complementary.

1. Hospitalization: a hospital stay of 4 weeks aimed at behavioral intervention with total reliance on the multidisciplinary team of nutrition management.
2. High-intensity outpatient pathway (2–3 times per week): this is also a multidisciplinary intervention, including the implementation of a physical activity program.

Medications should never be indicated as the sole treatment for BED; however, in some patients, association with drug therapy is imperative. Indeed, regardless of the presence of a concomitant depressive disorder, CBT can be associated with pharmacological treatment with antidepressants, which has been shown to enhance the effectiveness of the psychotherapeutic intervention. Numerous medications have been tested in the treatment of BED, including antidepressants (SSRIs, serotonin, and norepinephrine reuptake inhibitors, and bupropion), anticonvulsants (topiramate), weight-loss agents (sibutramine), and agents for the treatment of substance use disorders (naltrexone). Although some antidepressants may reduce the frequency of binge episodes, they are sometimes associated with weight gain. Topiramate, which has been shown to reduce both frequencies of binge episodes and weight, is no more recommended because it may worsen cognitive dysfunction.

In case the above-mentioned interventions are not sufficient, *bariatric surgery* can be considered if severe obesity needs to be addressed due to metabolic risks. The surgical approach should always, however, be integrated with psychological and nutritional support.

Types of Bariatric Surgery currently foreseen for BED

- Gastric bypass (Roux-en-Y).
  - Gastric banding.
  - Biliopancreatic diversion.
  - Sleeve gastrectomy.

13.5% of patients who undergo bariatric surgery are BED. However, approximately 20% of patients who undergo these procedures do not benefit from them. Within 2 years, a high proportion of BED patients who have undergone surgery relapse, with at least 25% experiencing binge eating and 12% vomiting for reasons related to shape and weight.

Risks associated with the surgical procedure may include excessive bleeding, infection, adverse reactions to anesthesia, deep vein thrombosis, pulmonary or respiratory problems, leakage into the gastrointestinal system (i.e., fistulas), and death (rare). However, malabsorption syndromes are a frequent complication, resulting in nutritional deficiencies with consequent anemia (iron, vitamin B12, folate deficiencies), neurological disorders such as Wernicke's Encephalopathy (B12 deficiency), beri-beri (B1 deficiency) alterations of vision (vitamin A and vitamin E deficiencies), skin alterations (vitamin A deficiency), osteoporosis, and secondary hyperparathyroidism. Finally, the surgical intervention can be followed by the so-called dumping syndrome (also called "accelerated gastric emptying syndrome"). An early form and a late form of this disorder may occur. The early dumping syndrome is caused by an accelerated emptying of the stomach, that is a too-rapid passage of food from the stomach to the intestine through the communication created by the surgery, which, thanks to the gastric resection, reduces the volume of the stomach and alters to a greater or lesser extent the normal digestive processes. The passage in the intestine of food not yet fully processed by the gastric digestive phase determines the release of vasodilator substances and stimulators on nerve endings, which cause the classic symptoms of the syndrome (abdominal cramps, nausea, and diarrhea).

The late dumping syndrome is instead caused by hypoglycemia that follows the production of an excessive amount of insulin.

Finally, often as a complication of surgery, we can observe the development of a secondary eating disorder, for example, the appearance of BN.

8.7 Other Eating Disorders

8.7.1 Other Specified Feeding or Eating Disorder

The Other Specified Feeding or Eating Disorder (OSFED) corresponds to the diagnostic category of the DSM-5 that has replaced the Eating Disorder Not Otherwise Specified (EDNOS) present in the DSM-IV. This category includes five types of disorders: atypical anorexia nervosa, atypical bulimia nervosa of low frequency and/or limited duration, binge eating disorder of low frequency and/or limited duration, purging disorder, and night eating syndrome (NES). These disorders are quite common and account for 32–53% of all individuals with eating disorders.

In the DSM-5, these individuals present an eating disorder that has similarities with AN or BN and causes clinically significant impairment, but the diagnostic criteria listed for those disorders are not fully met.

For example, atypical disorders are those in which the patient's weight is just above the diagnostic threshold for AN, or binge eating and purging are less frequent than required for the diagnosis of BN.

Although the diagnostic criteria are not met, these forms can also be very severe and incapacitating, and the treatment and prognosis can be virtually superimposable on those for AN and BN.

The following disorders are included among OSFED:

- *Atypical anorexia nervosa*: All criteria for AN are met, with the exception of significant weight loss, since the patient's weight is within or above the normal range.
- *Binge eating disorder (of low frequency and/or limited duration):* All of the criteria for BED are met, but at a lower frequency and/or for less than 3 months.
- *Bulimia nervosa (of low frequency and/or limited duration):* All of the criteria for BN are met, but the binge eating and inappropriate compensatory behavior occurs at a lower frequency and/or for less than 3 months.
- *Purging disorder:* It is characterized by recurrent purging behaviors that influence weight or shape (e.g., self-induced vomiting, abuse of laxatives, diuretics, or other medications) in the absence of binge eating.
- *Night eating syndrome (NES):* It is characterized by recurrent episodes of nocturnal food consumption (eating after waking from sleep or excessive food consumption after the evening meal). The patient has a conscious memory of eating. Nocturnal food consumption is not explained by external influences with changes in the sleep–wake cycle or local social norms and causes significant difficulties and/or problems with functioning. The prevalence of NES in the general population is 1.5%, while in obese subjects is 6–16%. NES is characterized by the following symptomatology.
  - Evening hyperphagia, that is, consumption of at least 25% of the daily intake after the evening meal and\or > 2 nocturnal food intakes (defined as waking up at night to eat) per week.
  - At least 3 of the following 5 characteristics.

    "Morning anorexia" (defined as no appetite in the morning).

    A strong urge to eat between dinner and the onset of sleep or during nighttime awakenings.

    Insomnia at least 4–5 times a week.

    The belief that eating is necessary to initiate or resume sleep.

    Depressed mood that worsens during nighttime hours.
  - Awareness and ability to remember the nightly or evening food intake. This criterion is necessary to differentiate NES from sleep-related eating disorder (SRED), a disorder in which the nightly food intake occurs without awareness and is not remembered by the subject.

Several studies show that the serotonergic system, involved in the regulation of appetite and circadian rhythms, has a key role in the pathophysiology of NES. Antidepressant treatments with sertraline and psychological therapies can be used for the optimal administration of patients with NES. The pharmacological treatment with sertraline shows a significant reduction in the number of awakenings every week, night intake of the week, and percentage of calorie consumption after the evening meal. Other treatment options such as melatonergic medications, light therapy, and topiramate represent promising treatment options.

8.7.2 PICA

It is an eating disorder characterized by the consumption of nonnutritious substances that are not typically thought of as food, such as hair, dirt, and paint chips.

It is characterized by the following diagnostic criteria

- The consumption is continued for at least 1 month.
- The consumption of nonnutritious substances is inappropriate for the individual's developmental level. In children under the age of two, putting objects in their mouth is a normal part of development that allows the child to explore the senses. Mouthing can sometimes lead to swallowing. Children under 2 years of age should not be diagnosed with pica.
- The eating behavior is not part of a culturally or socially supported practice.
- If the eating behavior occurs in the context of another mental disorder (e.g., cognitive disability, autism spectrum disorder) or medical condition (e.g., pregnancy), it is severe enough to require additional clinical attention to avoid conditions of intoxication, poisoning, or bezoar formation.

The prevalence of PICA is difficult to establish due to differences in the definition. A high incidence of PICA (more than 50%) is reported in patients with intellectual disabilities (about 10%).

Mood disorders, anxiety disorders, and obesity are important predictors of PICA in adults. In children, autism spectrum disorder (ASD) is the most significant PICA predictor.

Iron deficiency, anemia, and malnutrition are two of the most common causes of PICA, and some women can develop PICA during pregnancy because of the nutrient deficiency. In these cases, the disorder is a response of the body to a significant nutrient deficiency. The first-line treatment for PICA is thus to test for and correct vitamin or nutrient deficiencies. In many cases, PICA disappears once the deficiencies are corrected. If the disorder is not caused by malnutrition, several behavioral interventions are available.

8.7.3 Rumination Disturbance

It involves the repeated regurgitation of food, which may be rechewed, reswallowed, or spit out.

It is characterized by the following diagnostic criteria:

- The repeated regurgitation of food over a period of at least 1 month.
- The repeated regurgitation is not attributable to gastrointestinal or other associated medical conditions (e.g., gastrointestinal reflux).
- The disorder does not occur exclusively in the course of anorexia nervosa, bulimia nervosa, uncontrolled eating disorder, or avoidant/restrictive eating disorder.
- If symptoms occur in the context of another mental disorder (e.g., intellectual disability), they are severe enough to warrant additional clinical attention.

Some of the oldest observations of rumination disturbance suggested that the syndrome occurs primarily in children and adults with developmental delays but later studies have demonstrated that most patients with rumination are of normal-level intellect. Due to the lack of awareness of the many health care professionals, it is conceivable that the true occurrence of the rumination syndrome is underestimated.

8.7.4 Avoidant/Restrictive Food Intake Disorder (ARFID)

ARFID involves reductions in the amount and/or types of food consumed, but unlike anorexia, it does not involve any distress about body shape or size, or fears of fatness.

It is characterized by the following diagnostic criteria

- A food and nutrition abnormality (e.g., lack of interest in food or eating; avoidance based on sensory characteristics of food) manifested by a persistent inability to take in adequate nutritional and/or energy intake associated with one or more of the following.
  - Significant weight loss or in children inability to achieve growth-related weight.
  - Significant nutritional deficiency.
  - Dependence on enteral nutrition or oral nutritional supplements.
  - Marked interference with psychosocial functioning.
- The disorder is not related to food deficiency or associated with cultural practices.
- The disorder does not occur exclusively in the course of anorexia or bulimia nervosa and there is no evidence of abnormality in the way one's weight and body shape are perceived.
- The abnormality is not best attributable to a medical condition or other mental disorder. If the eating disorder occurs in the course of another disorder, its significance exceeds that of the underlying disorder and requires specific clinical attention.

From the epidemiological point of view, the average age of onset is about 12 years, and in 30% of cases it affects male subjects. It is a clinical entity not to be underestimated since about 15% of patients who require treatment in centers for eating disorders meet the criteria of ARFID.

Children with autism spectrum conditions, ADHD, intellectual disabilities and anxiety disorders are much more likely to develop ARFID. Moreover, children who do not outgrow normal picky eating appear to be at higher risk to develop ARFID.

8.7.5 Orthorexia

Orthorexia nervosa (ON) is a disorder of very recent origin, which has developed simultaneously to the birth and spread of health philosophies of life (e.g., vegetarianism, veganism, and organic food), and the increasing attention of our society toward healthy eating. This condition has always been difficult to define in clinical terms; in fact, it has not yet been clinically recognized as one of the disorders of nutrition and eating, as validated diagnostic criteria for ON have not yet been established. The term orthorexia nervosa was introduced by Steven Bratman, in 1997, to indicate a pathological preoccupation on healthy food consumption.

Nor DSM-IV and the DSM-5 recognize the ON as a proper disorder independent of others: it is placed together with "inverse anorexia" within the area of the avoidant/restrictive disorder of food intake. The obsession with "healthy eating" typical of orthorexia negatively affects the individual's relational, emotional, and physical spheres of life. According to the literature, the onset of ON is characterized by a desire to eat better in order to have a better physical shape. The pathological core of orthorexia lies in a series of distorted beliefs about what is healthy and a sense of personal superiority resulting from dietary restrictions.

Many subjects who develop AN or BN exhibit orthorexic behaviors at early stages of their clinical history.

Orthorexia nervosa is characterized by the following:

1. Devoting more than 3 h a day to preparing and searching for food.
2. Feeling a sense of superiority over those with other eating habits.
3. Following a rigid regimen that is qualitatively controlled and putting in place compensatory restrictions in case of transgressions.
4. Associate one's self-esteem to the adherence to the diet.
5. Putting healthy nutrition at the center of one's life.
6. Overlooking values, relationships, previous interests and activities, and even physical health to achieve the goals described.

According to data released by the Italian Ministry of Health, orthorexics would be 300,000 in Italy (11.3% males vs. 3.9% females).

For the assessment of orthorexic behaviors, it is possible to use the Bratman Test (see Table 8.2), a popular self-report questionnaire. If the answer is affirmative to at least 4–5 questions above, the subject is considered at risk. If the answer is

Table 8.2 Bratman Test

Bratman Test
1. Do you spend more than 3 h a day thinking about your diet?
2. Do you plan your meals several days ahead?
3. Is the nutritional value of your meal more important than the pleasure of eating it?
4. Has the quality of your life decreased as the quality of your diet has increased?
5. Have you become stricter with yourself lately?
6. Does your self-esteem get a boost from eating healthily?
7. Have you given up foods you used to enjoy in order to eat the "right" foods?
8. Does your diet make it difficult for you to eat out, distancing you from family and friends?
9. Do you feel guilty when you stray from your diet?
10. Do you feel at peace with yourself and in total control when you eat healthily?

affirmative to all questions, the attention to "healthy" food could configure a real obsession and increase the risk for the development of a proper eating disorder.

Finally, abnormal eating behaviors can be present as a "symptom" of other psychiatric disorders:

- *Mood disorders:* the depressed subject often shows appetite and weight loss, while a subject with a manic episode can sometimes appear so busy that he does not feel the sense of hunger. On the contrary, in some cases, hyperphagia with weight gain can occur.
- *Schizophrenia:* altered eating behaviors can be found also in psychotic disorders, both related to the psychopathological symptoms and as a side effect of antipsychotic treatment. For example, a patient with a delusion of poisoning can refuse to eat, while a subject treated with antipsychotics can experience an increased appetite.
- *Anxiety disorders:* sometimes, a panic attack can occur while the subject is feeding, leading to the appearance of an intense fear of choking while eating.
- *Somatoform disorders:* in this case, the patient may fear that certain foods may lead to gastralgia or other gastrointestinal disorders.
- *Substance abuse:* the continuous intake of alcohol or drugs can be associated with a reduction in food intake, resulting in malnutrition.
- *Mental organic disorders:* the behavioral disorganization of the subject may lead to the inability to provide for their nutrition correctly.
- *Personality disorders:* Borderline personality disorder is often characterized by abnormalities of eating behavior (binge eating or food restriction), which usually have a short course in time and are the expression of the emotional dysregulation of the subject.

Further Reading

Attia E, Becker AE, Bryant-Waugh R, Hoek HW, Kreipe RE, Marcus MD, Mitchell JE, Striegel RH, Walsh BT, Wilson GT, Wolfe BE, Wonderlich S. Feeding and eating disorders in DSM-5. Am J Psychiatry. 2013;170(11):1237–9. https://doi.org/10.1176/appi.ajp.2013.13030326.

Castillo M, Weiselberg E. Bulimia nervosa/purging disorder. Curr Probl Pediatr Adolesc Health Care. 2017;47(4):85–94. https://doi.org/10.1016/j.cppeds.2017.02.004.

Fairburn CG, Cooper Z, Shafran R. Cognitive behaviour therapy for eating disorders: a "transdiagnostic" theory and treatment. Behav Res Ther. 2003;41(5):509–28. https://doi.org/10.1016/s0005-7967(02)00088-8.

Favaro A, Tenconi E, Santonastaso P. Perinatal factors and the risk of developing anorexia nervosa and bulimia nervosa. Arch Gen Psychiatry. 2006;63(1):82–8. https://doi.org/10.1001/archpsyc.63.1.82.

Gibson EL. The psychobiology of comfort eating: implications for neuropharmacological interventions. Behav Pharmacol. 2012;23(5–6):442–60. https://doi.org/10.1097/FBP.0b013e328357bd4e.

Godier LR, Park RJ. Compulsivity in anorexia nervosa: a transdiagnostic concept. Front Psychol. 2014;17(5):778. https://doi.org/10.3389/fpsyg.2014.00778.

Hilbert A, Pike KM, Goldschmidt AB, Wilfley DE, Fairburn CG, Dohm FA, Walsh BT, Striegel WR. Risk factors across the eating disorders. Psychiatry Res. 2014;220(1–2):500–6. https://doi.org/10.1016/j.psychres.2014.05.054.

Klump KL, Suisman JL, Burt SA, McGue M, Iacono WG. Genetic and environmental influences on disordered eating: an adoption study. J Abnorm Psychol. 2009;118(4):797–805. https://doi.org/10.1037/a0017204.

Lampard AM, Byrne SM, McLean N, Fursland A. Avoidance of affect in the eating disorders. Eat Behav. 2011;12(1):90–3. https://doi.org/10.1016/j.eatbeh.2010.11.004.

Le Grange D, Lock J. Eating bulimia in adolescents: a family based approach. New York: The Guilford Press; 2007.

Macht M. How emotions affect eating: a five-way model. Appetite. 2008;50(1):1–11. https://doi.org/10.1016/j.appet.2007.07.002.

Meule A, Richard A, Schnepper R, Reichenberger J, Georgii C, Naab S, Voderholzer U, Blechert J. Emotion regulation and emotional eating in anorexia nervosa and bulimia nervosa. Eat Disord. 2019;25:1–17. https://doi.org/10.1080/10640266.2019.1642036.

Pérez S, Marco JH, Cañabate M. Non-suicidal self-injury in patients with eating disorders: prevalence, forms, functions, and body image correlates. Compr Psychiatry. 2018;84:32–8. https://doi.org/10.1016/j.comppsych.2018.04.003.

Racine SE, Horvath SA. Emotion dysregulation across the spectrum of pathological eating: comparisons among women with binge eating, overeating, and loss of control eating. Eat Disord. 2018;26(1):13–25. https://doi.org/10.1080/10640266.2018.1418381.

Safer DL, Telch CF, Agras WS. Dialectical behavior therapy for bulimia nervosa. Am J Psychiatry. 2001;158(4):632–4. https://doi.org/10.1176/appi.ajp.158.4.632.

Stice E, Gau JM, Rohde P, Shaw H. Risk factors that predict future onset of each DSM-5 eating disorder: predictive specificity in high-risk adolescent females. J Abnorm Psychol. 2017;126(1):38–51. https://doi.org/10.1037/abn0000219.

Tozzi F, Thornton LM, Klump KL, Fichter MM, Halmi KA, Kaplan AS, Strober M, Woodside DB, Crow S, Mitchell J, Rotondo A, Mauri M, Cassano G, Keel P, Plotnicov KH, Pollice C, Lilenfeld LR, Berrettini WH, Bulik CM, Kaye WH. Symptom fluctuation in eating disorders: correlates of diagnostic crossover. Am J Psychiatry. 2005;162(4):732–40. https://doi.org/10.1176/appi.ajp.162.4.732.

Tremblay L, Lariviere M. The influence of puberty onset, body mass index, and pressure to be thin on disordered eating behaviors in children and adolescents. Eat Behav. 2009;10(2):75–83. https://doi.org/10.1016/j.eatbeh.2008.12.001.

Udo T, Grilo CM. Prevalence and correlates of DSM-5-defined eating disorders in a nationally representative sample of U.S. adults. Biol Psychiatry. 2018;84(5):345–54. https://doi.org/10.1016/j.biopsych.2018.03.014.

Substance-Related Disorders

9

F. Martini, L. Fregna, M. Bosia, G. Perrozzi, and R. Cavallaro

9.1 Introduction

Historically, the practice of using consciousness-altering substances has been followed by humankind for millennia, mostly with therapeutic, cultural and religious purposes. The first reference to nicotine use dates back to 40,000 years ago in two different sources of Australian aborigines and American indigenous. Moreover, archaeological evidence of coca use dates back to at least 3000 BCE, while the earliest report of marijuana use was found in China 12,000 years ago.

In an evolutionary biology perspective, one of the most ancient and important circuits in the brain, the reward system, evolved to increase the adaptive fitness of animals. Survival for most species depends on maximizing contact with advantageous stimuli (positive reinforcers, such as food and sex) and minimizing contact with noxious stimuli (negative reinforcers, such as pain and environmental threats).

Thanks to their ingenuity, humans have managed to extract and refine highly reinforcing stimuli (e.g., high content alcohol, high potency drugs) for

F. Martini (✉) · L. Fregna (✉)
IRCCS San Raffaele Scientific Institute, Milan, Italy
e-mail: francesca.martini1@hsr.it; fregna.lorenzo@hsr.it

M. Bosia · R. Cavallaro
IRCCS San Raffaele Scientific Institute, Milan, Italy

Vita-Salute San Raffaele University, Milan, Italy
e-mail: bosia.marta@hsr.it; cavallaro.roberto@hsr.it

G. Perrozzi
Vita-Salute San Raffaele University, Milan, Italy
e-mail: perrozzi.giulia@hsr.it

R. Cavallaro, C. Colombo (eds.), *Fundamentals of Psychiatry for Health Care Professionals*, https://doi.org/10.1007/978-3-031-07715-9_9

recreational purposes. However, in a subset of individuals with high vulnerability (i.e., preexisting mental illness, genetic predisposition, young age), easy access to these highly reinforcing substances, combined with promotive environments (i.e., chronic stress, peer pressure), can dramatically increase the risk of developing a substance use disorder (SUD). Repeated exposure to drugs overactivates the reward system and triggers neuroplastic changes in several neuronal circuits of the brain. As a consequence, the addicted individual experiences enhanced brain's reactivity to drug cues, decreased motivation for non-drug rewards, increased sensitivity to stressful and negative emotional stimuli, and weakened ability of self-regulation and decision-making. These brain adaptations result in compulsive drug-seeking and reduced ability to control the urge to take the drug despite severe harm.

Thanks to the increasing evidence on the neurobiological, genetic and epigenetic mechanisms underlying it, addiction, though not without resistance, is no longer considered a "moral deficiency," but as both a complex brain disorder and a severe psychiatric illness.

9.2 Definition

The term addiction derives from the Latin verb ad-dicere, which means to deliver. It originally referred to a penalty provided by the Roman law consisting in selling a debtor to his creditor until his work had extinguished the debt. Only in the twentieth century, the term addict became associated with someone whose inclinations toward a drug/habit had led to dependence upon it. A new word was needed to describe a growing phenomenon leading to serious medical consequences.

Today addiction is defined as a chronic relapsing disorder characterized by compulsive drug seeking, difficulties in controlling drug use, persistent drug taking despite harmful consequences and long-lasting changes in the brain.

In the most recent edition of the Diagnostic and Statistical Manual of Mental Disorders (DSM-5) the terms "Substance Dependence" and "Substance Abuse" have been replaced with "Substance Use Disorders" in order to include the frequent nondependent but still maladaptive pattern of substance use causing clinically and functionally significant impairment, regardless of the presence of physical dependence. According to the number of symptoms identified, the DSM-5 allows clinicians to specify how severe an individual's substance use disorder is. The disorder is then classified as mild (presence of 2–3 symptoms), moderate (presence of 4–5 symptoms), or severe (presence of 6+ symptoms).

The term "addiction", still in widespread use, refers to the long-term, neurobiological disease that overlaps with severe substance use disorders as defined in the DSM-5 (see Box 9.2 for DSM-5 Criteria). Clinicians may also add specifiers ("in early remission," "in sustained remission," "on maintenance therapy" for certain substances, and "in a controlled environment") to further describe the current state of the SUD.

Addiction is characterized by three peculiar phenomena: craving, tolerance, and physical dependence (Box 9.1).

Box 9.1: Which Phenomena Underlie Addiction?

Craving: An intense, urgent or abnormal desire to use the substance, persisting for months and even years after drug use cessation and triggered by emotional stimuli or drug-related cues in the environment. Craving is a crucial factor in maintenance of drug addiction and probability of relapse.

Tolerance: Repeated exposure to a drug induces a state of adaptation characterized by "drug-opposite changes" that result in a reduction of the drug's effect over time with a consequent need for increased dosages of the substance to reach intoxication.

Physical Dependence: A state of adaptation which can occur with the regular use of a substance. It is due to the biochemical and structural changes taking place in the brain after repeated use. It results in drug class–specific withdrawal syndromes, that is, a constellation of symptoms and signs produced by abrupt discontinuation or rapid dose reduction, decreased blood levels of the drug and/or administration of an antagonist. Withdrawal's signs and symptoms are usually opposite to the main effect of the substance.

Box 9.2: DSM-5 Diagnostic Criteria for SUD

SUD, as defined by DSM-5, is a problematic pattern of substance use leading to clinically significant impairment or distress as manifested by at least two of the following occurring in a 12-month period.

1. The substance is often taken in larger amounts or over a longer period of time than was intended.
2. Persistent desire or unsuccessful efforts to cut down or control substance use.
3. A great deal of time is spent in activities necessary to obtain/use the drug, or recover from its effects.
4. Craving or strong desire to use the substance.
5. Recurrent use resulting in failure to fulfill major role obligations at work, school, and home.
6. Continued substance use despite having persistent or recurrent social or interpersonal problems.
7. Important social, occupational, or recreational activities are given up/reduced because of drug use.
8. Recurrent substance use in situations in which it is physically hazardous.
9. Substance use is continued despite knowledge of having a persistent or recurrent physical or psychological problem that is likely to have been caused or exacerbated by the substance.
10. Tolerance.
11. Withdrawal.

Together with SUDs, the DSM-5 includes in the broader diagnostic entity of substance-related disorders also substance-induced disorders, which comprise a variety of behavioral or psychological anomalies resulting from exposure to a drug of abuse. This subgroup includes intoxication, withdrawal and other mental disorders secondary to substance use (e.g., mood disorders, psychotic disorders, anxiety disorders, neurocognitive disorders, delirium, sleep disorders, and others).

The DSM-5 recognizes substance-related disorders resulting from the use of ten separate classes of drugs: alcohol; caffeine; cannabis; hallucinogens; inhalants; opioids; sedatives, hypnotics, or anxiolytics; stimulants; tobacco; and other or unknown substances.

9.3 Epidemiology

Addiction is a major contemporary public health issue. Over the last decade, the number of people consuming drugs has notably increased and SUD is expected to become more prevalent over time. Polydrug use is frequent among drug users and individual patterns of use range from recreational to habitual and dependent. Experience of drug use is usually higher in males. Cannabis is the most frequently tried drug in the lifetime, with much lower estimates reported for the use of cocaine, MDMA, and amphetamines. While use of heroin and other opioids remains limited, these drugs are still associated with the more harmful forms of use, including injecting, a mode of administration that is increasingly common also for stimulants. In addition to traditional plant-based substances (cannabis, cocaine and heroin), the past decade has witnessed a considerable growth in production and consumption of synthetic drugs and in recreational use of prescription medicines, in particular opiates.

In 2018, an estimated 269 million people worldwide (5.4% of the global population aged 15–64) had used illicit drugs at least once in the previous year and more than 35 million people were estimated to suffer from a SUD. Among illicit substances, cannabis is the most widely used drug globally, with an estimated 192 million people having used it in 2018 (3.9% of the global population aged 15–64).

Concerning legal substances, alcohol use disorder (AUD) is the most prevalent SUD. The 2018 World Health Organization's (WHO) Global Status Report on Alcohol and Health determined that in 2016 harmful use of alcohol was responsible of around 3 million deaths, more than hypertension and diabetes combined. In 2016, an estimated 2.3 billion people were current alcohol users and 283 million people aged more than 15 years suffered from an AUD (5.1% of adults).

European data are not more reassuring. An estimated 20 million young adults (aged 15–34) used drugs in 2019 (16.6%) in the European Union. Alcohol intake in the European area is the highest in the world, with over one fifth of the population aged more than 15 reporting heavy episodic drinking (five or more drinks on a single occasion) at least once a week. Tobacco consumption is the biggest preventable health risk and the leading cause of premature death in the EU.

Finally, as previously mentioned, abuse of prescription medicines is a growing and alarming phenomenon. The most used drugs are central nervous depressants (benzodiazepines, nonbenzodiazepines, and barbiturates), prescription stimulants (dextroamphetamine, dextroamphetamine/amphetamine combination product, and methylphenidate), and prescription opioids (hydrocodone, oxycodone, fentanyl, and codeine). The number of deaths from prescription opioid overdoses was markedly higher in 2016 than in 1999. In 2012, opioid painkillers caused more "drug poisoning deaths" than heroin and cocaine.

9.4 Etiopathogenesis

9.4.1 Etiology

Similarly to other chronic diseases, substance use disorder is a complex and multifactorial pathology.

Drug use is a dynamic process that starts differently for each person. Many people try substances capable of inducing addictive behavior, often for specific reasons. For instance, people suffering from insomnia or anxiety may try to treat the condition by using sedative substances, such as alcohol, marijuana or benzodiazepines, while people with social anxiety can use alcohol to be more at ease in social situations. Moreover, patients with pain problems may be prescribed opioids and people with attention deficit-hyperactivity disorder may be prescribed or use stimulants to deal with the condition. Adolescents frequently experiment with substances out of curiosity, rebelliousness, and peer pressure.

However, not every person using a substance for recreational or medical purposes develops a substance use disorder. Many genetic, environmental, and social factors contribute to determinate a person's individual vulnerability to start using drugs, sustain drug use, and develop addiction.

9.4.1.1 Genetic Factors

Scientific literature has long recognized that alcohol and drug addiction appear in clusters in families. Approximately half of the risk for addiction is attributable to genetic factors, as shown by twin and adoption studies. Genes that influence psychological traits as reactivity to stress, risk-taking, impulsivity, and novelty-seeking can influence the willingness to try drugs as well as the susceptibility to develop a substance use disorder. Moreover, several polymorphisms have been identified as candidates for genetic susceptibility, including polymorphisms in the alcohol-metabolizing enzyme genes, opioidergic, dopamine, gamma-aminobutyric acid (GABA), and serotonin receptor genes, catechol-*O*-methyltransferase enzyme genes, and others.

9.4.1.2 Environmental Factors

Environmental stressors, especially in early life, may increase the risk for SUD as well. For example, social deprivation during infancy has been associated with altered brain maturation and connectivity, which could underlie the increased impulsivity observed in these children. Moreover, stressful environments with social isolation and poor familial cohesion during adolescence have been documented to have detrimental effects on addiction vulnerability. In addition, parents' habits or attitudes toward substance consumption (e.g., parents' substance social use, substance/alcohol easy access, excessive permissiveness) could increase the risk to experiment with drugs. This increased susceptibility is partly due to the shared genetic vulnerability, partly to role modeling and social learning processes (adolescents may adopt the excessive substance use habits of parents and siblings even if they perceive their negative consequences).

9.4.1.3 Epigenetic Factors

Traumatic experiences, stress, or substance exposure in infancy or early adolescence can influence biological drivers of substance use and addiction through epigenetic mechanisms. Epigenetics refers to the processes through which gene expression is altered by environmental events. It relies on three main processes—DNA methylation, histone acetylation, noncoding RNA—and the induced alterations become heritable despite no specific mutations in DNA sequences. For instance, early life stress may affect the development of the major component of the stress system (hypothalamic–pituitary–adrenal axis), leading to increased reactivity to stress and vulnerability to addiction. Interestingly, transgenerational epigenetic effects of parental drug-taking before conception on addiction-related behavior in the progeny have been identified.

Age of Onset

Early use of drugs or alcohol is one of the most relevant risk factors for the later development of a substance use disorder. During adolescence, the brain undergoes significant maturational and remodeling processes, including synaptic pruning and myelination. This increased neuroplasticity and the greater sensitivity of adolescents to environmental stimuli, such as stress and peer pressure, influencing drug-taking, make adolescents at greater risk both to experiment with drugs and develop substance use disorders. The human brain continues to develop until the early to mid-20s, and, while the development occurs faster for reward/motivation and emotional circuits, the prefrontal top-down control circuits develop years later. Therefore, during adolescence, the reward/motivation and emotional circuits are hyperactive and lead to marked emotional reactivity and reward-seeking behaviors. Meanwhile, the prefrontal cortex (PFC) is still immature and unable to self-regulate, leading to greater impulsivity and risk-taking. Exposure to drugs of abuse in an early phase of life, when the development of some brain structures is not yet complete, may further prejudice the development of the PFC, increasing the long-term risk for addiction.

Substance-Specific Characteristics

Each substance has a different propensity to induce dependent behavior due to its specific characteristics. One of the most determinant aspects of drug liability to induce dependence is the rapidity of its effect. Substances that can be injected intravenously, absorbed through the nasal mucosa, or smoked exert their effects on the brain within seconds. Heroin, crack cocaine, powder cocaine, tobacco, and cannabis are all taken by one or other of these rapid routes. Drugs can be particularly addictive also because of their tendency to induce adaptive changes responsible for physical dependence: short half-life of the drug, high pharmacodynamic efficacy, and liability to produce tolerance. Addictive drugs are generally used repeatedly and frequently, partly because of the power of craving and partly to avoid unpleasant withdrawal symptoms. Finally, some drugs can lead to repetitive use that seems to rest more on psychological dependence and craving than tolerance or physical symptoms directly related to drug withdrawal.

Other Psychiatric Disorders

Another important source of vulnerability to develop drug addiction is the comorbidity with psychiatric disorders. Substance use disorder has a higher prevalence in several psychiatric illnesses such as anxiety disorders, mood disorders, schizophrenia, personality disorders, and attention deficit hyperactivity disorder (ADHD), compared to the general population. This increased vulnerability may be ascribed, on the one hand, to genetic or environmental variables that could represent common risk factors for both substance use and psychiatric disorders. On the other hand, preexisting psychiatric disorders or symptoms could lead the person to "self-medicate" with alcohol or other substances. The co-occurrence of substance use and mental health disorders is of great relevance not only because of its high prevalence but also due to the difficulty of recognizing and managing it. Comorbidity of psychiatric conditions and SUDs is associated with lower rates of treatment success, a higher rate of psychiatric hospitalizations, and a higher prevalence of suicide than the two conditions alone.

9.4.2 Neurobiological Basis of Addiction

From a neurobiological point of view, addiction can be conceptualized as a complex, three-stage disease characterized by a disturbance in three major neurocircuits:

- Basal ganglia-driven binge/intoxication stage,
- Extended amygdala-driven withdrawal/negative affect stage,
- Prefrontal cortex-driven preoccupation/anticipation stage.

9.4.2.1 Binge/Intoxication Stage

Every substance of abuse strongly activates a brain circuit called "reward system", mainly composed of a group of subcortical nuclei called basal ganglia. Intoxicating doses of alcohol and drugs cause an increase in dopamine release in this area, leading to intensely positive feelings that act as a reinforcement and promote the repetition of the behavior. Repeated consumption, hence repeated reward experiences, thus become associated with the drug-related stimuli through a phenomenon of associative learning called conditioned reinforcement. Stimuli, including contextual or environmental, associated with the drug (e.g., places where the subject used the drug, people the subject used the drug with) become conditioned and able to increase dopamine release before drug exposure, thus triggering craving for a drug even long after use has stopped.

With repeated use, ordinary, healthy rewards, such as food, sex, or social interactions, lose their former motivational drive, making the brain's reward system much less sensitive to stimulation by both drug-related and non–drug-related rewards (desensitization process) and resulting in the need to increase the substance intake.

9.4.2.2 Withdrawal/Negative Affect Stage

In addition to resetting the brain's reward system, repeated exposure leads to neuroadaptations in another brain circuitry involved in stress and emotional response, including the extended amygdala and the basal forebrain. These changes result in an increase in a person's reactivity to stress and the emergence of the negative emotional state associated with withdrawal.

At this stage, substance-taking aims not only to pursue the rewarding effect, lessened by the desensitization process, but also to escape the discomfort associated with the aftereffects of use. Unfortunately, although drug administration temporarily alleviates the distress, repetition of this pattern worsens the withdrawal syndrome in the long term, triggering a vicious cycle.

9.4.2.3 Preoccupation/Anticipation Stage

Lastly, the neuroadaptations of these two stages are accompanied by a disruption of the brain region, called prefrontal cortex (PFC), responsible for executive processes, including the ability to make decisions, regulate one's own actions and emotions, organize thoughts and activities, prioritize tasks, manage time and inhibit impulses, including the urge to use the substance.

The combination of increased motivation for the drug-related reward (binge/intoxication stage), reduced reward sensitivity and hyperactivity of the stress circuit (withdrawal/negative affect stage), and impaired executive functions (preoccupation/anticipation stage) often results in the inability to balance the urge to take the drug with the will to abstain.

9.5 Assessment

Detecting alcohol/substance hazardous use, abuse, or dependence problems has continued to challenge healthcare professionals, especially in light of the fact that most of the subjects with a SUD do not actively seek out treatment in the substance abuse treatment system. Screening is therefore a vital first step and primary care providers are recommended to routinely ask patients about alcohol and illicit substance use at initial assessment.

Patients may present symptoms suggestive of substance intoxication (e.g., sedation, slurred speech), withdrawal (e.g., unexplained tachycardia or hypertension, restlessness), or medical/psychiatric complications of substance use such as cardiovascular or pulmonary diseases, anxiety, insomnia, or depressed mood. Otherwise, alcohol/substance use disorders could be identified only through routine screening questions asked by physicians or other staff.

When a substance use disorder is suspected, the health care professional should perform a detailed clinical assessment, using a nonjudgmental and empathetic attitude.

Healthcare staff should preferably use structured assessment tools such as *Alcohol Use Disorders Identification Test (AUDIT)* or *CAGE Alcohol Abuse Screening Tool* and *CAGE-AID Substance Abuse Screening Tool.* Key aspects of the assessment should include anyway the following.

- Frequency, amount, and route of administration of the substance.
- Past or current use of other psychoactive substances (including alcohol and tobacco).
- Prior treatments and response to treatment.
- Overdose history.
- Risky behaviors, such as driving while intoxicated or sharing needles.
- Medical and psychiatric history.
- Family history of substance use disorders.
- Current social situation and supports.
- Readiness for change.

The type and route of administration of each substance may be relevant in identifying a subject's probability of contracting an infectious disease, risk of mental and physical health consequences, and adherence to treatment recommendations. A careful physical examination should look for complications, such as abscesses, cirrhosis, or oral health problems. Patients who report a recent interruption of use of a highly addictive drug should be evaluated for withdrawal symptoms.

In order to establish an appropriate treatment plan for the patient, it is crucial to know what type of substances the patient is consuming. Substance classes used include stimulants, hallucinogens, and depressants. Most substances have their own profiles of effects but they could share similar signs and symptoms with other drugs of the same class. Several factors, such as social desirability bias, legal implications, and implicit cognitive process may lead a patient to deny or minimize use of a substance. Therefore, a urinary drug screen should be performed to confirm the drug of use and laboratory investigations to identify possible health consequences.

Once established the problem, health care providers should promote patients' awareness concerning their substance use, increase their motivation to reduce or stop the use of harmful substances and offer options for referrals or treatment recommendations.

9.6 Clinical Presentation

The clinical presentation of SUDs is highly heterogeneous due to both the wide range of disorders covered by the definition and the drug-specific effects and consequences.

Substance-related disorders can be subdivided conceptually in conditions linked to the actual use of the substance and its immediate effects (intoxication and withdrawal), and those which reflect the complications of substance use such as disease processes in the brain and the body (substance-induced mental disorders and substance-related physical health problems).

Moreover, drug classes differ in terms of physiological and metabolic effects, psychological and behavioral consequences, and cognitive effects. It is thus essential to separately address the clinical pictures related to the most frequently used class of substances.

9.6.1 Alcohol

Alcohol use disorders are among the most prevalent mental disorders globally, affecting 8.6% of men and 1.7% of women in 2016, with evidence of the gender gap narrowing over time. Ethanol has several complex effects on the brain, due to its ability to cross biological membranes and to interact with multiple molecular targets. The effects of ethanol are mediated by the activation of GABA-A receptors, leading to an increase in GABA function, which eventually results in a dopamine increase.

Ethanol is highly addictive. Its legal status and perception as a harmless socializing drug lead consumers to frequently adopt unhealthy use patterns, encompassing a spectrum from hazardous use to alcohol use disorders. Risky use has been defined as more than 4 drinks per occasion or more than 14 drinks per week. Risky alcohol use progresses into a disorder when the individuals experience negative consequences and/or loss of control around their drinking, as defined in the DSM-5.

9.6.1.1 Alcohol Acute Intoxication

It is generally easily recognized, being characterized by alcoholic breath, dysarthria (slurred speech with poor articulation of words), hypotension, ataxia (impaired coordination and unsteady gait), impaired attention and memory, emotional lability, behavioral disinhibition with altered critical capacity. Presentations depend on the blood alcohol concentration but are also individual-dependent. It is a serious medical emergency and can lead to respiratory failure, coma and death for pulmonary aspiration of gastric content or respiratory arrest.

9.6.1.2 Alcohol Withdrawal

It occurs in subjects with a physical dependence on alcohol as a consequence of the neuroadaptive compensatory changes that occur during protracted exposure to alcohol's effects. It is defined by the appearance, within 4–8 h from last drink, of at least two symptoms among autonomic instability, increased hand tremor, insomnia, nausea or vomiting, transient hallucinations, psychomotor agitation, anxiety, and seizures. The time course and symptoms' severity of alcohol withdrawal should be closely monitored to establish adequate treatment and avoid complications. One of the most widely used rating instruments is the *Clinical Institute Withdrawal Assessment of Alcohol* Scale, revised (CIWA-Ar).

In some cases, the clinical picture may evolve in two severe and potentially fatal complications, that is, delirium tremens and seizures

- *Seizures*: They occur within 48 h, either as a single generalized tonic–clonic seizure or as a brief episode of multiple seizures.
- *Delirium tremens*: It occurs in about 5% of patients with alcohol withdrawal within 1–4 days from the last drink. It is characterized by fluctuating disturbances in consciousness (confusion), changes in cognition (memory, orientation, language, visuospatial ability), exacerbation of autonomic symptoms (sweating, fever, hypertension, tachycardia, nausea, palpitations, and severe tremor), visual, auditory or tactile hallucinations. Typical alcohol withdrawal hallucinations are defined as microzooptic or lilliputian and feature frightful miniature individuals, animals, objects, or fantasy figures. These symptoms get typically worse at night and on the fourth or fifth day after the stopping of heavy drinking. Delirium tremens usually lasts for 2–3 days, but symptoms may linger for as long as a week. In up to 4% of hospitalized patients, withdrawal delirium results in death, mainly due to hyperthermia, cardiac arrhythmias, complications of seizures, or concomitant medical conditions.

9.6.1.3 Physical Health Consequences

High chronic alcohol use damages the human body. The incidence and severity of tissue injury are influenced by several factors such as total volume of alcohol consumed, pattern of consumption, duration of alcohol misuse, age, gender and diet of the drinker. In particular, adolescents and women are more vulnerable to the effects of alcohol both in the brain and the body.

- *Gastrointestinal diseases:* Alcohol is one of the most frequent causes of liver disease including alcoholic hepatitis, fatty liver degeneration, fibrosis and cirrhosis. Moreover, excessive alcohol intake is linked with reflux esophagitis, acute and chronic gastritis and acute and chronic pancreatitis.
- *Cardiovascular diseases*: Heavy alcohol consumption has been shown to increase the risk of hypertensive heart disease, cardiomyopathy, cardiac arrhythmias such as atrial fibrillation, and hemorrhagic strokes.
- *Cancer:* There is an established causal link between alcohol use and cancer development in the oropharynx, larynx, esophagus, liver, colon, rectum and female breast.
- *Others:* Chronic heavy use has been linked to a disruption of glucose homeostasis and to the development of insulin resistance, resulting in a higher risk of diabetes mellitus. Moreover, chronic alcohol intake has been linked with hematopoietic

system changes including anemia and thrombocytopenia (abnormally low levels of platelets). Nutritional deficiencies, in particular thiamine and vitamin B12, are often associated with alcoholism and can result in peripheral neuropathies, bone marrow toxicity, cognitive changes, and intracranial hemorrhages.

9.6.1.4 Mental Health Consequences

- *Wernike encephalopathy and Korsakoff syndrome:* Wernicke encephalopathy results from thiamine inadequate intake or absorption from the gastrointestinal tract and hepatic storage often associated with alcoholism. It consists of psychomotor slowing or apathy, ocular disturbances, ataxia, impaired consciousness, and, if untreated, it may lead to coma and death. Wernike encephalopathy may fully remit, persist, or degenerate into Korsakoff syndrome, a late complication characterized by cognitive dysfunction, loss of recent memory and inability to learn new information, confusion, and behavioral changes.
- *Alcoholic hallucinosis*: It is a rare alcohol-induced psychotic disorder almost exclusively seen in people with severe and long-lasting alcohol use disorder. It occurs after 7–48 h from the last drink, in 2–3% of alcohol-dependent patients. It is characterized by sudden onset of auditory hallucinations and, often, persecutory delusions, but in clear consciousness, correct orientation, intact attention, and memory. It usually has a favorable prognosis with symptoms typically improving within weeks and hallucinations clearing within one to 6 months.
- *Cognitive impairment*: Long-term excessive alcohol intake has been associated with a decrease in gray and white matter (brain atrophy) and several degrees of cognitive impairments. Mechanisms involved are multiple, including direct neurotoxic effects, severe vitamin deficiencies, dietary neglect and malnutrition, traumatic brain injury, cerebrovascular events, alcoholic liver disease. They can manifest with generalized cognitive deficits, psychotic symptoms such as delusions, and behavioral changes (emotional lability, irritability, and disinhibition).
- *Depression and anxiety*: Repeated heavy drinking is associated with an increased risk of transient depressive episodes, characterized by severe anxiety, insomnia, low mood with guilt thoughts, and suicidal ideas and attempts. However, the majority of these conditions are substance-induced and likely to improve within 2–4 weeks of abstinence without antidepressant treatment. However, after the acute withdrawal phase, people may develop a protracted abstinence syndrome, lasting up to several months. This condition must be promptly recognized and treated with education and cognitive-behavioral approaches. If depressive symptoms continue to occur daily and almost all day long after four or more weeks of abstinence, the clinician should hypothesize that a major independent depressive episode is present and treat it accordingly. Since the clinical features of an alcohol-induced depressive episode are very similar to those of independent mood disorders, family history is an important aspect of differential diagnosis.

Box 9.3: Binge Drinking

Binge drinking is a pattern of alcohol intake that brings blood alcohol concentration to 0.8 g/dL and typically occurs after the intake of 5 or more alcoholic drinks by men and 4 or more by women over around 2 h. Interest in binge drinking behaviors has increased in recent decades due to its notable growth, especially among adolescents and young adults. Binge drinking is estimated to affect around 7.5% of the population of 15+ years worldwide. It commonly occurs on weekends, with moderate or no alcohol intake on most other weekdays.

Binge drinking seems particularly common among people with certain personality traits like impulsivity, disinhibition, and sensation/novelty-seeking. However, it can also represent a dysfunctional strategy to alleviate worries or reduce social anxiety, especially among females. Other major risk factors for binge drinking are peer use (having friends who drink) and binge drinking behavior of parents.

This drinking pattern is linked to an increased risk of acute consequences, including alcohol poisoning, involvement in car accidents and fatalities, alcohol-related blackouts and injury, physical and sexual violence, and impaired scholastic or professional performance. Chronic binge alcohol consumption has systemic effects on several organs. It has been associated with gastritis and gastric mucosal atrophy, gastric and esophageal dysmotility, hepatic steatosis and fibrosis, acute hepatitis, pancreas inflammation, impaired intestinal nutrient absorption. It can lead to increased risk of cardiovascular comorbidities, including hypertension, stroke, myocardial infarction, and atrial fibrillation. It has been linked to impairments in multiple aspects of lung function, skeletal muscle dysfunction and acute muscle injury.

Binge drinking could also lead to potentially long-term brain functional and structural changes, especially in the context of the ongoing brain maturation that occurs during adolescence. The specific effects of binge drinking during adolescence on the brain and cognition are influenced by the timing, dose, and duration of alcohol exposure. Brain alterations can result in behavioral and cognitive impairment (difficulties in decision-making, impulse control, motor skills) but also increase the risk of developing alcohol or other substances addiction later in life.

9.6.2 Cannabis

Cannabis is the most widely used illicit psychoactive substance worldwide. Cannabinoids target cannabinoid receptors, CB1 and CB2, which are located in the central nervous system (basal ganglia, hippocampus, cerebellum, and cortex) and systemically (immune cells and spleen) and are usually activated by endogenous cannabinoids. Although cannabis has long been considered a "soft" drug, studies have demonstrated the harmful addictive effects and mental health consequences associated with its use. Cannabis addiction affects at least 13 million people worldwide, mostly young adults, men, and people in high-income countries. Increased vulnerability to cannabis addiction is due to a multiplicity of variables, including environmental and genetic factors, level of use (high doses and frequent use), potency of the substance used, and age at onset. The commencement of usage during adolescence has been associated with an increased risk of developing dependence, other substance use and psychiatric disorders and poorer cognitive and educational outcomes in adulthood.

Box 9.4: THC and CBD: What Is the Difference?

Marijuana is derived from the cannabis Sativa, Indica, hybrid species, and/or Ruderalis plants, which contain more than 140 compounds. Among them, the most abundant are δ-9-tetrahydrocannabinol (THC), responsible for the psychotropic effects (psychotomimetic and euphoric), and cannabidiol (CBD), responsible for the anxiolytic and therapeutic (anti-inflammatory, immunomodulator, antioxidant, analgesic) effects. Over the past decades, cannabis potency (i.e., the ratio between THC and CBD) has radically increased worldwide, leading to more serious complications.

Furthermore, of great concern are synthetic cannabinoids (e.g., so-called Spice), which have frequently been associated with serious adverse reactions, including death.

9.6.2.1 Cannabis Intoxication

It occurs within a 2-h period of using cannabis and is associated with physical symptoms such as increased appetite, conjunctival injection, xerostomia (dry mouth), and tachycardia. Moreover, it may be characterized by a range of behavioral alterations, including impaired motor coordination, euphoria, perception of slowed time, anxiety, and, on some occasions, persecutory ideas.

9.6.2.2 Cannabis Withdrawal

It occurs within 1 week following discontinuation of frequent and prolonged cannabis usage (usually daily over a period of months). It is characterized by the appearance of irritability, anxiety, nervousness, restlessness, depressed mood, sleep difficulties (i.e., insomnia, vivid and disturbing dreams), reduced appetite or weight loss, physical symptoms causing discomfort (i.e., abdominal pain, nausea, tremors, fever, sweating, headache), and cannabis craving. Symptoms of withdrawal may persist for weeks or months after cessation of cannabis use.

9.6.2.3 Physical Health Consequences

Chronic cannabis use has been associated with harm in several systems

- *Respiratory diseases*: Marijuana smoking increases the risk of several respiratory problems, such as spontaneous pneumothorax (collapsed lung) and chronic obstructive pulmonary disease.

- *Cancer*: Evidence supports that cannabis use has an impact on testicular germ cell tumor and lung cancer development.
- *Cardiovascular diseases*: Cannabis use has been associated with an increased risk for cardiovascular diseases, particularly ischemic strokes, hemorrhagic strokes, and ischemic heart disease.
- *Gastrointestinal diseases*: The use of cannabis has been associated with hyperemesis syndrome, a condition characterized by persistent vomiting and diarrhea.

9.6.2.4 Mental Health Consequences

- *Other substance use disorders*: Cannabis use, especially if daily and started during adolescence, has been considered a risk factor for the development of other substance use disorders. Indeed, together with alcohol and tobacco, cannabis has been defined as a "gateway drug". The co-occurrence of different SUDs could reflect either overlapping genetic vulnerability or the use of other drugs to deal with the effects of the first. These comorbidities can make treatment more difficult.
- *Depression and anxiety:* Several studies have suggested that cannabis use, especially if long-term and heavy, may represent a risk factor for the emergence of anxiety disorders (panic attacks and social anxiety) and mood disorders (depression and bipolar disorder). Moreover, heavy cannabis use is linked to a higher risk of suicide ideation and attempt.
- *Cannabis-induced psychosis:* Cannabis use is considered one of the major preventable risk factors for psychosis. THC, especially at high doses, can precipitate acute, transitory, dose-dependent psychosis. Moreover, evidence suggests that cannabis use may promote the emergence of lasting psychotic disorders in some individuals, especially those with a genetic vulnerability (past history or family history of psychotic symptoms). Indeed, the use of cannabis is estimated to increase the risk of schizophrenia by almost twofold, accounting for 8–14% of cases. The link between cannabis use and schizophrenia is complex and still controversial. It could originate from direct causality, shared etiology, gene–environment interactions, or self-medication for premorbid symptoms. Factors increasing the risk of transitioning to a first episode of psychosis are male gender, high doses and long duration of abuse, early onset age, use of high potency substances, and synthetic cannabinoids. Finally, cannabis consumption is particularly harmful to patients with a preexisting psychotic disorder.
- *Amotivational syndrome*: Heavy long-term cannabis use has been associated with apathy, defined as reduced initiative and motivation, diminished ability to concentrate, reduced social engagement and emotional expression, lack of energy, decreased social and professional functioning. Cannabis chronic use can decrease dopamine levels in the brain's striatum, an area associated with motivation and novelty-related decision-making. Consistently, cannabis consumers show reduced striatal dopamine synthesis capacity, inversely correlated to amotivation.
- *Cognitive impairment:* Evidence suggests that cannabis use has detrimental effects on cognition, with the magnitude of neuropsychological impairment

being related to frequency and duration of use and age at onset. Once again, adolescence represents a period of greater vulnerability to drug exposure. Cannabis use reduces educational attainment, leading to memory and learning deficits. In particular, short-term use of cannabis has been linked to impairment of memory, psychomotor coordination, altered decision-making skills, attention, and concentration, while long-term/heavy use has been associated with a significant decline in cognitive development and deficits in executive functioning, memory, learning, and processing speed. Since THC can disrupt reward-based learning processes, reduced motivation is likely to be one of the pathways leading to impaired learning.

9.6.3 Psychostimulants

The psychostimulant class comprises methamphetamine, amphetamine, cocaine, methylenedioxymethamphetamine (MDMA), and several prescribed stimulants, such as dextroamphetamine and methylphenidate. Stimulants increase noradrenaline and dopamine neurotransmitter activity and sympathetic arousal. Route of administration are multiple: they can be ingested, snorted, injected, and smoked (in the form of crack cocaine and crystalline methamphetamine). Compared with amphetamine/methamphetamine and cocaine, patterns of consumption are generally more occasional for MDMA, with lower rates of severe complications including dependence and death.

9.6.3.1 Stimulants Intoxication

Stimulant drugs are used worldwide for their ability to produce euphoria, boost self-confidence, increase sociability, energy, and alertness, and reduce need for sleep, fatigue, and appetite. On some occasions, their use may increase libido and enhance sexual pleasure. The incidence of undesired side effects increases with increasing dose and duration of use. They include dysphoric mood, hyperarousal, anxiety, panic attacks, altered judgment, interpersonal susceptibility, and psychotic symptoms such as paranoid thoughts and hallucinations. In particular, cocaine use is sometimes associated with the appearance of a characteristic phenomenon called "formication" or "cocaine bugs," that is, tactile hallucinations consisting of bugs crawling beneath or upon the skin. Signs associated with stimulant intoxication are tachycardia, hyper/hypotension, dilated pupils (mydriasis), nausea or vomit, repetitive or stereotyped behaviors, tremor and dyskinesia. Overdoses can be fatal, usually for arrhythmias, respiratory depression, and epileptic status.

9.6.3.2 Stimulants Withdrawal

The withdrawal syndrome is prominently psychological and characterized by fatigue, anxiety, reduced ability to feel pleasure (anhedonia), psychomotor agitation or retardation, concentration impairment, increased appetite, sleep disturbances, and vivid unpleasant dreams. The associated physical signs may include tremor, shivers, involuntary movements, and musculoskeletal pain. During the first days, stimulant withdrawal syndrome can be associated with depressed mood with an increased risk for suicidal ideation and attempt (which has been termed the "crash").

9.6.3.3 Physical Health Consequences

- *Cardiovascular diseases:* The physical harms are primarily related to cardiovascular accidents, resulting in deaths from acute (e.g., acute coronary syndrome, aortic dissection, and cardiac arrhythmias) and chronic (e.g., cardiomyopathy and coronary disease) cardiovascular pathology, even in young healthy individuals.
- *Cerebrovascular diseases:* Stimulants increase the risk of ischemic or hemorrhagic stroke, in particular among young (aged < 45 years) users.
- *Respiratory diseases:* Respiratory problems are of course higher among users who smoke the substance, and can range from moderately decreased pulmonary diffusion capacity, pneumothorax ("crack lung"), and bronchospasm to pulmonary edema, hypertension, and infarction.
- *Others:* Other acute complications include seizures that can occur even at the first time of use and without a prior history, hyperthermia, acute muscular damage (rhabdomyolysis), and acute renal failure. Chronic cocaine snoring can result in nasal septum perforation, osteolytic sinusitis and oropharyngeal ulcers. The gastrointestinal chronic consequences of stimulants' use are mainly gastric and duodenal ulcers. Other consequences of stimulants' use depend on the route of administration: as with other substances, injection and sharing equipment can lead to infectious complications (e.g., blood-borne viruses, abscess, endocarditis).

9.6.3.4 Mental Health Consequences

Chronic stimulant use can also lead to neuropsychiatric consequences:

- *Excited delirium:* It is a condition mainly associated with the use of stimulants, but also hallucinogens, cannabinoids, opioids, and alcohol. Excited delirium syndrome is a condition characterized by delirium, agitation, and hyperadrenergic autonomic dysfunction. The etiology is multifactorial and involves a complex interaction of concurrent medical conditions, including neurologic, toxicologic, psychiatric and metabolic disturbances. It is an infrequent but life-threatening condition, which affects heavy and habitual abusers and evolves rapidly in 2–4 h. Its clinical presentation may include paranoid symptoms, hallucinations, bizarre or aggressive behavior, hyperactivity, unusual pain tolerance, hyperthermia, tachycardia, sweating, and pupil dilation. Morbidity and mortality related to excited delirium have been ascribed to acute myocardial infarction, hypoventilation, pulmonary edema, and cardiopulmonary arrest.
- *Depression and anxiety*: The use of stimulants is associated with an increased risk of depression. Depressive syndromes are frequent in subjects with stimulant dependence and withdrawal can trigger or worsen depression. The mood-elevating effects of stimulant use can lead to a vicious cycle of stimulant self-medication of depressive symptoms. The association between stimulant use and anxiety is still uncertain although panic attacks can occur during acute intoxication.

- *Cognitive impairment:* A significant harm associated with psychostimulants use is cognitive impairment. Chronic stimulants consumers exhibit a range of short-term deficits across multiple cognitive functions, in particular learning, memory, processing speed, executive functions and motor and language skills, persisting for at least some weeks after interrupting the use. Moreover, some evidence suggest long-term impairment, particularly in memory and higher cognition. Several factors such as lifetime exposure, pattern of use and dosage may influence the potential for long-term impairments. Neurotoxicity, particularly to dopaminergic and serotoninergic neurons, has been proposed as the underlying mechanism for the cognitive impairment associated with psychostimulant consumption.
- *Parkinsonism:* Evidence supports a relationship between chronic methamphetamine use and the development of Parkinsonism. Long-term users can develop subtle Parkinsonian features such as abnormalities in hand movements, speech, gait, movement speed and coordination. Possible involved mechanisms are acute reduction in striatal dopamine levels and dopamine synthesis and, after repeated and high-dose exposure, lasting dopamine transporter alterations and structural degeneration of dopaminergic neurons.
- *Psychotic disorders:* Transitory psychotic symptoms such as hallucinations (commonly auditory, visual, or tactile) and paranoid ideation are frequent among recreational stimulants users in particular with frequent, high-dose, and dependent use or use of high-potency forms such as crystal methamphetamine. On some occasions, psychotic symptoms persist beyond intoxication, leading to the emergence of brief drug-induced psychoses. Stimulant-induced psychoses usually have better outcomes than cannabis-related psychoses, with lower rates of transition to chronic psychotic disorders. In approximately 30% of these patients, however, the appearance of psychotic symptoms may suggest an underlying susceptibility to lasting psychotic illnesses such as schizophrenia. Subjects with a family history of psychosis or premorbid schizoid or schizotypal personality traits are at a particularly high risk. Such patients may also exhibit characteristic clinical symptoms such as formal and content thought disorders, and negative symptoms including apathy, avolition, affective flattening, and social withdrawal, as well as impairments in several cognitive and functional domains.

Box 9.5: MDMA

MDMA is a somewhat unique substance in that it is officially classified as a hallucinogen in the DSM-5, but actually has strong stimulant properties and also acts as an amphetamine. It exerts its effects by increasing the activity of three neurotransmitters (serotonin, dopamine, and norepinephrine, and by binding to 5-HT2A and alpha-2 receptors. Its acute effects include increased energy, enhanced sense of well-being, increased extroversion, distortions in time and perception, and enhanced enjoyment from sensory experiences. It has also been defined empathogenic or entactogen since it can increase self-awareness, emotional warmth, and empathy toward others. MDMA can also cause a number of acute adverse reactions, including increase in blood pressure, rise in body temperature with renal failure or fatal cerebral edema, involuntary jaw clenching, lack of appetite, nausea, restless legs, headache, sweating. Psychological symptoms can include depersonalization, illogical or disorganized thoughts, anxiety, and panic attacks, and in severe cases, a loss of consciousness and seizures.

Some people report signs and symptoms of dependence, including withdrawal symptoms such as fatigue, apathy, depression, loss of appetite and concentration deficits.

Long-term exposure to MDMA has been associated to degeneration in serotonergic neurons and widespread reductions in grey matter volume. Repetitive use can lead to the appearance of sleep disturbances, depression, paranoia, increased impulsivity, and deficits in several domains of cognition, including attention, memory, learning, and visual processing.

9.6.4 Opioids

The prevalence of opioid use disorder has increased significantly with the still ongoing opioid epidemic in many Western nations. A wide range of opioids are used for medical and recreational purposes, and include substances directly derived from the opium poppy (e.g., morphine and codeine), semisynthetic (e.g., heroin and hydromorphone), and synthetic opioids (e.g., methadone and fentanyl). Opioids enter the brain rapidly and bind to opioid receptors including $\mu 1$, $\mu 2$, δ, κ on cells located in several areas, mainly in the central and peripheral nervous system, involved in feelings of pain and pleasure and in regulating heart rate, breathing, sleeping and in the gastrointestinal tract. Common routes of administration include injection, nasal insufflation, and inhalation. Some people practice the so-called speedballing, that is, the consumption of a mix of heroin and crack cocaine.

9.6.4.1 Opioid Intoxication

People using opioids report feeling a "rush," or a sense of euphoria. Opioid intoxication symptoms also include dry mouth, itching, confusion, impaired judgment, psychomotor agitation or depression, pupillary constriction ("pinpoint pupils"), nausea, vomiting, constipation, and decreased pain perception. At high doses, opioid use can lead to respiratory depression as these molecules suppress respiratory rhythm generation and reduce the physiological responsiveness of central and peripheral chemoreceptors. Opioid overdose should be considered in subjects with decreased respiratory rate or chest wall rise, diminished consciousness, and miotic pupils.

9.6.4.2 Opioid Withdrawal

The severity of withdrawal syndrome depend on the route of administration, dose and duration of use. The time to onset and duration of the withdrawal symptoms are related to the half-life of the substance. Withdrawal from methadone does not begin until 36 h after last use, whereas withdrawal from heroin may start 4–6 h after the last use, may reach its peak within 36–72 h, and may last for 7–14 days. For both long-acting and short-acting opioids, less acute withdrawal symptoms (e.g., anxiety, dysphoria, anhedonia, and insomnia) can persist for weeks to months. Other signs and symptoms include sweating, dilated pupils, lacrimation, rhinorrhea, yawning, nausea, stomach cramps, diarrhea ("flu-like" symptoms), piloerection, thermoregulation disturbances (sweating, fever). Moreover, patients report subjective symptoms such as muscle and joint aches, anxiety, restlessness, insomnia, and irritability. Opioid withdrawal can produce severe discomfort but it is usually not fatal (unlike alcohol withdrawal).

9.6.4.3 Physical Health Consequences

Factors influencing opioid side effects can be both subject-related (age, genetic variations, renal and liver dysfunction) and drug-related (half-life, route of administration).

- *Gastrointestinal diseases:* Opioid-induced constipation is the most common side effect of long-term opioid use. However, the adverse events on the gastrointestinal tract result in a more generalized and complex condition, defined as the "opioid-induced bowel dysfunction", which includes a constellation of symptoms such as xerostomia, vomiting, gastroesophageal reflux, abdominal pain, loss of appetite, biliary colic and epigastric discomfort and pain.
- *Endocrinopathy:* Medical and recreational opioid use has been associated with alterations in several hormonal systems. The use of opioids decreases the blood levels of several hormones as sex hormones and stress hormones, and increases the serum levels of others. The hormonal adverse effects of chronic use of opioids differ between sexes: men usually experience sexual dysfunction, depression, and decreased energy level, while women more frequently exhibit menstrual dysfunction. Opioids can also cause hyperglycemia, worsening diabetes, and increase the risk of metabolic syndrome. Finally, opioid-related altered hormonal production has been associated with osteoporosis and increased risk of pathologic fracture.
- *Immunologic disorders:* Long-term use of opioids may alter several processes involving the immune system, such as the response to stress, infection, and malignant transformations. For instance, opioid use in HIV-infected individuals has been linked to exacerbation of the infection and increases in viral load. Moreover, several studies have reported an association between chronic opioid use and various cancer types, including laryngeal, esophageal, gastric, lung, bladder, and pancreatic cancers.
- *Cardiovascular diseases:* Opioid use can be associated with hypotension and bradycardia. Methadone, especially at high doses, has also been reported to cause QT prolongation, possibly resulting in torsade de pointes ventricular tachyarrhythmia.
- *Infectious diseases:* Opioids are often injected, leading to an increased risk of getting HIV or hepatitis C, as well as bacterial infections of skin, soft tissues, bloodstream, and heart, transmitted by sharing needles or other drug equipment.

9.6.4.4 Mental Health Consequences

- *Delirium:* Opioid-induced delirium is most likely to happen with high doses, with coadministration of other psychoactive substances or medical compounds, or in individuals with preexisting brain damage or risk factors.
- *Sleep disorders:* In the long term, chronic use of opioids could lead to sleep dysfunction, including excessive daytime sleepiness, daytime fatigue, depression, and notably, respiratory depression during sleep and sleep-disordered breathing.
- *Cognitive impairment:* Opioid consumption has been linked to structural brain alteration with a reduction in grey and white matter. Consistently, heroin users and methadone or buprenorphine maintenance patients commonly show a variety of cognitive deficits, including impaired visuospatial and working memory, executive functions, and attention, even after prolonged abstinence. The prevalence of opioid-induced cognitive impairment is highly variable, depending on individual vulnerability, opioid doses, route of administration, and duration of use. The effects may be transient in some individuals but persistent in others.
- *Neurotoxicity:* Opioid-induced neurotoxicity refers to an array of neurologic symptoms due to the accumulation of toxic opioid metabolites, including confusion, hallucinations, sudden and involuntary muscle spasms, and hyperalgesia. Opioid-induced hallucinations (visual, auditory, or rarely tactile), often defined as "waking dreams", are an uncommon phenomenon experienced especially with high dosages and in patients with comorbidities that may predispose to hallucinations.

Box 9.6: Prescription Medicines

Misuse of prescription medicines is a rapidly growing public health problem. In 2017, around 18 million people had abused such drugs at least once in the previous year. The reasons for the high prevalence vary according to age, gender, and other factors, but they include easy access to the substances, misinformation about their addictive properties, and the perception that prescription medicines are less dangerous than illicit drugs. The most commonly misused classes of prescription medicines are opioids, central depressants, and stimulants.

Prescription opioids are often prescribed after surgery or injury, or for conditions characterized by moderate to severe pain such as terminal cancer. However, the last decades have witnessed a dramatic increase in the use of prescription opioids for the management of chronic, noncancer pain, such as osteoarticular pain, despite the potential development of tolerance and dependence, possible severe consequences and the lack of evidence about their long-term effectiveness. Moreover, due to their euphoric effect, prescription opioids are frequently used for recreational purposes, with a high risk of overdoses and death.

Central nervous system (CNS) depressants, including benzodiazepines, are usually prescribed for sedative and hypnotic purposes in the treatment of anxiety, panic, acute stress reactions, and sleep disturbances. Most of these drugs exert their action by potentiating the activity of GABA, a neurotransmitter that inhibits brain activity. This action is responsible for the drowsy and calming effects. Misuse of these molecules is manifested by slurred speech, confusion, dizziness, xerostomia, loss of memory, poor concentration, incoordination, hypotension, and slowed breathing. Benzodiazepines are highly addictive and their abrupt reduction or discontinuation can lead to withdrawal syndromes and severe complications such as seizures. Concomitant use of these drugs and other CNS depressants (as opioids or alcohol) can cause respiratory depression.

Prescription stimulants are used for the treatment of attention-deficit hyperactivity disorder (ADHD) and narcolepsy. Prescribed amphetamines as a treatment for ADHD in adolescents and young adults are increasingly common, especially in the United States. Moreover, these drugs have been used more and more even in the absence of clinical indication with the aim of enhancing cognitive performance, especially among high school and college students. Nonmedical use of stimulants to enhance mental performance carries potential physical and mental health risks, including dependence, cardiovascular events, and psychosis.

9.6.5 Hallucinogens

Hallucinogens are also defined as psychedelics or psychotomimetics because, besides inducing hallucinations, they cause an experience of expanded and heightened consciousness. They are usually divided into two categories: classic hallucinogens (such as lysergic acid diethylamide or LSD) and dissociative drugs (such as phencyclidine or PCP).

9.6.5.1 Classic Hallucinogens

Classic hallucinogens, including LSD, psilocybin, mescaline, and dimethyltryptamine (or DMT), are agents that act primarily on the 5-HT2A receptors. Classic hallucinogens exert their action, at least in part, by temporarily disrupting communication between neurotransmitter systems that regulate mood, perception, hunger, sleep, body temperature, sexual behavior, and muscle control. They are mainly consumed in the oral form.

Classical hallucinogen intoxication: Hallucinogens effects are often unpredictable and influenced by several factors, such as the amount ingested, the individual's personality traits, mood and expectations, and environmental and contextual factors. Physical signs of intoxication may include sweating, tachycardia, hypertension, pupillary dilatation, blurred vision, nausea, loss of appetite, tremors, and incoordination. Psychological effects include euphoria, intensified or distorted sensory experiences (brighter colors, sharper sounds, synesthesia, that is "seeing" sounds or "hearing" colors), hallucinations, rapid emotional shifts, distorted sense of time/space, depersonalization/derealization, dreamlike feelings. These experiences can lead to severe behavioral abnormalities. On some occasions, hallucinogens consumption can produce acute adverse or unpleasant experiences, defined by users as "bad trips." They include terrifying thoughts and feelings of anxiety, confusion, agitation, and disorientation. In serious cases, acute psychotic episodes, characterized by frightening hallucinations, severe anxiety, paranoid ideation, and total loss of contact with reality, may occur. These symptoms may lead to accidents, injury, or suicide attempts. A bad trip is generally followed by sadness, depression, apathy and paranoid thoughts, which may persist even for months.

Classical hallucinogen withdrawal: Psychedelic drugs are associated with no proper withdrawal syndrome. However, discontinuation after protracted use can precipitate a dysphoric state associated with irritability, restlessness and sleep disturbances.

9.6.5.2 Physical Health Consequences

The physiological adverse effects are usually not severe and consist of dizziness, nausea, weakness, myalgia, shivering, abdominal pain, and pupillary dilatation. Fatal intoxications are rare and mainly due to the concomitant use of other drugs, mostly alcohol. Accidental fatalities, suicide, violent behavior and risky acts due to impairment in judgment are more often the causes of hallucinogen-related deaths than the physical symptoms caused by the drugs themselves.

9.6.5.3 Mental Health Consequences

- *Flashbacks:* Flashback is a generally short-term, nondistressing, benign, and reversible state occurring in the absence of recent substance use and characterized by a total or partial reappearance of perceptual alterations experienced during a previous intoxication. Recurrent visual flashbacks include false perception of motion in the peripheral visual fields, geometric hallucinations, flashes of color, intensified colors, image traces of moving objects, halos of light around objects, and altered perception of object sizes (macropsia/micropsia). Affected subjects may experience the repeated occurrence of the same flashback or different ones.
- *Hallucinogen persisting perception disorder:* On some occasions, flashbacks can occur as a part of a lasting, distressing, pervasive, sometimes irreversible condition. In these cases, the recurrence of one or more perceptual disturbances may cause distress and anguish or impair personal, familial, social, and professional functioning. Affected individuals are aware of these symptoms and usually actively seek medical or psychiatric assistance.
- *Psychosis and depression:* Hallucinogens have been associated with persistent psychosis characterized by visual disturbances, disorganized thinking, paranoid thinking, and mood changes. Moreover, in predisposed individuals, their use can favor the onset of a chronic psychosis or depression, sometimes difficult to distinguish from a nonorganic psychotic or mood disorder.

9.6.5.4 Dissociative Drugs

This class includes dissociative anesthetics (ketamine and phencyclidine) that have some hallucinogenic effects. They can produce hallucinations and a dissociated state characterized by an altered sense of the self and the body and reduced concern for the environment. Laboratory studies suggest that dissociative drugs cause their effects by disrupting the actions of the neurotransmitter glutamate at *N*-methyl-D-aspartate (NMDA) receptors on neurons throughout the brain. Glutamate plays a crucial role in cognition (including learning and memory), perception of pain and emotion. Phencyclidine also alters the actions of dopamine, the neurotransmitter responsible for the euphoric and rewarding effects of many substances of abuse.

Dissociative drug intoxication: At lower doses, these substances can induce euphoria, numbness in the extremities, confusion, disorientation, loss of coordination, and altered perceptions (with visual colorful hallucinations), condition known among ketamine users as "*K-land*" status. Physical associated symptoms can be increase in heart rate, blood pressure, breathing rate, and body temperature, dizziness, nausea, and vomiting. At high doses, ketamine induces a more severe dissociative state usually referred to as a "*K-hole*", wherein the individual experiences an inability to move followed by a total disconnection from physical sensations and the appearance of vivid hallucinations, a condition often compared to an out-of-body or near-death experience. The inability to move accounts for the classification of ketamine as a date rape drug. Cognitive deficits (attention, memory, concentration) and marked psychological distress, including feelings of panic, terror, anxiety, depersonalization/derealization, and paranoia can also occur when using high doses of dissociative anesthetics.

Dissociative drug withdrawal: A specific dissociative drugs withdrawal syndrome has not yet been identified, but discontinuing a protracted use has been associated with craving for the drug, headaches, anxiety, depression, shaking, sweating, and palpitations.

9.6.5.5 Physical Health Consequences

- *Cardiovascular diseases:* Dissociative drugs stimulate the cardiovascular system producing an increase in heart rate, cardiac output, and blood pressure. As a consequence, their use can increase acute cardiovascular and cerebrovascular risk especially for people with preexisting risk factors, such as hypertension and severe cardiac disease. Acute cardiac risk is notably increased when these substances are combined with stimulant drugs.
- *Urinary diseases:* Frequent ketamine use has been associated with ulcerative cystitis characterized by frequency and urgency of urination, painful or difficult urination, urge incontinence, and occasionally hematuria. This condition can have a severe and long-term impact on the individual. Finally, frequent, high-dose ketamine use has been linked to the appearance of hydronephrosis (swelling of the kidney) due to urinary tract problems.
- *Others:* At high doses, dissociative drugs can lead to the appearance of seizures, respiratory failure, acute damage to skeletal muscles, coma and death.

9.6.5.6 Mental Health Consequences

- *Psychosis:* Dissociative anesthetics use can induce transient psychotic states characterized by confusion, agitation, aggression, hallucinations, and delusions of grandeur and possibly evoking aggressive or self-mutilating behavior. In this context, a sense of enhanced strength and invulnerability, together with the inability to feel pain and impaired judgment, can result in serious injury. Acute intoxication may also favor the onset of psychotic episodes similar to schizophrenia, with positive symptoms and deficient and cognitive symptoms. However, differential diagnosis is assisted by features such as the limited duration over time and the reversibility of symptoms. Finally, on some occasions, the onset of long-term "schizophrenia-similar" psychosis may occur, particularly in repeated and/or high-dose use.
- *Cognitive impairment:* Frequent use can lead to persistent speech difficulties, impairments in episodic and working memory, visual recognition, executive functions.
- *Others:* Dissociative drugs use can lead to the emergence of depressive symptoms, suicidal ideation, anxiety, and social isolation that may persist for months or years after chronic use stops.

Box 9.7: New Psychoactive Substances

The acronym NPS refers to the heterogeneous array of psychoactive substances that rapidly emerged in the 2000s. Although NPS have been designed as legal alternatives to well-known illicit drugs, such as cannabis, cocaine, MDMA, and LSD, most of them are now considered illegal. Due to the lack of information on their contents, their different potency and risk profiles, and the lack of knowledge of their pharmacological properties and toxicology, their use can produce unexpected effects and acute harm as evidenced by the numerous hospital emergency admissions and deaths associated with their use.

NPS can be classified based on the illicit substances they are designed to mimic.

Synthetic cannabinoids are a group of synthesized compounds with effects similar to those resulting from cannabis use, such as euphoria and drowsiness, but usually more severe.

Stimulant NPS are drugs with effects comparable to those of amphetamine, cocaine, and MDMA, including increased alertness, self-confidence, sociability, and energy, and reduction of appetite, need for sleep and fatigue (e.g., mephedrone).

Hallucinogen NPS include dissociative and classic hallucinogens. Dissociative NPS (e.g., methoxetamine) have effects similar to those produced by ketamine; psychedelic NPS (e.g., 2C drugs, NBOME drugs) have effects similar to those resulting from LSD and psilocybin. However, these effects may differ in duration and intensity.

Depressant NPS comprise two types of substances: opioids, causing euphoria, analgesia, sedation, and drowsiness (e.g., fentanyl analogues), and novel benzodiazepines, which are used for their sedative-hypnotic, anxiolytic, muscle-relaxant, and anticonvulsant effects.

Use of NPS has been associated with confusion, cardiovascular and cerebrovascular effects (from tachycardia and hypertension to myocardial infarction, cardiac arrest, and stroke), gastrointestinal effects (liver dysfunction, nausea or vomiting), hyperthermia, renal failure, and psychiatric symptoms/syndromes (anxiety and agitation, insomnia, delirium, psychosis, suicidal ideation, hallucinations, depression, mania).

Table 9.1 Summary of signs and symptoms of intoxication and withdrawal of the most used drugs

Substance	Intoxication	Withdrawal
Alcohol	Disinhibition, euphoria, mood lability, incoordination, dysarthria, blackout	Tachycardia, hypertension, tremor, nausea/vomiting, sweating, anxiety, insomnia
Cannabis	Tachycardia, conjunctival hyperemia, incoordination, increased appetite; euphoria, slowed time perception, paranoid ideation	Irritability, insomnia, anxiety, apathy, nausea
Cocaine and amphetamine-type stimulants	*Pupillary dilation,* hypertension, tachycardia, insomnia, anorexia, tremor; euphoria, anxiety, hallucinations, paranoid ideation	"Crash" includes depression, irritability, craving, fatigue, increased appetite
Opioids	*Pinpoint pupils*, respiratory depression, drowsiness, euphoria	Pupillary dilation, nausea/vomiting, diarrhea, yawning, sweating, rhinorrhea, anxiety, muscular pain
Hallucinogens	Tachycardia, hypertension, ataxia, pupillary dilation; euphoria, dissociation, hallucinations/illusions	Depression, restlessness, anxiety, fatigue, anorexia, sleep disturbances
Benzodiazepines	Sedation, disinhibition, disorientation, respiratory depression	Anxiety, insomnia, seizures, tremors

Table 9.1 shows a summarizes signs and symptoms of intoxication and withdrawal of the main categories of substances of abuse.

9.7 Treatment

9.7.1 Assessment and Management of Substance-Related Emergencies

Emergency drug reactions may be seen at home, in the emergency room, or in the office of any practitioner of medicine or health care professional. Intoxicated patients present unique challenges. They can be agitated and disruptive, sedated, or even unconscious. Patients may present with single substance intoxication or after intake of a combination of drugs.

In addition to direct consequences of substance use, such as intoxication, withdrawal and overdoses, medical complications, as well as injury and trauma, including motor vehicle accidents, falls, drownings, harm of others, and suicide can represent severe emergencies.

Assessment and management include the following steps

1. Supporting vital functions and maintenance of airway patency, ventilation, and circulation.
2. Performing serum and urine toxicological screening to identify the compounds used and detect associated substances not reported (high rate of polyabuse). Unfortunately, no systems of tracing are available for serum and urinary metabolites of most NPS.
3. Reviewing all potential medications the subject may have access to.
4. Monitoring blood pressure, heart rate, oxygen saturation, body temperature, respiratory rate, and diuresis; obtain venous access; performing an electrocardiogram and an accurate clinical examination (pay special attention to pupils and look for transdermal patches and signs of injection, such as "track marks").
5. Performing blood analyses (including blood count, glucose level, electrolytes, liver enzymes, cardiac enzymes, coagulation, and muscle enzymes) to review kidney and liver function and to exclude other conditions such as infections.
6. Getting a brain scan in patients with known or suspected trauma, unexplained focal deficits, or high risk of intracranial lesions (e.g., anticoagulant/antiplatelet therapy); performing chest X-ray in patients with cardiopulmonary signs or symptoms and chest trauma.
7. Setting up an appropriate pharmacological treatment if necessary.

9.7.2 Pharmacological Therapy

A variety of medications and tools can be employed to handle acute intoxication, withdrawal, overdose and manage concurrent conditions. Alcohol, benzodiazepines, and opioids are particularly at risk of inducing physical dependence and withdrawal symptoms. Therefore, their treatment will be discussed more in detail.

9.7.2.1 Intoxication

Treatment of intoxication and acute complications is often merely supportive and symptomatic

- Hypertension: benzodiazepines, if severe: intravenous antihypertensive drugs, nitrates, clonidine. Beta-blockers are contraindicated due to the risk of unopposed alpha-adrenergic effects.
- Hypotension: intravenous crystalloids.

- Dehydration and renal failure: intravenous fluids.
- Nausea and vomiting: antiemetics.
- Anxiety, psychomotor agitation: intravenous or intramuscular benzodiazepines (e.g., lorazepam, diazepam). Restraints should be avoided if possible because they may worsen agitation and increase the risk of sudden death, especially during stimulant drug use.
- Agitation with psychotic symptoms: benzodiazepines, antipsychotics (e.g., risperidone, olanzapine, quetiapine, haloperidol).
- Seizures: intravenous or intramuscular benzodiazepines, barbiturates.
- Severe hyperthermia: physical cooling (cold ablutions). If resistant, intubation and paralysis with curaric agents.
- Excited delirium: intravenous or intramuscular (e.g., diazepam, midazolam, lorazepam), antipsychotics (e.g., haloperidol, chlorpromazine, risperidone).
- Serotoninergic syndrome: benzodiazepines; if refractory cyproheptadine (antihistamine with additional anticholinergic, antiserotonergic, and local anesthetic properties).

9.7.2.2 Overdose

Overdose may be suspected from history collected with patients, family members, or caregivers, or the identification of overdose signs or symptoms.

- *Opioids:* intravenous, intramuscular, or subcutaneous naloxone (opioid receptor antagonist). It may cause symptoms of opioid withdrawal, including nausea, vomiting, restlessness, agitation, increased heart rate, and sweating.
- *Benzodiazepines*: flumazenil (selective GABA-A receptor antagonist) administered via injection, otic insertion, or intranasally. It can lead to complications, such as withdrawal symptoms, agitation and anxiety, lowered seizure threshold.

9.7.2.3 Withdrawal

General principles regarding the management of substance withdrawal syndromes include the use of symptomatic treatment, prescription of opioid substitution treatment and establishment of a plan for long-term management of the underlying SUD.

It is important to remark that detoxification alone is only the first step of the process: patients who do not undergo any additional treatment are likely to relapse.

The patient should be supported through the stage of acute detoxification and withdrawal to be medically stable and drug-free. This may be carried out at an inpatient treatment facility, or an outpatient program under medical supervision. Involving the individual's family is crucial to support the treatment process.

Besides medications and counseling, treatment generally includes social supports such as linkage to community recovery groups depending on the patient's individual needs and level of existing family and social support.

The goals of SUD treatment are to minimize the symptoms of the disease, improve physical and mental health and social functioning, educate and motivate patients to manage situations possibly triggering a relapse.

9.7.2.4 Opioids

Opioid withdrawal syndrome is a potentially fatal condition resulting from opioid dependence. Pharmacologic options for its management include opioid agonist therapy (methadone and buprenorphine), alpha-adrenergic drugs, and supportive treatments that can provide symptom relief. Methadone and buprenorphine may be used both during the withdrawal phase, and for the long-term management of the disorder.

When a patient shows signs of opioid withdrawal, the alternatives for pharmacological management of the latter are the gradual suspension of a full opioid agonist (methadone), the short-term use of a partial μ-opioid agonist (buprenorphine), or the detoxification using opioid antagonists (naltrexone and naloxone).

- *Methadone:* It is a long-acting μ agonist that prevents other opioids to bind μ receptors and suppresses craving. Methadone should be started at 10–30 mg daily and slowly up titrated to 20–40 mg. These dosages are usually sufficient to manage withdrawal symptoms. It should be avoided in patients with signs of intoxication, especially due to depressant drugs, for the risk of fatal overdose.
- *Buprenorphine:* It is an opioid partial agonist with a very high affinity for μ receptors, giving adequate opioid effects while decreasing the danger of respiratory depression. It can be coformulated with naloxone as an abuse-deterrent. It is not significantly different from methadone in terms of completion of detoxification treatment but it can control more quickly withdrawal symptoms.
- *Naltrexone:* It is an opioid receptor antagonist which can be used as a treatment for relapse prevention. It should be used only in patients who are highly motivated to remain on treatment and have already been detoxified due to the potential to cause fatal overdoses and severe withdrawal.

9.7.2.5 Alcohol

Withdrawal from alcohol is associated with significant morbidity and mortality when improperly managed.

Alcohol withdrawal may be simulated or accompanied by other conditions such as infection (e.g., meningitis), trauma (e.g., intracranial bleeding), metabolic derangements, drug overdose, hepatic failure, and gastrointestinal hemorrhage.

Benzodiazepines (e.g., diazepam, lorazepam, and chlordiazepoxide) are the gold standard for the management of alcohol detoxification. Long-acting benzodiazepines with active metabolites (e.g., diazepam) are particularly useful to achieve a smooth clinical course with a lower chance of recurrent withdrawal or seizures. On the other hand, lorazepam should be used to treat patients with advanced cirrhosis or acute alcoholic hepatitis. This molecule may prevent prolonged effects if oversedation occurs due to its shorter half-life and the absence of active metabolites.

Two different approaches are possible: front-loading therapy consists of administering higher initial doses of benzodiazepines to accomplish a faster control of alcohol withdrawal symptoms. This approach is most suited for patients who are at higher risk of undergoing serious complications in the case of severe withdrawal. For front loading, a variety of dosing schedules is available.

- *Diazepam*, 5–10 mg IV every 5–10 min until the desired level of sedation is achieved.
- *Lorazepam*, 2–4 mg IV every 15–20 min, is a valid alternative.

A symptom-triggered approach is suggested for most patients to handle alcohol withdrawal when pharmacotherapy is indicated. It consists on the administration of pharmacological treatment on the basis of symptom severity and subsequent titration of the dose required, in order to minimize the risk of over-sedation and reduce the overall quantity and duration of treatment. To apply this method, a systematic evaluation of the patient's conditions should be performed using a validated tool to measure withdrawal severity, such as the Clinical Institute Withdrawal Assessment for Alcohol Scale (CIWA-Ar). When the score is high (any score of 8 or greater on the CIWA-Ar), further medication is given. For acute withdrawal symptoms, the following is recommended: *diazepam* 5–10 mg IV or *lorazepam* 2–4 mg IV in individuals with severe liver dysfunction.

Supportive care, such as intravenous fluids, correction of electrolyte imbalance, nutritional supplementation, and frequent clinical reassessment, including vital signs, is essential. Thiamine and glucose should be administered to prevent or manage Wernicke encephalopathy. Supplements of folate should be administered routinely, and deficiencies of potassium, magnesium, and phosphate should be corrected. In the case of Delirium Tremens, besides the above treatment, antipsychotics such as haloperidol are indicated to manage agitation and hallucinations, carefully monitoring the occurrence of side effects (e.g., hypotension, QTc prolongation).

After the medically monitored detoxification phase, some pharmacological treatments could be offered in combination with psychosocial treatment to prevent relapse.

- *Naltrexone*: Blockade of μ-opioid receptors involved in the rewarding effects of alcohol and the craving reduces the risk of relapse and should be offered to moderately to severely dependent drinkers.
- *Acamprosate:* It acts as a functional glutamatergic antagonist and increases GABA-ergic activity, reducing craving for alcohol. It may reduce symptoms of prolonged withdrawal such as insomnia, restlessness, anxiety and dysphoria. It may be more effective in subjects with severe dependence.
- *Disulfiram:* It is an aversive agent that inhibits ethanol's metabolism, leading to the accumulation of a toxic metabolite (acetaldehyde) and resulting in an intolerance reaction to alcohol characterized by headache, flushing, vomit and other unpleasant symptoms. Disulfiram should be offered to highly motivated individuals who are participating in structured treatment programs.

9.7.3 Behavioral Treatment and Psychosocial Interventions

It is well known that social environments play a crucial role in the evolution, as well as the recovery, of addiction problems. The connections between the use of substances and the quality of relationships and social bonds are two-way and very strong. SUDs contributes to psychiatric disorders, disability, death caused by accidents, medical illnesses caused or exacerbated by drug use, and are associated with high rates of suicide. There are other problems associated with SUD, such as housing and homelessness, unlawful behavior (as a victim or perpetrator) and incarceration, infectious diseases associated with intravenous injection or risky sexual behavior, and unemployment. A thorough approach to SUDs includes a combination of professional addiction and addiction-related services (e.g., assessment, detoxification, pharmacological treatment, rehabilitation, psychotherapy, substance use monitoring, and mutual support programs). SUDs deeply affect the social functioning of individuals. Because of that, there are many useful interventions and peer support programs to assist subjects affected and their families. It is essential to rebuild a social network around the person with SUD in order to promote reintegration and reconnection of the person with others.

Many psychosocial interventions are empirically recommended for the treatment of SUD, although there is no definite evidence that any of them is to be preferred or remarkably effective for a specific patient's features. The crucial factors for a successful outcome of all psychosocial interventions are empathetic and nonjudgmental attitude, and use of positive reinforcement methods to promote positive changes in patients' lives.

Clinicians should acknowledge the following interventions

- *Cognitive-behavioral therapy:* This form of psychotherapy applies the principles of conditioning and learning. CBT encompasses a range of interventions, including relapse prevention and coping skills therapy, used to provide skills useful in supporting subjects to reduce or interrupt substance use and sustain abstinence.
- *Motivational enhancement therapy:* It is a set of techniques focused on promoting subjects' insight into their disease and encouraging constructive changes in their lives.
- *Contingency management/motivational incentives therapy:* It is based on the principles of positive reinforcement of behaviors consistent with a reduction or discontinuation of substance use.
- *Family and couples therapy:* It uses a combination of psychoeducation and cognitive-behavioral techniques to provide those affected and their family members with skills valuable in reducing the maladaptive communication patterns that can trigger relapse to substance use. Moreover, family approaches may also have the benefit of decreasing the emotional burden for family members and allow them to deal more adequately with the patient.
- *Self-help groups:* These groups (e.g., Alcoholic Anonymous) aim at supporting patients to achieve and maintain abstinence from substances.

More in general, all interventions should stress some key points such as the following

- Avoiding or reducing contact with persons, places, or situations at high risk.
- Changing dysfunctional relationship dynamics.
- Promoting new social bonds.

- Forming a network of sober and supportive peers.
- Resisting social influences to use drugs.
- Developing communication or assertiveness abilities.
- Actively seeking help and assistance.
- Involving in drug-free social and recreational activities.
- Joining peer support programs.

Addiction is a complex disorder strongly influenced by individual factors such as genetic vulnerability, developmental, and social, and interpersonal factors. Moreover, it is an evolving condition that changes through time and across the lifespan of the individual. This heterogeneity makes it crucial to personalize interventions to accommodate the unique neurobiological characteristics and social environments of individual patients.

Box 9.8: Brain Stimulation Techniques

Evidence has shown that modulating the activities of brain areas involved in addiction through brain stimulation techniques may represent a promising treatment option for SUD. These techniques are effective in reducing subjectively rated substance craving, which is assumed to be an important risk factor for relapse.

Repetitive Transcranial Magnetic Stimulation (rTMS). rTMS is a technique requiring the use of an electromagnetic coil placed on the patient scalp; the coil produces repeated trains of magnetic pulses, resulting in a transient magnetic field. This process generates transitory electrical currents in the brain areas underneath the coil, which are able to modulate cortical excitability. rTMS has been demonstrated to induce significant neuroplasticity changes in the targeted brain regions, lasting beyond active stimulation periods. Excitatory rTMS of the dorsolateral prefrontal cortex has been shown to have an acute effect on reducing craving and substance abuse in SUD patients.

Transcranial Direct Current Stimulation (tDCS). tDCS is a technique involving the application of a weak direct electrical current (e.g., 1–2 mA) through electrodes placed on the scalp; tDCS modulates resting membrane potential and cortical excitability in targeted brain regions through depolarization or hyperpolarization depending on stimulation parameters. Cortical excitability is enhanced by the anodal electrode and reduced by the cathodal electrode. Although its action mechanisms are not fully understood, tDCS may induce enduring neuroplasticity changes in the targeted brain areas. tDCS has shown in several studies positive effects on alcohol/substance craving and consumption.

Deep Brain Stimulation (DBS). DBS is performed via invasive electrodes inserted in specific brain regions. Such systems additionally include a subcutaneous pulse generator. Deep Brain Stimulation for SUD has been applied to a small number of severe and otherwise treatment-resistant cases. DBS of the nucleus accumbens has been shown to successfully decrease craving and enable long-term abstinence.

Further Reading

American Psychiatric Association. Diagnostic and statistical manual of mental disorders. 5th ed. Washington, DC: American Psychiatric Association; 2013.

Campeny E, López-Pelayo H, Nutt D, et al. The blind men and the elephant: systematic review of systematic reviews of cannabis use related health harms. Eur Neuropsychopharmacol. 2020;33:1–35. https://doi.org/10.1016/j.euroneuro.2020.02.003.

Carvalho AF, Heilig M, Perez A, Probst C, Rehm J. Alcohol use disorders. Lancet. 2019;394(10200):781–92. https://doi.org/10.1016/S0140-6736(19)31775-1.

Center for Health Information and Analysis. Access to substance use disorder treatment in Massachusetts. Boston: Center for Health Information and Analysis; 2015.

Daley D. Family and social aspects of substance use disorders and treatment. J Food Drug Anal. 2013;21(4):S73–6.

Donroe JH, Tetrault JM. Substance use, intoxication, and withdrawal in the critical care setting. Crit Care Clin. 2017;33(3):543–58. https://doi.org/10.1016/j.ccc.2017.03.003.

European Monitoring Centre for Drugs and Drug Addiction. European drug report 2020: trends and developments. Luxembourg: Publications Office of the European Union; 2020.

Khademi H, Kamangar F, Brennan P, Malekzadeh R. Opioid therapy and its side effects: a review. Arch Iran Med. 2016;19(12):870–6.

Klein WJ. Pharmacotherapy for substance use disorders. Med Clin N Am. 2016;100(4):891–910.

Kuntsche E, Kuntsche S, Thrul J, Gmel G. Binge drinking: health impact, prevalence, correlates and interventions. Psychol Health. 2017;32(8):976–1017. https://doi.org/10.1080/08870446.2017.1325889.

Lappin JM, Sara GE. Psychostimulant use and the brain. Addiction. 2019;114(11):2065–77. https://doi.org/10.1111/add.14708.

Morgan CJ, Curran HV, Independent Scientific Committee on Drugs. Ketamine use: a review. Addiction. 2012;107(1):27–38. https://doi.org/10.1111/j.1360-0443.2011.03576.x.

Ouzir M, Errami M. Etiological theories of addiction: a comprehensive update on neurobiological, genetic and behavioural vulnerability. Pharmacol Biochem Behav. 2016;148:59–68. https://doi.org/10.1016/j.pbb.2016.06.005.

Peacock A, Bruno R, Gisev N, et al. New psychoactive substances: challenges for drug surveillance, control, and public health responses. Lancet. 2019;394(10209):1668–84. https://doi.org/10.1016/S0140-6736(19)32231-7.

Tenegra JC, Leebold B. Substance abuse screening and treatment. Prim Care. 2016;43(2):217–27. https://doi.org/10.1016/j.pop.2016.01.008.

Volkow ND. Personalizing the treatment of substance use disorders. Am J Psychiatry. 2020;177(2):113–6. https://doi.org/10.1176/appi.ajp.2019.19121284.

Volkow ND, Boyle M. Neuroscience of addiction: relevance to prevention and treatment. Am J Psychiatry. 2018;175(8):729–40. https://doi.org/10.1176/appi.ajp.2018.17101174.

Volkow ND, Koob GF, McLellan AT. Neurobiologic advances from the brain disease model of addiction. N Engl J Med. 2016a;374(4):363–71. https://doi.org/10.1056/NEJMra1511480.

Volkow ND, Swanson JM, Evins AE, et al. Effects of cannabis use on human behavior, including cognition, motivation, and psychosis: a review. JAMA Psychiatry. 2016b;73(3):292–7. https://doi.org/10.1001/jamapsychiatry.2015.3278.

World Drug Report. World Drug Report. 2020. (United Nations publication, Sales No. E.20.XI.6).

World Health Organization. Global status report on alcohol and health 2018. Geneva: World Health Organization; 2018.

Organic Mental Disorders and Psychiatric Issues in the Elderly

10

M. Bosia, F. Cuoco, G. Bigai, F. Martini, L. Fregna, C. Congedo, and R. Cavallaro

> *"Dementia cannot recognize itself, in the same way that blindness cannot be seen"*
>
> *—Apuleio*

Key Points

- Cognitive disorders are more frequent in older adults who are often affected by other comorbid pathologies and take pharmacological polytherapy.
- Although delirium manifests as a psychiatric syndrome, it is caused by organic medical conditions, objectifiable by laboratory and instrumental examinations.
- Dementia is not an inevitable condition due to aging, but a disorder that may be due to various etiological causes and that includes behavioral and psychiatric symptomatology.

M. Bosia · R. Cavallaro
IRCCS San Raffaele Scientific Institute, Milan, Italy

Vita-Salute San Raffaele University, Milan, Italy
e-mail: bosia.marta@hsr.it; cavallaro.roberto@hsr.it

F. Cuoco · F. Martini (✉) · L. Fregna (✉)
IRCCS San Raffaele Scientific Institute, Milan, Italy
e-mail: cuoco.federica@hsr.it; francesca.martini1@hsr.it; fregna.lorenzo@hsr.it

G. Bigai · C. Congedo
Vita-Salute San Raffaele University, Milan, Italy
e-mail: bigai.giorgia@hsr.it; congedo.carlo@hsr.it

R. Cavallaro, C. Colombo (eds.), *Fundamentals of Psychiatry for Health Care Professionals*, https://doi.org/10.1007/978-3-031-07715-9_10

10.1 Delirium

10.1.1 Introduction

Delirium is one of the most common and challenging conditions in older patients, although it can affect people of any age. Despite its high prevalence in a variety of care settings, it remains under-recognized and inadequately managed. Without timely diagnosis and appropriate treatment, it can lead to significant consequences such as increased morbidity, mortality, long-term cognitive impairment, and loss of autonomy, frequently leading to institutionalization. Although delirium is the direct pathophysiological consequence of an organic condition, it lies at the interface between psychiatry and medicine due to its frequent emotional and behavioral manifestations. The psychiatrist is therefore often required to play a central role in its identification and management.

10.1.2 Definition

Delirium, also known as acute confusional state, represents an organically caused condition characterized by a decline of consciousness and cognitive functioning. It is a transient, rapid-onset (over hours or few days) and usually reversible clinical condition characterized by the coexistence of various symptoms. Besides disturbances of consciousness and cognitive impairment, it includes psychomotor disturbances, alteration of the circadian rhythm and the sleep–wake cycle, and psychopathological manifestations such as emotional and behavioral changes, hallucinations, and delusions. The symptoms tend to fluctuate throughout the day, typically worsening during the night.

10.1.3 Epidemiology

Delirium affects about 1–2% of the general population, with equal prevalence between genders. In clinical practice, older people are particularly vulnerable to this disorder (it is estimated that more than 10% of people over 85 years old experience delirium). Furthermore, patients with dementia show a higher risk of developing delirium. About 10–30% of medical and surgical patients develop delirium during hospitalization; up to 50–70% of them stay in intensive care units (ICU). Despite its clinical significance, delirium is greatly underdiagnosed.

10.1.4 Etiopathogenesis

The onset of the disorder may depend on the primary underlying illness, on the development of overlapping medical conditions (e.g., sepsis, low oxygen levels), and on the use of sedatives or analgesics commonly used in ICU, emergency wards, and nursing homes. The etiology is usually multifactorial, with main causes including neurological disorders, drug use (or discontinuation), and systemic illnesses (see Box 10.1).

Box 10.1: Most Frequent Causes of Delirium

Neurological disorders	*Drugs (use or discontinuation)*
• Space-occupying lesions: tumors, abscesses, hematomas • Head injuries (concussion) • Infections (meningitis, encephalitis) • Cerebrovascular accidents: transient ischemic attack (TIA), ischemia or hemorrhage, hypertension, vasculitis (e.g., systemic lupus erythematosus) • Neurodegeneration • Epilepsy	• Substances of abuse: alcohol, amphetamines, cocaine, opiates, cannabis • Medications: sedative and hypnotic medications, anticholinergics, steroids, opiates (analgesics), benzodiazepines, antiparkinsonian agents

Systemic conditions

- Metabolic encephalopathies: organ failure (liver, kidney), hypoxia, malnutrition (thiamine deficit which leads to Wernicke's encephalopathy, B12 vitamin, folic acid or niacin deficiency), endocrine disorders, dehydration, infections and sepsis, electrolytes imbalances, cardiovascular disease, uremia, hypoglycemia, hypo/hyperthyroidism, hypo/hyperparathyroidism, Cushing's syndrome, Addison's syndrome
- Intoxication with poisons: carbon monoxide, heavy metals (lead, mercury, manganese)

The risk of developing delirium depends on the interaction between different predisposing and precipitating factors. Subjects with multiple significant predisposing factors are considered at high risk of developing delirium, usually triggered by a single or mild precipitating factor. On the contrary, in healthy individuals, delirium may arise if serious or multiple precipitating factors are concomitantly present. Notably, the factors presented may vary over time, making an individual's risk of delirium dynamic. The most significant risk factor for delirium is a preexisting neurocognitive disorder. See Box 10.2 for an overview on main delirium predisposing and precipitating factors.

Box 10.2: Delirium Predisposing and Precipitating Factors

Predisposing factors	Precipitating factors
• Older age (>65 years) • Male sex • Pre-existing cognitive impairment (neurocognitive disorders) • Physical comorbidities (e.g., single or multiple organ failure, cancer, cerebrovascular accidents) • Sensory impairment (vision, hearing) • Compromised global functioning, requiring assistance for self-care • Dehydration/malnutrition • Drugs misuse, alcohol and drug-dependence	• Prolonged sleep deprivation • Environmental and emotional stress factors • Inadequately managed pain • ICU admission • Immobilization • Severe constipation with fecal retention • Medications: sedatives such as benzodiazepines and opioids, anticholinergics, dopaminergics, corticosteroids, polypharmacy, general anesthetics • Substance intoxication or withdrawal • Primary neurological diseases • Cerebrovascular accidents: cerebral hypoperfusion due to severe blood pressure drop, ischemic stroke and transient ischemic attack (TIA), intracranial hemorrhage • Cerebral infections (meningitis, encephalitis), respiratory infections (pneumonia) and urinary tract infections • Iatrogenic complications, surgery, use of bladder catheter • Malnutrition, dehydration, electrolyte imbalances, hypoglycemia, hypoxia, hypercapnia, anemia • Shock, heart attacks, heart failure • Metabolic disorders (e.g., SIADH, Addison's disease, hyperthyroidism) • Chronic/terminal diseases • Traumatic event (e.g., brain injury, concussion)

The pathophysiological mechanisms remain unclear and they may vary depending on underlying causes. Abnormal stress response (neurotoxic effects of excess of glucocorticoids), disrupted blood–brain barrier (which allows the entry of toxins and cytokines to the brain) and impaired cholinergic neurotransmission may be involved.

10.1.5 Clinical Presentation

The core symptoms of delirium include consciousness alterations, global impairment of cognitive functions (e.g., attention, memory, perception, and orientation), psychopathological manifestations such as emotional and behavioral changes, hallucinations and delusions, psychomotor disturbances, and sleep–wake cycle alterations (see Table 10.1 for further details).

Table 10.1 Symptomatology of delirium

Alterations of consciousness (all patients):	• Range from clouding of consciousness and stupor to hypervigilance. • Fluctuation of the awareness status within the day, with typical nocturnal worsening.
Inattention (all patients):	• Distractability, inability to maintain attentional focus. • Reduction or loss of spatial orientation.
Motor activity changes (almost all patients):	• According to DSM-5, a distinguish classification into hypoactive, hyperactive, and mixed psychomotor subtypes must be done (see further DSM-5 diagnostic criteria).
Memory impairment (65-100% of patients):	• Strictly connected to inattention and to diffuse cognitive impairment. • Impaired acquisition of new long-term memories rather than short-term memories. confabulations occur in 8–15% of patients. • Long-term memories stored before the onset of delirium are generally preserved.
Mood lability (50-90% of patients):	• Fluctuation of mood and affect, which appear incongruous and present rapid changes between fear, sadness and joking. • Irritability and nervousness are often associated .
Disorganized thinking (50-90% of patients):	• Various signs of formal thought disorders (e.g., limited coherence and poverty of speech, loss of associations, perseveration, tangentiality).
Disorientation (45-100% of patients):	• Loss of awareness of the surroundings (including time, place, self and environment).
Language disturbances (40-90% of patients):	• Impairment of linguistic information processing (e.g., anomic aphasia, paraphasia, impaired comprehension, agraphia, and word-finding difficulties).
Sleep changes (30-90% of patients):	• Connected to alterations of circadian rhythm regulation • Often precedes the onset of delirium • Characterized by fragmented sleep or even sleep-wake cycle reversal
Psychotic symptoms (20%-70% of patients):	• Delusions: they are typically poor structured and non-stereotyped; the content relates to persecutory and paranoid themes, suspiciousness • Perceptual Disorders: illusions and hallucination, which may involve one or more senses but are more frequently visual

A particular form of delirium, which manifests in patients with alcohol use disorder, is *Delirium Tremens*, an acute confusional state caused by alcohol withdrawal (see also Chap. 9). In this case, the clinical presentation includes physical symptoms such as shaking, shivering, irregular heart rate, sweating and, occasionally, hyperthermia or seizures, which may even result in death. Psychiatric

Box 10.3: Clinical Case: Postsurgical Delirium in Patient with Mild Cognitive Impairment

Mr. G. is a 75-year-old man who was admitted to the hospital for surgical treatment for hip fracture. Mr. G lived with his wife. He was able to perform daily activities with minimal help, although he received a diagnosis of mild cognitive impairment 2 years before. He had no personal or family history of psychiatric disorders, but he has been taking low dose benzodiazepines for sleep difficulties for the past few months. He had a history of major tobacco consumption (2 cigarette packs/day for 37 years) but denied alcohol consumption. He had moderate hypertension and had a heart attack treated with percutaneous coronary intervention 4 months ago. His chronic medications consisted of clopidogrel, aspirin, valsartan, and atenolol. Postoperative analgesia consisted of intravenous paracetamol and morphine for the first 48 postoperative hours.

Forty-eight hours after the intervention, the nurses reported that Mr. G. was restless and agitated, would not follow commands, kept trying to undress and get out of bed. He refused to cooperate with the staff, repeating that nurses have tried to kill him. Moreover, he was found to speak loudly when no one else was in the room. He appeared disoriented and confused and appeared to have much more severe memory deficits than before the surgery. He could not sleep at night, when these symptoms usually got worse.

A Mini-Mental State Examination (MMSE) could not be obtained due to Mr. G.'s uncooperativeness. The Confusion Assessment Method (CAM) was administered and resulted positive. The psychiatrist consultant suggested the possibility of delirium during the recovery period after surgery. The patient gradually recovered after symptomatic and supportive treatment, with a clear mind and accurate answers and was discharged 7 days after surgery.

symptoms can vary: typically the patient appears irritable, confused, fearful, and may manifest visual and somatic hallucinations called microzoopsies (vision and feeling of little, creeping animals moving on or under patient's skin, but also animals like snakes, bats, and spiders or even courtyard animals or elves in the surrounding environment).

10.1.5.1 Diagnostic Criteria

The early identification of signs and symptoms, an accurate history collection, along with specific clinical investigations, can help in establishing a prompt diagnosis. A previous assessment of the patient's baseline level of cognitive functioning is important to make a correct diagnosis. The core symptoms of Delirium are summarized in DSM-5 diagnostic criteria (see Box 10.4).

Box 10.4: Diagnostic Criteria for Delirium According to DSM-5

A. Lack of attention and awareness.
B. Acute onset (from hours to days), characterized by fluctuant changes from baseline of cognitive functioning throughout the day.
C. There must be at least one additional cognitive disturbance involving memory, orientation, language, visuospatial ability, or perception.
D. A pre-existent neurocognitive disorder or a primer consciousness alteration (e.g., coma) cannot explain disturbances indicated in criteria A and C.
E. There is evidence that the disturbances are a "direct physiological consequence" of an underlying medical condition or a combination of causes, objectifiable by clinical history, physical examination, laboratory or instrumental examinations.

Delirium diagnosis is possible only *if all five criteria are satisfied.*
Specify if:

- *Acute*: Lasting a few hours or days.
- *Persistent*: Lasting weeks or months.

In addition to etiology, in DSM-V is necessary to specify the level of patient's psychomotor activity

(a) *Hyperactive Delirium*: increased psychomotor activity which may associate with hypervigilance, restlessness, fast or loud speech, irritability, combativeness, impatience, swearing, singing, laughing, uncooperativeness, euphoria, anger, wandering, easy startling, fast motor responses, distractibility, tangentiality, nightmares, and persistent thoughts (hyperactive subtyping is defined with at least three of the above).
(b) *Hypoactive Delirium*: reduced psychomotor activity, usually associated with slowness of physical movements and low consciousness levels, from lethargy to stupor. It may include unawareness, decreased alertness, sparse or slow speech, lethargy, slowed movements, staring, and apathy (hypoactive subtyping is defined with at least four of the above).
(c) *Mixed Delirium*: thus present in case of nonaltered psychomotor activity or when it swings between its two opposite polarities.

10.1.5.2 Differential Diagnosis

Various psychiatric conditions might have similar clinical presentations to those seen in delirium. The more frequent pathologies that must be excluded are the following.

Dementia: this group of disorders (discussed below in this chapter) is associated with cognitive and psychosocial functioning decline that are typically progressive, not acute-onset, and irreversible, and the level of consciousness is intact and stable until the latest stadium. Nevertheless, a delirium my manifest within the frame of a dementia.

Depression: symptoms of depression may be present in a delirium state, especially in the hypoactive subtype. However, depression is usually characterized by normal level of consciousness and intact orientation, similar previous episodes, the absence of fluctuations during the day and the dominance of depressive feelings and thoughts. Gathering the patient's clinical history from the caregivers can help the differential diagnosis.

Psychosis: in some types of acute psychosis, consciousness and cognition may result impaired. However, in psychotic disorders visual hallucinations are rare.

Other mental illnesses: delusional manic episodes (confused mania) or dissociative disorders can present as delirium-like states, characterized by a rapidly fluctuating impairment of cognitive function and ability to maintain attentional focus.

10.1.5.3 Course and Prognosis

The duration of a delirium episode depends strictly on the timing of diagnosis (the faster it is, the better is the outcome) and the access to adequate medical assistance. Nevertheless, more complications and worse outcomes are more frequent in older people.

Evidence shows that delirium results in poor long-term outcome and higher mortality (up to 60% increase) in the elderly admitted to hospital, compared to the general population.

Many patients still meet the criteria for delirium for a prolonged period after hospital discharge, with up to 21% of individuals showing persistence of some symptoms at 6 months postdischarge.

According to literature, after an episode of delirium in older people, functional dependence probability may even double, thus increasing the possibility of institutionalization.

Older people are also reported to have an increased risk of developing dementia or some cognitive decline. Indeed, as previously mentioned, the connection between delirium and dementia is complex: dementia is the leading risk factor for delirium, and delirium is an independent risk factor for subsequent dementia.

10.1.6 Assessment

Detection of delirium can be supported by the use of validated delirium screening scales, which differ in duration, complexity, and need for training.

Examples of tools for the assessment of delirium used in clinical practice are as follows:

- Mini Mental State-Examination (MMSE): used to evaluate cognitive functioning. A baseline level functional assessment, done before the onset of delirium, may be helpful to measure the global deterioration of the patient's cognition.
- Confusion Assessment Method (CAM).
- Richmond Agitation and Sedation Scale (RASS): highly sensitive and specific for the identification of delirium in the elderly.
- Observational Scale of Level of Arousal (OSLA): highly sensitive and specific for the identification of delirium in the elderly.
- Delirium Observation Screening Scale (DOS).
- Nursing Delirium Screening Scale (Nu-DESC).
- Recognizing Acute Delirium As part of your Routine (RADAR).
- 4AT (4 A's Test).
- Delirium Diagnostic Tool-Provisional (DDT-Pro).

ICU patients have a significantly higher risk of developing delirium. In this population, delirium is associated with increased duration of mechanical ventilation, prolonged hospitalization, and increased mortality. International guidelines recommend checking these patients for delirium every day (usually twice or more a day) using a validated screening scale. The most commonly used are the Confusion Assessment Method for the ICU (CAM-ICU) and the Intensive Care Delirium Screening Checklist (ICDSC). If these tools are not used, delirium can remain undiagnosed by the healthcare team in 75% of ICU, leaving the patient without prompt and proper intervention.

10.1.6.1 Evaluations

In order to make a prompt recognition and to conduct a thorough differential diagnosis, it is mandatory to perform a complete and systematic evaluation of the patient, including an accurate history taking, a careful physical examination, blood exams screening, as well as possible further instrumental examinations, as detailed below.

Clinical history: Since the patient is usually confused and disoriented, the clinical history should be obtained by questioning the patient's relatives/caregivers or the medical staff under whose supervision the patient was. A detailed history is useful for understanding the baseline cognitive and functional level of the patient and for identifying conditions that could contribute to the onset, progression, and persistence of delirium (e.g., medical illnesses, drug treatments, and alcohol use).

Physical Examination: It includes monitoring of vital parameters (pulse, blood pressure, temperature, blood oxygen level, and respiratory rate), evaluation of movement and motor abilities, mimics, reflexes, and autonomic features.

Laboratory: It includes blood count with cell definition, electrolytes, renal and hepatic indexes, glucose, thyroid function, nutritional status (B12 and folic concentrations), urine analysis, blood and urine drug screening, serological test for syphilis, HIV antibody test.

Chest radiograph: It is used to exclude the presence of infections, such as pneumonia, or cardiac diseases, such as congestive heart failure.

EEG: In delirium, it typically shows a generalized slowing of resting-state activity (versus focal areas of hyperactivity, usually related to epilepsy), with abnormally decreased background alpha power and increased theta and delta frequency activity.

Neuroimaging: The role of neuroimaging has not yet been established, and its feasibility is questionable. However, in patients with focal neurological symptoms, a computed tomography (CT) scan or magnetic resonance imaging (MRI) scan of the head may be recommended to exclude the presence of space-occupying lesions, like tumors, abscesses, and hemorrhages.

Cerebrospinal fluid: Examination of cerebrospinal fluid is rarely indicated.

10.1.7 Treatment

10.1.7.1 Nonpharmacological Treatment

Identifying hospitalized people at risk (e.g., patients over age 65, those with baseline cognitive deficits, ICU patients) and possible precipitating factors (e.g., constipation,, hypoxia, dehydration, immobility, sleep deprivation, functional decline, visual or hearing deficits), and implementing close observation and adequate patient screening through the use of proper clinical tools (tests and assessment scales), are essential to prevent delirium occurrence.

Once delirium has been diagnosed, its management and treatment are based on the following steps.

- *Step 1*: Identifying and treating the underlying causes (e.g., reducing the dose of psychotropic, analgesic, or anticholinergic drugs, restoring hydro–electrolyte balance).
- *Step 2*: Managing and treating symptoms (see further "medications").
- *Step 3*: Reducing risks of complications (e.g., ensuring adequate nutrition and hydration, supporting motor needs, preventing aspiration, falls, and bedsores).
- *Step 4*: Providing an ideal therapeutic environment, with particular attention to sensory stimuli from the surrounding (e.g., single and silent room, low light exposure, assistance from the same few caregivers to favor familiarity of faces and their association with care).

10.1.7.2 Pharmacological Treatment

- *Antipsychotics*: haloperidol is considered the first choice in treating delirium symptoms, and it is usually used low-dose and short-term (1 week or less). Antipsychotic drugs should be avoided or used with extreme caution in people who suffer from Parkinson's disease or dementia with Lewy bodies due to a higher risk of worsening of motor and cognitive status. Moreover, patients at risk for delirium also appear to be more vulnerable to several antipsychotic-related adverse effects, such as EPS. Low-dose atypical antipsychotics (i.e., risperidone, olanzapine, and quetiapine) may be also used since they have shown comparable efficacy. However, caution is recommended, as atypical antipsychotics have been associated with increased risk of cerebrovascular events in elderly patients, and increased all-cause mortality in elderly patients with dementia.

- *Benzodiazepines*: this type of drug should be avoided or used with caution, as it may trigger or worsen delirium. Benzodiazepines are, instead, recommended if the delirium is caused by alcohol withdrawal (Delirium Tremens), benzodiazepine withdrawal, or in case of contraindications to antipsychotics (e.g., in Parkinson's disease or neuroleptic malignant syndrome).
- *ICU sedatives*: patients who stay in ICU and need respiratory assistance may benefit from using dexmedetomidine, which seems to shorten the length of the delirium.

Box 10.5: Delirium Nursing Assessment and Management

The patient's examination in the course of delirium should follow a fairly precise plan that allows evaluating the development of symptoms quickly and effectively. We propose a possible scheme of evaluation and management of the patient with delirium.

Assessment

- Psychiatric interview: it should include an assessment of behaviour and mental state, with particular attention to the state of consciousness, formal aspects of language (speed, continuity, and organization), content of thought, cognitive status and judgment. The patient evaluation must take particular account of possible risk factors for the safety of the patient.
- Periodic re-evaluation: the assessments listed above must be repeated at regular daily intervals in order to capture the fluctuations of symptoms over time.

Management of patients with Delirium

The main nursing care goal to achieve is to avoid worsening of symptoms:

- Assessing the anxiety level and preventing its increase.
- Providing an adequate therapeutic environment (low-stimuli setting).
- Ensuring patient safety by removing potentially dangerous and harmful objects from the surrounding.
- Maintaining a soft and gentle approach toward the patient (talk slowly and not loudly, frequently reassure the patient, answer calmly and clearly to patient's questions, avoiding possible misinterpretations).
- During agitation, constant one-to-one observation is essential to prevent accidental self-harming.
- Involving family members can be useful as maintaining visual, verbal, and tactile contact with relatives reduces patients' behavioral disorders.

10.2 Minor and Major Neurocognitive Disorders

10.2.1 Introduction

Dementia is a contemporary public health issue, severely impacting not only affected individuals and their families but also society and health care services. The hallmark feature of dementia is progressive cognitive decline, usually resulting in daily functioning impairment. However, although often perceived as a mere "cognitive" disorder, dementia is frequently associated also with behavioral and psychological symptoms, which occur in clusters or syndromes identified as psychosis, depression, anxiety, and disinhibition.

In many patients, the cognitive impairment is prominent at the onset and the underlying etiology is easily recognized (e.g., Parkinson's disease, traumatic brain injury, HIV infection, or stroke). In other patients, however, behavioral and psychopathological symptoms may be the first hint of underlying dementia and the direct reason why patients seek medical attention, with the causative disorder being revealed much later in the course of illness (e.g., Alzheimer's disease, cerebrovascular disease, frontotemporal dementia, and Lewy body disease). Moreover, neuropsychiatric symptoms do not always arise from the dementia phenotype but may result from psychiatric disorders or psychological vulnerability preceding dementia.

Although NCDs fall within the area of neurological conditions, the psychiatrist's intervention is often required in their assessment, recognition, diagnosis, and treatment.

10.2.2 Definition

Neurocognitive disorder (NCD) refers to a syndrome that progresses with an impairment in one or more areas of cognition (including memory, speech, reasoning, judgment, and orientation) compared to previous level of cognitive performance and may also be associated with behavior alterations.

The fifth edition of DSM (DSM-5) includes in this category two subtypes: mild and major NCD.

Major NCD, previously called dementia, refers to a gradually progressive brain illness characterized by a cognitive decline and changes in behavior to such an extent that daily functioning is impaired. On the contrary, Mild NCD is described as a less severe form of cognitive impairment that exceeds what would be expected for normal aging but does not interfere with the capacity for independence in everyday activities. In this subchapter, we will focus primarily on the major NCD, often referred to as dementia, due to its higher prevalence and impact.

Dementia may result from a primary disorder (degenerative) or may be secondary to other organic conditions. While primary dementias are usually permanent and progressive, secondary dementias can be prevented and treated.

10.2.3 Epidemiology

The prevalence of moderate to severe dementia is rising with the aging population and varies in different population groups. See Box 10.6 on the prevalence of dementia in different population groups.

Box 10.6: Prevalence of Dementia in Different Population Groups

- General population older than 65 years of age: 5%.
- General population older than 85 years of age: 20–40%.
- Outpatient general medical practice: 15–20%.
- Chronic care: 50%.

The most common type of dementia is the Alzheimer's type that occurs in about 50–60% of all subjects with dementia and increases in prevalence with age. Indeed, in people aged 65 years old, the prevalence is 0.6% in men and 0.8% in women, while at age 90 about 1/5 of people have the disorder. Moreover, in about 40–60% of these people, dementia is moderate–severe.

Individuals with Alzheimer disease occupy more than 50% of nursing home beds, while more than two million persons with dementia live at home. Current predictions suggest that from 2015 to 2050, the number of people affected by dementia will have increased around twofold in Europe and North America, threefold in Asia, and fourfold in Latin America and Africa. The second most common type of dementia is vascular dementia, secondary to cerebrovascular diseases, representing about 15–30% of all dementia cases. Vascular dementia is most common in people 60–70 years old and is more frequent in men than in women. Finally, approximately 10–15% of people have coexisting vascular and Alzheimer's type Dementia. Other relatively common types of dementia are Lewy body dementia and Frontotemporal dementia.

10.2.4 Etiopathogenesis

The most common causes of dementia in individuals older than 65 years of age, as mentioned, are Alzheimer's disease, vascular dementia, and a mix of these two. Beyond neurodegenerative causes, several other conditions, including different medical disturbances, traumatic brain injury, drugs, and toxins can determine the onset of dementia. There are also many types of secondary dementia induced by reversible causes (i.e., metabolic abnormalities or nutritional deficiencies). Early detection of these latter is thus of the utmost importance when considering a diagnosis of NCD.

See Fig. 10.1 for an overview of the main causes of dementia.

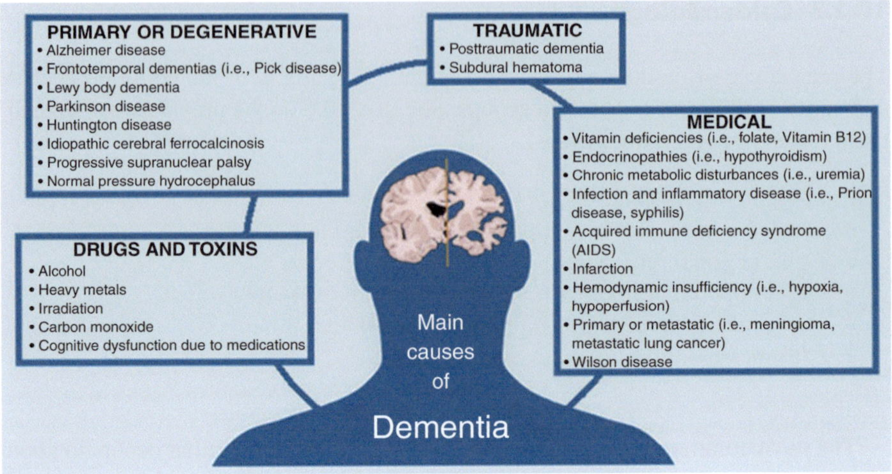

Fig. 10.1 Main causes of dementia

10.2.5 Assessment

The diagnosis of dementia is mainly clinical, based on anamnesis, physical examination, evaluation of cognitive and functional abilities. The critical clinical points are the identification of the disease and the clinical workup of its cause. The underlying etiology of Major and Mild NCD is similar. Laboratory tests and neuroimaging are essential for etiological and differential diagnosis.

10.2.5.1 Diagnostic Criteria

DSM-5 distinguishes between Major and Mild Neurocognitive Disorders, based on levels of functioning, specifying a variety of types of dementia according to the main underlying process. Diagnostic criteria are reported in Table 10.2.

10.2.5.2 Evaluations

When a NCD is suspected, a comprehensive evaluation should be performed in order to confirm the diagnosis and clarify the underlying cause. The assessment should include a global clinical evaluation (clinical history taking, mental status examination), cognitive testing, functional assessment, and neuroimaging.

From a practical point of view, it is important to understand the pattern of cognitive impairment and preserved skills, to help care planning.

Clinical history: Whenever possible, the clinical history should be obtained both from the patient and a relative, caregiver, or other knowledgeable informants. Commonly, caregivers and family members are more aware and concerned about cognitive and functional impairment than people with dementia themselves and, therefore, more informative.

Table 10.2 DSM-5 diagnostic criteria for major and mild neurocognitive disorders

Mild neurocognitive disorder	Major neurocognitive disorder
A. Evidence of *modest cognitive decline* in one or more cognitive domains (attention, executive function, learning/memory, language, perceptual–motor, or social cognition) based on: 1. Concern of the individual, an informant, or the clinician and 2. A *modest impairment* in cognitive performance, documented by neuropsychological testing or quantified clinical assessment B. The cognitive deficits *do not interfere with capacity for independence* in everyday activities (complex instrumental activities of daily living such as paying bills or managing medications are preserved) C. The cognitive impairment do not occur just in the context of a delirium D. The cognitive impairment is not better explained by another mental disorder	A. Evidence of *significant cognitive decline* in one or more cognitive domains (complex attention, executive function, learning/memory, language, perceptual–motor, or social cognition) based on: 1. Concern of the individual, an informant, or the clinician and 2. A *substantial impairment* in cognitive performance, documented by neuropsychological testing or quantified clinical assessment B. The cognitive deficits *interfere with capacity for independence* in everyday activities (complex instrumental activities of daily living such as paying bills or managing medications are impaired) C. The cognitive impairment do not occur just in the context of a delirium D. The cognitive deficits are not better explained by another mental disorder
Specify: 1. Due to Alzheimer's disease, frontotemporal lobar degeneration, Lewy body disease, vascular disease, traumatic brain injury, substance/medication use, HIV infection, prion disease, Parkinson's disease, Huntington's disease, other medical condition, multiple etiologies, or unspecified 2. Accompanied by a clinically significant behavioral disturbance (e.g., psychotic symptoms, mood disturbance, agitation, apathy, or other behavioral symptoms)	Specify: 1. Due to Alzheimer's disease, frontotemporal lobar degeneration, Lewy body disease, vascular disease, traumatic brain injury, substance/medication use, HIV infection, prion disease, Parkinson's disease, Huntington's disease, other medical condition, multiple etiologies, or unspecified 2. Accompanied by a clinically significant behavioral disturbance (e.g., psychotic symptoms, mood disturbance, agitation, apathy, or other behavioral symptoms) 3. Current severity: *Mild*: Difficulties with instrumental activities of daily living (e.g., housework, managing money) *Moderate*: Difficulties with basic activities of daily living (e.g., feeding, dressing) *Severe*: Fully dependent

Inquiring about the past life of people provides an invaluable source of baseline data regarding cognitive functioning, intellectual abilities, personality, motor skills, mood, and perception. Moreover, a comprehensive history, including comorbid diseases and medications, may provide information that could help elucidate the etiology of the disorder (e.g., cardiovascular diseases, obstructive sleep apnea, diabetes may suggest a vascular cause).

The clinician should then obtain a detailed description of changes in the subject's functioning, addressing all activities of daily routine. Early cognitive impairments are often observed in the subject's ability to handle finances, manage medications, deal with unexpected situations and solve problems of everyday life.

Look at Box 10.7 on the daily routine to investigate in the assessment of dementia.

Box 10.7: Daily Routine and Activities to Investigate in the Assessment of Dementia

- Self-care.
- Personal autonomy in daily activities.
- Job responsibilities, work performance, work habits.
- Shopping and personal support.
- Interactions with family and friends, social activities, relationships.
- Recreational activities, hobbies, sports.
- Reading interests.
- Ability to maintain personal finances.

Moreover, features such as change onset, trajectory over time, and cognitive domain affected should be investigated in order to clarify the cause of NCD.

Look at Box 10.8 on key points of clinical history.

Box 10.8: Key Points of Clinical History

The time of the change

- Sudden onset (Vascular etiology?).
- Gradual change: few years (Degenerative disease?).
- Progressive change: days, weeks or months (Tumor? Space-occupying lesion? Medical conditions?).

The skills changed

- Memory.
- Language.
- Visuospatial skills.
- Personality.

The cognitive decline should then be investigated accurately. The clinician should inquire about change over time in memory skills (short-term memory versus long-term memory), language skills, executive functioning and visuospatial abilities, as well as alterations in the subject's judgment and personality.

Look at Box 10.9 on the main cognitive and behavioral changes to investigate in the assessment of dementia.

Box 10.9: Clinical Eye

Characteristic symptomatology of the onset of dementia:

- Difficulty learning and remembering new information.
- Difficulty performing complex tasks.
- Difficulty in reasoning and judgment.
- Difficulty in problem solving skills.
- Spatial-temporal orientation difficulties: people get lost, do not remember the current date.
- Worsening difficulty in finding the words that express what one wants to say.
- Difficulty following conversations.
- Behavioral changes: inappropriate behavior and reactions, passivity, irritability, suspiciousness.

Mental status examination: The mental status examination of patients with NCD follows the same format as for younger adults. However, in the case of older adults, the clinician should pay attention to features, signs, and symptoms that are particularly common in this population.See the following section "Psychiatric issues in the elderly" for a more detailed description.

Cognitive testing and functional assessment: Neuropsychological testing to measure cognitive skills is an essential complement to the clinical history. On the other hand, clinical history must guide the interpretation of cognitive tests because several factors can alter cognitive test performance (i.e., IQ, premorbid intellect, educational status, cultural differences, learning disability, vision and hearing disturbances, cooperation). Cognitive testing allows standardized, reproducible evaluation of people's cognitive abilities. Thus, the clinicians can evaluate memory, executive functions, visuospatial and constructional abilities, both at the initial assessment and also during the follow-up to monitor the progression of the cognitive impairment. Cognitive tests vary depending on their purpose and also on the severity of the NCD. Moreover, in order to precisely quantify the level of functional impairment of patients, specific assessment scales may be used as a complement to the information provided by caregivers and family members. Look at Box 10.10 for the main cognitive tests and functional assessment tools.

Box 10.10: Main Cognitive Testing and Functional Assessment Tools

Cognitive Tests

- Mini Mental State Examination (MMSE) and Modified Mini Mental State (MMMS).
- Clock Drawing Test (CDT).
- Frontal Assessment Battery (FAB).
- The Montreal Cognitive Assessment (MOCA).

- General Practitioner Test of Cognitive Function (GP COG).
- Addenbrookes Cognitive Assessment (ACE).
- Repeatable Battery for the Assessment of Neuropsychological Status (RBANS).

Functional Assessment Tools

- Instrumental Activities of Daily Living (IADL).
- Activity of Daily Living (ADL).

Laboratory tests: Blood tests are essential to exclude rare but possible medical causes of NCD, sometimes treatable. They should include, besides routine tests, thyroid-stimulating hormone, vitamin B12 levels, and other specific tests (e.g., for HIV and syphilis) required to complement previous clinical findings. In selected cases (e.g., when Alzheimer's disease is suspected), lumbar puncture for cerebrospinal fluid analysis could be considered.

Neuroimaging: Brain imaging is recommended as a component of the diagnostic assessment, giving a screenshot of the brain's structural appearance and perfusion. CT or MRI scans can detect potentially reversible structural disorders such as subdural hematoma normal-pressure hydrocephalus, brain tumors, or subdural hematoma. Moreover, they may identify patterns of regional brain atrophy suggestive of a specific neurodegenerative disorder (e.g., Alzheimer's disease, Frontotemporal Dementia) or diffuse cerebral microinfarcts indicative of Vascular Dementia. PET and SPECT can identify cerebral perfusion patterns and further help with differential diagnosis.

10.2.5.3 Differential Diagnosis

Several conditions might mimic the clinical presentations of NCDs. In particular, the clinician should exclude:

Physiological aging: Aging is not necessarily associated with any significant cognitive decline, although some minor memory deficits can occur. These deficits can be distinguished from those of dementia because they are less severe and have little impact on people's behaviour and daily functioning.

Delirium: Generally, delirium has a rapid onset, brief duration, cognitive impairment fluctuation during the day, nocturnal exacerbation of symptoms, marked disturbance of the sleep–wake cycle, and prominent disturbances in attention and perception.

Depression: Some elderly people with depression may present symptoms of cognitive impairment that are challenging to distinguish from symptoms of dementia. The clinical picture is sometimes referred to as "pseudo-dementia" but using the term "depression-related cognitive dysfunction" is preferable. People with depression-related cognitive dysfunction often have a clinical history of depressive episodes. Moreover, these subjects generally have prominent depressive symptoms and more insight into their symptoms than people with dementia. Furthermore, memory impairment from depression usually responds to antidepressant medication.

Factitious disorder: In factitious disorder, people attempt to simulate memory loss, doing so erratically and inconsistently. On the contrary, in the course of dementia, one loses memory for time and place before memory for persons, and recent memory before remote memory.

Other disorders: Intellectual disability, which does not include memory impairment, occurs in childhood. Amnestic disorder is characterized by a circumscribed loss of memory and no functional deterioration.

10.2.6 Clinical Presentation

Box 10.11: Clinical Case

Mr. M., a 69-year-old retired university professor of history, was accompanied to the psychiatric service by his wife, on the recommendation of the family physician. Mr. M. retired at age 65 years. In the 18 months prior to retirement, Mr. M. often was irritable and aggressive with his colleagues. He also forgot some important faculty meetings, student exams, or graduation sessions. Furthermore, his wife realized that he has not renewed his theater subscription and has not played tennis, his favorite sport, for some time. Mr. M. became so forgetful that his family was afraid to leave him alone, even at home. His wife told the doctor that last month, while Mr. M. was walking in the historic center of his city, an area he knew well, he could not find his way home. From then on, his memory failure began to worsen. Mr. M. failed to recognize neighbors and colleague she knew for many years. The morning of the psychiatric visit, his wife had to start dressing him because he was trying to put his leg in his shirt sleeve. On the psychiatric visit, Mr. M. had a hard time remembering the current date and year but was able to recall his name and date of birth. During the interview, when the doctor asked to recall words and numbers, Mr. M. at first appeared tense and distressed, then he railed at the doctor and his wife. His general medical condition was good; neuroimaging showed diffuse atrophy on the CT scan.

10.2.6.1 Cognitive Impairment

Cognitive impairment is the clinical hallmark of NCD and may involve different cognitive domains.

Patients with NCD often complain about *memory loss*, specifically reporting difficulties in learning and retaining new information, defined as "short-term memory". Memory impairment occurs early in the course of the disease, is usually mild at the onset, and is more marked for recent events. As the disease advances, memory deficit will progress, preserving mostly remote memories. Poor recall of old memories will finally occur, as well as a loss of general knowledge. Since memory is vital for orienting in space, time, self, and others, orientation will be progressively affected.

Forgetting is usually associated with perseveration, which must be distinguished from that that may accompany other psychiatric disorders, such as depression or anxiety. In these last, repetitiveness is related to the presence of ruminations about past experiences and worrying about future dangers, resulting in continuous requests for reassurance.

Changes in language skills are also common in people with dementia and may cause great distress. Difficulties in finding words (anomia) may result in interrupted speech or increased use of paraphasias or circumlocutions. Moreover, word substitutions are described, with the wrong word slipping out, either a similar-sounding word or a word from a similar semantic category.

Impairment in executive functions could manifest with changes in the ability to organize and plan complex tasks, difficulty in making decisions, and struggle in adjusting to new situations.

Changes in visuospatial abilities may have a significant impact on daily living skills, like, for example, driving the car, moving in familiar or unfamiliar places, putting things in the right place, or using ordinary objects.

Finally, it is possible to detect changes in *social cognition*. Family members may complain about socially inappropriateness, impulsivity, inflexibility and loss of empathy.

10.2.6.2 Behavioral and Psychological Symptoms of Dementia (BPSD)

As previously mentioned, although cognitive impairment is the clinical hallmark of dementia, behavioral or psychological symptoms will affect nearly all subjects with dementia over the course of illness, sometimes dominating the presentation.

BPSD include alterations in perception, thought content, mood, or behavior, as defined by the International Psychogeriatric Association. These symptoms are frequent, impacting 60% of people living in the community and 80% of those residing in long-term care facilities. Moreover, BPSD are responsible for premature institutionalizations, increased direct and indirect costs, and, above all, considerable suffering and diminished quality of life on the part of persons with dementia and their families. Indeed, people suffering from BPSD present more severe functional deficits, emotional distress, abuse, and neglect, and have higher mortality rates. Thus, clinicians should screen for BPSD during all visits with people with dementia. BPSD can be distinguished into those with psychotic and nonpsychotic features. Look at Box 10.12 for BPSD in course of dementia.

Box 10.12: BPSD in Course of Dementia

Psychotic Features

- **Hallucinations**—Most commonly auditory and visual.
- **Delusions**.
- **Delusional misidentification syndrome**.
- Misidentification of persons, places, objects, or events.

Nonpsychotic Features

- Agitation and irritability.
- Apathy.
- Anxiety.
- Depressive syndrome.
- Emotion lability.
- Sleep disturbances (night wandering, sundowning, sleep–wake cycle reversal).
- Wandering (in particular nocturnal).
- Physical aggression.
- Repetitive vocalizations and movements.
- Inappropriate sexual behavior.
- Inappropriate eating behaviors and hyperorality.

Although BPSD are seen almost universally in dementia, regardless of the underlying etiology, specific types of dementia are more frequently associated with certain symptoms. For example, vascular dementia and the prodromal phase of Alzheimer's disease typically presents with apathy, depression, anxiety and irritability.

In some types of dementia, neuropsychiatric symptoms are defining elements of the illness, integrated into the diagnostic criteria. Lewy body dementia, for instance, frequently presents with visual illusions and hallucinations, well-formed and detailed, which later on become florid, persistent, and, in many cases, are associated with misidentification, paranoia, delusions. Rapid eye movement (REM) sleep disturbance is also typically seen in these conditions. Finally, the behavioral variant of frontotemporal dementia typically manifests with stable personality changes. In particular, patients may manifest inappropriate or offensive social behaviors, with indifference, impatience, distractibility insensitivity, jocularity, and disinhibition. Moreover, ritualistic or stereotyped behaviors may appear, as well as hyperorality and dietary changes. Apart from social cognition decline, many of these patients do not have other noticeable cognitive deficits until the illness is established. Frontotemporal dementia with prominent dysexecutive deficit may present at onset with psychotic symptoms. Differential diagnosis between primary psychotic disorders and frontotemporal dementia must be kept seriously into account when psychotic symptoms appear in patients with a silent psychiatric history and a late onset in comparison to the usual range of age of onset of the psychiatric diagnoses.

10.2.6.3 Onset and Course of Illness

Generally, the course of dementia is characterized by onset in people older than 60 years of age with progressive decline over 5–10 years, leading eventually to death. The age of onset and the speed of decline varies among the different types of dementia. Usually, the course of illness begins with several subtle signs that, at first, may be overlooked by people affected and their families. Data suggest that the course of the disease is usually more rapid in people with an early onset of dementia or a family history of dementia. Insidious onset of symptomatology is most common in vascular dementia, Alzheimer's disease, endocrine and metabolic disorders, or brain tumors. Instead, the onset of a Neurocognitive Disorder secondary to head trauma, cardiac arrest with cerebral hypoxia, or encephalitis can be sudden. The symptomatology becomes conspicuous as the illness progresses, and family members may then bring people affected to the physician. After diagnosing the Neurocognitive Disorder, people must have a complete medical and neurological workup because 10–15% of all individuals with dementia, as stated above, have a potentially reversible condition if the treatment begins before permanent brain damage occurs.

Individuals with dementia may be more sensitive to some ongoing pharmacological treatment, such as benzodiazepines, or habitual alcohol use, which can precipitate aggressive or agitated behavior.

In the terminal stages of the disease, people become amnestic, profoundly disoriented, incoherent, and incontinent of urine and feces.

The course of the illness range from a steady progression to a rapidly worsening dementia to stable dementia. Psychosocial factors can modify the severity and course of the disease. For example, higher education and premorbid intelligence (cognitive reserve) may better compensate the cognitive deficits in particular at the beginning of illness. On the other hand, anxiety and depression can intensify and aggravate the symptomatology. Moreover, the symptoms of dementia may progress more slowly after psychosocial and pharmacological treatment, when they are possible.

10.2.7 Treatment

The treatment of dementia must follow a person-centered care approach. The first step consists of the verification of the diagnosis, the evaluation of the progression of illness, the implementation of preventive measures (i.e., control of diabetes and hypertension), and the beginning of appropriate treatment, including the treatment of underlying medical conditions through specific pharmacotherapies. The second step consists of identifying the different stages of dementia, as they have distinctive and decisive requirements for planning a personalized care plan. The general treatment approach to people suffering from dementia includes the following.

- Supportive medical assistance.
- Cognitive and psychological interventions.
- Psychosocial support for those affected and their families.
- Pharmacological treatment for specific symptoms, including BPSD.

10.2.7.1 Nonpharmacological Treatment

Cognitive impairment has a significant psychological meaning for people with dementia and their families. Memory is essential to maintain a sense of continuity of the self over time. Memory impairment is thus usually very distressing: emotional reactions range from depression and anxiety to a catastrophic terror for the realization that the sense of self is dissolving. People with dementia often benefit from supportive and educational interventions aimed at explaining the nature and course of illness. They may also benefit from assistance in grieving and accepting the extent of their disability and from attention to self-esteem issues. Therefore, on the one hand, it is useful to help people find ways to support the impaired functions, for example, keeping calendars for orientation problems and taking notes for memory difficulties. On the other hand, it is important to maximize any areas of intact functioning by assisting people in the identification of activities in which successful functioning is possible.

Since the individuals who take care of people with dementia struggle with feelings of guilt, grief, anger, and exhaustion as they watch their relatives gradually deteriorate, they should also be offered supportive and educational interventions.

Due to the limited efficacy and significant side effects of pharmacological treatments, nonpharmacological treatments represent the first line of therapy for people with dementia. These treatments include supportive and rehabilitative interventions aimed, on the one side, at helping those affected "manage change", contain disability, and reduce the discomfort due to cognitive symptoms, and, on the other, at developing greater knowledge in caregivers and family members. Several other approaches have been proposed to improve cognitive outcomes in dementia, including cognitive and psychological interventions. Finally, some evidence suggests that physical exercise may improve dementia's symptomatology, particularly in the early stages.

Look at Box 10.13 on the nonpharmacological treatment of people with dementia.

Box 10.13: Nonpharmacological Treatment in People with Dementia

Cognitive Interventions

- Cognitive stimulation (CS).
- Cognitive training (CT).
- Cognitive rehabilitation (CR).
- Reality orientation therapy (ROT).

Psychological Interventions

- *Emotion-Oriented Approaches*
  Psychotherapy.
  Reminiscence therapy.
  Validation therapy.
  Simulated presence therapy.

- *Sensory Stimulation-Oriented Approaches*
  Art therapy.
  Music therapy.
  Snoezelen multisensory stimulation.

Other Psychosocial Interventions

- PET therapy.
- Physical exercise.

10.2.7.2 Nonpharmacological Management of BPSD

The strongest evidence-based approach for the nonpharmacological management of BPSD (i.e., the DICE approach) emphasizes the importance of the following steps

- Describe the problematic behavior.
- Explore possible causes.
- Establish a treatment plan.
- Evaluate the outcome of the plan.

First, a detailed description of similar behaviors that occurred in the past should be obtained, including information about their context, timing, precipitants, consequences, and, if any, effective interventions. Second, the clinicians should investigate possible underlying and reversible causes of the behavior, including possible undetected medical conditions or pain, conflictive patient–caregiver relationship, and problematic environmental factors.

Clinicians should work with the caregiver and, when possible, the individual affected, to develop a personalized management plan that includes measures to improve the environment and the caregiver skills, as well as targeted interventions. Environmental interventions include reducing confusion and loud noises, setting clear lighting, and using simple visual reminders to help the patients with common activities (i.e., picture of a toilet on the bathroom door). Caregivers should be instructed to use a caring tone of voice, maintain a daily routine, provide simple instructions, and avoid conflicts. Moreover, caregivers will benefit from education regarding BPSD, including strategies to improve communication with those affected and understanding that behaviors are not intentional. Finally, targeted interventions include specific strategies for people with dementia aimed at reducing BPSD, such as cognitive and emotion-oriented interventions, behavioral management techniques, exercise, and sensory stimulation interventions. Among these, cognitive stimulation therapy and reminiscence therapy are the most frequently applied interventions. Cognitive stimulation therapy include several activities (i.e., art, dancing, gardening, etc.) to stimulate concentration, thinking, and memory. This treatment may improve memory, communicative skills, and quality of life; moreover, it may reduce BPSD, including depression and anxiety. On the other hand, reminiscence therapy uses memory aids (i.e., music, videos, etc.) to assist people with dementia remembering past events. It also may improve communication, retaining, and depressive symptomatology. Planned, repetitive physical activity is a good nonpharmacological intervention that should be encouraged.

10.2.7.3 Nursing Management of People with Dementia

Box 10.14: Nursing Interventions for Individuals with Dementia and Their Caregivers

- **Support reality and environment orientation**
  - Surround patients with familiar objects and use tools such as a clock, a calendar, or pictures to help maintain reality orientation.
  - Provide psychoeducational interventions for caregivers to inform them on how to orient people with dementia to time, persons, places, and circumstances, especially after discharge from the hospital.
- **Give positive feedback**
  Reinforce people with dementia when their behavior is appropriate and encourage caregivers to do the same. Positive feedback favors self-esteem and the probability to repeat appropriate behavior.
- **Simplify communication**
  Prefer face-to-face interactions, speak slowly, and use easily understandable terms and explanations. Encourage caregivers to do the same.
- **Discourage misgiving and suspiciousness**
  Display reasonable doubts when subjects with dementia show suspiciousness toward other people or incorrect interpretation based on delusional thinking.
- **Monitor and observe carefully**
  Close observation is indicated in case of aggressiveness, agitation, hallucinations, or delusional thinking.

Nursing staff plays an important role in the management and care planning of people with dementia, especially during hospitalization.

Nursing interventions should address individuals with dementia, as well as their families and caregivers. Look at Box 10.14 on nursing interventions.

10.2.7.4 Pharmacological Treatment

Clinicians may prescribe pharmacological therapy for specific symptomatology: anxiolytics and hypnotics for anxiety and insomnia, antipsychotics for hallucinations, delusions, agitation, and aggression, and antidepressants for depressive symptoms. However, particular attention should be paid to the possible drugs' adverse effects, which may be worse in older people (e.g., disinhibition, confusion, and oversedation), and the interactions with other ongoing pharmacotherapies, since this population often suffers from one or more chronic diseases. In general, clinicians should avoid drugs with high anticholinergic activity.

10.2.7.5 Antidementia Medications

Cholinesterase inhibitors, such as donepezil, rivastigmine, galantamine, and tacrine, are used to treat cognitive impairment in Alzheimer's disease. They reduce the inactivation of the neurotransmitter acetylcholine by the enzyme cholinesterase and, thus, potentiate the cholinergic system, which, in turn, improve cognition. These treatments are particularly useful for subjects with mild to moderate memory deficits, since a partial preservation of basal forebrain cholinergic neurons is essential to benefit from augmentation of cholinergic neurotransmission. Memantine, believed to be neuroprotective against excitotoxicity in the cortex and hippocampus, showed small beneficial effects on cognition in moderate to severe Alzheimer's disease.

Many other pharmacotherapies have been proposed to improve cognitive outcomes in people with dementia, including statins, nonsteroidal anti-inflammatory drugs, *Ginkgo biloba*, vitamin E, vitamin B12, folic acid, omega-3 fatty acids, ginseng, latrepirdine, hirudin, and cocoa, but all lack sufficient evidence to make any recommendation.

10.2.7.6 Pharmacological Treatment for BPSD

If nonpharmacological interventions have proven ineffective or patients' behaviors represent a risk to themselves or third parties, pharmacological treatment for BPSD should be considered.

SSRI antidepressants have proven useful in treating behavioral/psychiatric symptoms in frontotemporal dementia, although their use is not associated to improvements in cognition.

Antipsychotics should be prescribed with caution in dementia due to the risk of adverse cerebrovascular events and death in people with dementia. In clinical practice, these drugs are used not only to treat psychotic symptoms (i.e., delusions, hallucinations) but also to manage aggressiveness and agitation, symptoms that are extremely distressing for individuals with dementia and constitute a risk of harm for themselves and others. Atypical antipsychotics are preferable to typical ones, and risperidone has the largest evidence base in the treatment of agitation, aggression, and wandering in individuals with dementia at very low doses, not corresponding to those usually indicated in psychotic disorders. The main features of treatment should be "start low, go slow", i.e., a low start dosage with slow up-titration to therapeutic dose is recommended. Adverse effects may should be prevented by closely monitoring them and using the lowest effective dose. If severe side effects occur, the medication should be immediate down titrated or discontinued. Moreover, antipsychotics should be used for a maximum of 3 months, since evidence for efficacy beyond this period is limited, while the risk of adverse effects increases with the duration of treatment. If treatment with antipsychotics is required for longer periods, it is necessary to plan a follow-up to evaluate treatment response and tolerability. Moreover, discontinuation should be attempted at regular intervals.

Table 10.3 Antipsychotic treatment in older people with dementia

	Atypical antipsychotics		
Drug	Risperidone	Olanzapine	Quetiapine
Dosage	0.25–2 mg a day	2.5–10 mg a day	12.5–300 mg a day
Side effects	– Prolonged QTc – Extrapyramidal side effects	– Metabolic syndrome – Prolonged QTc – Higher mortality in people with dementia	– Prolonged QTc – More sedating

Look at Table 10.3 for usual dose range and main side effects of antipsychotics in older adults with dementia.

10.3 Psychiatric Issues in Older People

10.3.1 Introduction

Generally, old age, or late adulthood, refers to the stage of the life cycle that goes from age 65 to beyond.

The aging process, also called senescence, is characterized by a progressive deterioration in all of the body's systems' (cardiovascular, respiratory, genitourinary, endocrine, and immune, among others) functioning. However, the deterioration process is different for each organ and is strictly linked to the person's genetic predisposition and its overlap with environmental stressors or intentional misuse (e.g. smoking, drugs abuse, diet, type of job activity).

Physical decline is associated with changes in the subject's social and environmental functioning, which usually determine an increased need of assistance and result in changes in self-perception and ability to relate with the surrounding. These modifications in social activity, in association with ageism (discrimination toward older adults held by younger people), retirement, the possible loss of spouses and friends, and the acknowledgement of time limitation and the nearness of death, may predispose to the development of psychiatric symptoms such as depression and anxiety.

Moreover, older individuals typically have to deal with different illness and physical signs of aging, which often represent the organic substrate for the outbreak of psychiatric disorders.

Also drugs and medications use or discontinuation may lead to the development of psychiatric symptoms, which are, in this case, usually transient and reversible after the removal of the cause.

10.3.2 Psychiatric Examination of the Older Patient

Early recognition and management of signs and symptoms of psychiatric disturbances in older people increase the chances to obtain a good clinical outcome, which translates into better quality aging and prolonged life chances. When older individuals present with psychiatric symptoms, it is necessary to conduct a systematic investigation, in order to determine the cause of the onset and, consequently, to find the more appropriate treatment.

Psychiatric History: gathering both past physiological and pathological history is useful to individuate potential risk factors that may have taken part in the onset of current psychiatric disturbances. Additionally, past psychiatric disorders anamnesis is fundamental, since the current episode may be a manifestation of a previously diagnosed illness. Because of the high prevalence of cognitive disorders, in particular memory impairment, in older persons, anamnestic history might be taken with the help of caregivers and family members.

Mental Status Examination: as previously mentioned, this evaluation should follow the same format as for younger adults but take into account peculiar elements of this population. For example, indications of sensory impairments (e.g., hearing aid) should be noted, as well as the presence of *motor disturbances* (e.g., stooped posture, slowed movements, "pill rolling" movements of the fingers, tremors) which may be suggestive of Parkinson's disease or be iatrogenic.Mental Status Examination should include the assessment of the following

- *General functioning*: Particular attention should be paid to independence in daily life activities and social functioning.
- *Mood and affect*: An expansive mood with disinhibited behavior may indicate a manic episode or suggest a frontal lobe dysfunction. Depression in the elderly may manifest with atypical signs and symptoms, such as chronic pain, somatic complaints, or pseudodementia. Moreover, feelings of loneliness, hopelessness, and worthlessness are connected to a higher risk of committing suicide in older people.
- *Perceptual disturbances*: Hallucinations and illusions in older adults may be transitory phenomena resulting from the inability to interpret the significance of sensory impressions due to sensory impairment.
- *Thought disturbances*: Severe depression in the elderly may be accompanied by delusions of poverty and physical illness; while dementia is sometimes associated with simple, unelaborated delusions, such as the belief of being robbed by an intruder or that a spouse is an impostor.
- *Language form and content*: These are frequently linked to thought alterations and can vary from loss of fluency, using of passepartout words and slurred speech to different types of aphasia, which are usually related to organic brain disorders; the loss of the ability to appreciate metaphoric language (abstract thinking) may be an early sign of dementia.

Neuropsychological and Cognitive Evaluation: it involves the examination of *sensorium* (specific senses integrity is necessary to understand correctly external and internal stimuli and to elaborate appropriate reactions to them), *consciousness* (a rapid decline or fluctuations of awareness level are usually connected to the presence of an acute organic condition), *orientation* (disorientation may be associated both with organic and primary psychiatric disturbances), *memory* (in terms of immediate, recent and remote memory), *intellectual skills* (loss of problem-solving ability and reduction of global intellectual level from baseline typically associate with dementia) and *visuospatial capability* (a certain degree of decline is, in some ranges, normal with aging). To obtain a valid result, it is recommended to execute a comprehensive battery of neuropsychological tests. Some examples of the most used cognitive assessment tools are the Mini Mental State Examination (MMSE), assessing orientation, attention, calculation, immediate and short-term recall, language, and the ability to follow simple commands, and Montreal Cognitive Assessment (MoCA), evaluating visuospatial/executive, naming, memory, attention, language, abstraction, delayed recall, and orientation.

10.3.3 Mental Disorders of Old Age

Most of the psychiatric disorders of older people can be prevented, ameliorated, or even reversed. Nevertheless, if not detected and treated promptly, these conditions can progress to an irreversible state requiring a patient's institutionalization, leading to global worsening and reduction of the patient's life expectancy. Although in the elderly the onset of psychiatric symptoms may often be secondary to organic conditions, such as delirium and dementia, discussed in detail above, it is important to remember that any primary psychiatric disorder may manifest in this population. In some cases, a relapse of a previously diagnosed psychiatric disorder may occur, while, in other occasions, the episode may be the first of a late-onset psychiatric disturbance. As mentioned above, it is, therefore, important to recognize them in order to evaluate the most appropriate management plan. Below, the main primary psychiatric issues that can be found in older people are described.

10.3.3.1 Depressive Disorders

Depressive symptoms are the most common psychiatric symptoms shown in the older population, because of the many predisposing factors, such as a chronic medical illness, the loss of friends or spouse, social retirement, socioeconomic and housing status, and gradual loss of independence in everyday life. All these factors are associated with a higher vulnerability to depressive disorders.

An older individual may report a clinical history of previous depressive episodes (early-onset depression, EOD) or the onset of depressive symptoms after the age of 60 (late-onset depression, LOD). In both cases, the clinical manifestation of a depressive episode in the elderly usually differs from that of younger subjects. Indeed, episodes are characterized by peculiar features, such as the following.

- Increased emphasis on somatic problems rather than on low mood with a tendency to hypochondriasis. This condition is called "masked depression" because it often leads the physicians to focus on physical symptoms instead of considering the presence of a possible depressive episode.
- Reduce energy and low psychomotor activity.
- Decreased appetite associated with progressive weight loss.

- Sleep disorders, in particular early morning awakening and multiple nocturnal awakening rather than hypersomnia.
- Higher frequency of presentation of delusions whose contents relate to guilt, hypochondriasis, nihilism, persecution, and sometimes jealousy.
- Higher occurrence of melancholic features.
- Difficulties with concentration, speed of mental processing, and executive function (pseudodementia) which often improve, but do not completely resolve, after remission of late-life depression.
- Higher risk of suicidal ideation, seldom spontaneously verbalized.
- High relapse rate and high risk of chronicity.
- Worse prognosis of comorbid diseases.

The assessment of depressive episodes in older people may be done using general depression scales (e.g., Hamilton Rating Scale for Depression) or more specific tools, such as the Geriatric Depression Scale, a 15-item self-related scale that focuses on the most frequent depressive symptoms shown in elderly.

In order to exclude an organic caused depressive syndrome (e.g., initial dementia, cerebrovascular depression), the gathering of an accurate clinical history and pharmacological anamnesis, the execution of a physical examination and cognitive assessment, and the use of neuroimaging may be required.

10.3.3.2 Late Life Schizophrenia

Late-life schizophrenia includes subjects who were diagnosed with schizophrenia early in life (early-onset schizophrenia) and who are now over 50 years old; and those few who are diagnosed after age 45 (very rare). The clinical presentation of older persons with schizophrenia partially differs from that of younger persons. Compared to young adults, late-life schizophrenia patients show fewer and less severe positive symptoms, persisting negative symptoms, and generally stable neurocognitive functioning. Finally, older people with schizophrenia show more frequent and severe medical comorbidities such as diabetes and heart and lung diseases, often related to inadequate health care and resulting in higher levels of disability and higher mortality rates. Chronically institutionalized patients show a greater cognitive decline and poorer outcomes.

In rare cases, schizophrenia can manifest for the first time after age 65, with higher frequency in women. Late-onset schizophrenia (also called paraphrenia) is characterized by the prevalence of paranoid ideas, eccentric behavior, delusional thoughts, hallucinations, emotional blunting, and social retirement.

10.3.3.3 Delusional Disorders

Delusions usually represent associated symptoms of various psychiatric illnesses, such as schizophrenia, depressive disorders, and bipolar I disorder, or may be linked to an underlying organic medical condition, like, for example, dementing disorders, brain tumors or drug abuse. Nevertheless, an older age onset of the so-called delusional disorder may occur. In the elderly, delusional contents are usually related to somatic problems (e.g., the belief of having fatal illnesses) and persecutory ideas (e.g., fear of being poisoned by the caregivers), which may lead to aggressive defensive behaviors and oppositional attitudes toward caregivers. Physical or psychological stressors often represent precipitating factors for the onset of delusional syndromes (e.g., loss of spouse, retirement, social isolation, severe illnesses or surgery, sensory impairment).

10.3.3.4 Anxiety Disorders

Anxiety is one of the most occurrent symptoms in the older population and usually links to previous anxiety disorders (e.g., chronic anxiety disorder) or other psychiatric disturbances (e.g., depression, dementing disorders). More rarely, disturbances such as generalized anxiety disorder, panic attack disorder, posttraumatic stress disorder, and phobias can manifest for the first time in older age. In older people, anxiety frequently manifests with physical symptoms related to autonomic hyperactivation. Among these, the most common are gastrointestinal (e.g., dry mouth, dysphagia, borborygmi, and abdominal spasms), respiratory (e.g., chest tightness, shortness of breath), cardiovascular (e.g., precordial malaise, tachycardia, palpitations), and neurological symptoms (e.g., irritability, memory and attentional deficits, restlessness, tremor, headache blurred vision, instability in walking and posture, insomnia).

10.3.3.5 Obsessive-Compulsive Disorders

Obsessive-compulsive disorder usually develops in late adolescents and young adults, but a late-onset in the elderly population rarely may occur. Usually, in this case, individuals have shown obsessive symptoms in the past, which have exacerbated in older ages (e.g., being orderly, perfectionistic, punctual as a young adult and becoming more rigid and inflexible to changes as an elder). These disturbances often relate to anxiety and depressive disorders.

10.3.3.6 Sleep Disorders

Progressively with aging, physiological sleep changes occur, which may lead to alterations in sleep's quantity and quality. These include the following.

- Shifting in sleep phase (earlier sleep time and earlier wake time).
- Increasing of daytime napping.
- Reduced total sleep time (approximately 6–7 h).
- Overall decrease in sleep efficiency and restoration.
- Changes of sleep architecture, including longer sleep latency (time required from going to bed until falling asleep), decreased REM sleep and more night-time arousals.

These physiological modifications may represent the substrate of various type of sleep disturbances which can develop in older age. In addition, other factors may cause some sleep alterations, including general medical conditions (e.g., pain, nycturia, dyspnea), social and environmental factors, other mental disorders, and primary sleep disorders. Among the last, the most frequent are dyssomnias, such as primary insomnia, restless legs syndrome, sleep apnea, and nocturnal myoclonus. Typical of older age are parasomnias, defined as sleep behavioral and movement disorders which manifest during rapid eye movement (REM) sleep.

Adequate sleep hygiene and nonpharmacological treatments are recommended as the first choice, including guaranteeing an ideal sleep setting, limiting diurnal napping, and treating comorbid conditions. In addition, more specific pharmacological medications may be applied. Lower dosages and shorter half-life molecules should be preferred and benzodiazepine should be avoided (see Box 10.15 for further details).

10.3.3.7 Suicide Risk

Older age represents the most important risk factor connected with suicidal risk, since elderly population (above all males over the age of 65) have a rate of suicide which is up to five times higher than in younger subjects. Predisposing factors include the following.

- Feelings of loneliness and social retirement, widowhood.
- Feeling of hopelessness.
- Coexistence of more organic diseases that leads to poor medical health conditions.
- Mental illnesses (e.g., depressive disorders).
- Low access to medical health services.
- Loss of independence in everyday life activities.
- Financial problems.
- Access to lethal means, such as pharmacological treatments.
- Familial history of suicide.

In contrast, some factors were revealed to act as protectors and resulted in decreasing suicidal risk in the older population. These involve the maintaining of social relations, easy access to health-care services, rehabilitation programs, psychological support, specific pharmacological treatments (if needed, e.g., in depressive disorders), and the rapid detection and management of underlying potential risk factors. A valid clinical tool that can be used in order to evaluate the suicidal risk rate is the Geriatric Suicide Ideation Scale, a 30-item multidimensional scale that was developed specifically for older adults.

10.3.3.8 Alcohol Use Disorders

Alcohol use disorders (AUD) are usually under-recognized among older adults. Alterations in social, work, or recreational activities secondary to alcohol use may go undetected, especially if the individual lives alone, is retired, and is socially isolated. While two-thirds of older people with AUD started drinking at a young age, experiencing the medical and psychosocial consequences of early-onset AUD at an older age, a minority of them develop an AUD in later life. A late-onset AUD may be triggered by stressful life events, such as mourning, divorce, and retirement, or by the onset of a psychiatric disorder, such as anxiety or depressive disorder. Other risk factors are female sex, social isolation, and a painful or disabling somatic disease.

In the elderly, physiological age-related changes may affect the absorption, metabolism, and elimination of ethanol, increasing its toxic potential. Therefore, older people have higher blood levels of alcohol and are more sensitive to its harmful effects. Falls are common in intoxication and withdrawal states which easily result in fractures since alcohol use increases the risk of osteomalacia due to its inhibitory effect on vitamin D hydroxylation.

In patients with preexisting neurocognitive disorders, alcohol can worsen cognitive impairment, which, in turn, may affect their ability to monitor their alcohol consumption.

Interactions between alcohol and drugs, prescription and over-the-counter, can also be particularly dangerous in older people. For example, concomitant use of alcohol and antiplatelet or anticoagulants can increase the risk of gastrointestinal bleeding. Furthermore, since approximately 15% of older people with AUD concomitantly use or abuse benzodiazepines/sedatives, the interaction of these substances with alcohol can lead to decreased alertness, speed of reaction to stimuli, and motor coordination, increasing the risk of fractures or delirium.

Finally, another significant complication in older people with AUD is delirium tremens, a potentially life-threatening condition caused by alcohol withdrawal. In the elderly patient suffering from delirium, alcohol consumption habits should be actively investigated (often the patient is reluctant to speak or unaware of the severity of the problem) since differential diagnosis between delirium and delirium tremens is fundamental to set an adequate treatment (see Chap. 9 for further details).

Box 10.15: Psychopharmacology in Older People

General principles: the prescription of psychotropic drugs in the elderly patient must take into account various problems specific to the patient's age:

- *Pharmacokinetics and Pharmacodynamics:* physiological or pathological changes in hepatic and renal metabolism, cardiovascular function, gastrointestinal absorption, and the ratio of fat to lean mass can significantly alter the kinetics of the prescribed drug and, with it, the effects of the drug itself, including side effects. A general medical evaluation and ECG examination are always recommended.
- *Compliance*: there are several reasons for poor adherence to medical prescriptions. Difficulty drinking or swallowing the drug, cognitive impairment, reduced frequency of outpatient checks, side effects are all factors to consider.
- *Polypharmacy and comorbidities*: the elderly psychiatric patient (> 65 years) is the most frequent recipient of all pharmacological prescriptions. About 25% of all prescriptions involve this age group. This significantly increases the possibility of drug interactions. In addition, a greater

number of medications per individual patient makes adherence to therapy more difficult. Detailed drug history is always required.

All the considerations made previously have led to the formulation of an extremely widespread maxim in geriatric medicine and also valid for psychogeriatric prescription: "Start low and go slow."

Specific Medications

Antidepressants: the optimal antidepressant drug for the elderly should meet ideal requirements of efficacy, safety, and tolerability, should be manageable and relatively safe even if taken in overdose. This antidepressant molecule does not exist. However, of all the antidepressant categories, SSRIs are the ones that come closest to the ideal drug. The pharmacokinetics of SSRIs usually allows a single daily administration (better compliance), and their pharmacodynamics reduces side effects (pay attention to QT length!) even in case of overdose. It is always good to start with the lowest possible dosage and slowly increase it to the minimum effective dose, to reduce the side effects and the receptor upregulation effect.

Anxiolytics: benzodiazepines, the most widely used anxiolytics, should not be considered an elective treatment for anxiety in this population. They are a short-term treatment and should generally be prescribed only for a limited period while waiting for the main drug (i.e., SSRI) to take effect. All benzodiazepines, particularly in the elderly, are associated with some degree of worsening of cognitive function, psychomotor slowing, unsteady gait, and increased risk of falls and fractures. Long half-life benzodiazepines have an increased risk of accumulation in the elderly due to the reduced efficacy of renal and hepatic clearance. However, benzodiazepines are the first-line treatment of alcohol withdrawal and delirium tremens (always inquire about alcohol use habits!).

Antipsychotics: as for other drug classes, before prescribing an antipsychotic drug, it is necessary to consider the possible age-related pharmacokinetic alterations, current medical conditions and drug interactions due to polytherapy. Generally, second-generation antipsychotics are to be preferred over others due to the lower frequency of side effects. However, they should be prescribed with caution in elderly patients affected with dementia, since their use is associated with an increased risk of cerebrovascular events and of all-cause mortality. The use of atypical antipsychotics is reasonable both in the management of acute symptoms and in the long-term treatment of a specific psychiatric diseases, such as later life schizophrenia and mood disorders. The recommendation, as always, is to start with small dosages, also in consideration of the increased brain sensitivity. Risperidone should be started at 0.5–1 mg, while Quetiapine and Olanzapine should be prescribed at an initial dosage of 25 mg and 5 mg respectively. The dosage should than be increased with caution for possible cardiac and metabolic side effects. Clozapine is generally to be avoided in the elderly due to its cognitive, hematological, and sedative effects.

Further Reading

American Psychiatric Association. Diagnostic and statistical manual of mental disorders. 5th ed. Arlington: American Psychiatric Association; 2013.

Ames D, Chiu E, Lindesay J, Shulman KI. Guide to the psychiatry of old age. Cambridge: Cambridge University Press; 2010.

Boland R, Verduin ML. Kaplan and Sadock's synopsis of psychiatry. 12th ed. Baltimore: Lippincott Williams & Wilkins; 2021.

Folsom DP, Lebowitz BD, Lindamer LA, Palmer BW, Patterson TL, Jeste DV. Schizophrenia in late life: emerging issues. Dialogues Clin Neurosci. 2006;8(1):45–52. https://doi.org/10.31887/DCNS.2006.8.1/dfolsom.

Hugo J, Ganguli M. Dementia and cognitive impairment: epidemiology, diagnosis, and treatment. Clin Geriatr Med. 2014;30(3):421–42. https://doi.org/10.1016/j.cger.2014.04.001.

Inouye SK, Westendorp RGJ, Saczynski JS. Delirium in elderly people. Lancet. 2014;383:911–22.

International Psychogeriatric Association. IPA complete guides to behavioral and psychological symptoms of dementia (BPSD). Milwaukee: International Psychogeriatric Association; 2015.

Kafetz K. The Oxford textbook of old age psychiatry. Oxford: Oxford University Press; 2008.

Kales HC, Gitlin LN, Lyketsos CG. Assessment and management of behavioral and psychological symptoms of dementia. BMJ. 2015;350:h369.

Kirakosyan G, Frolova A. Delirium (clinical presentation, diagnosis and treatment): updates for neurologists and psychiatrists. Int J Health Sci Res. 2020;10(9):212–20.

Klotz U. Pharmacokinetics and drug metabolism in the elderly. Drug Metab Rev. 2009;41:67–76.

Onyike CU. Psychiatric aspects of dementia. Continuum (Minneap Minn). 2016;22(2 Dementia):600–14. https://doi.org/10.1212/CON.0000000000000302.

Salzman C. Clinical manual of geriatric psychopharmacology. 2nd ed. Baltimore: Lippincott Williams & Wilkins; 2015.

Taylor DM, et al. The Maudsley prescribing guidelines in psychiatry. 13th ed. New York: Wiley; 2018.

Todd OM, Teale EA. Delirium: a guide for the general physician. Clin Med (Lond). 2017;17:48–53.

Verdelho A, Gonçalves-Pereira M. Neuropsychiatric symptoms of cognitive impairment and dementia. Cham: Springer International Publishing; 2017.

Videbeck SL. Psychiatric-mental health nursing. Baltimore: Lippincott Williams & Wilkins; 2010.

Woods B, Spector A, Jones C, et al. Reminiscence therapy for dementia. Cochrane Database Syst Rev. 2005;2:CD001120.

Psychopharmacology

11

R. Zanardi, M. Spangaro, F. Attanasio, J. Sapienza,
F. Martini, L. Fregna, R. Cavallaro, and C. Colombo

11.1 Antidepressants

11.1.1 Introduction

Antidepressants, medicines originally used to treat depression, are nowadays FDA approved as a treatment in adults for a great variety of disorders such as depression, major depressive disorder, as adjunct therapies for bipolar I or II disorder, obsessive-compulsive disorder, bulimia nervosa, panic disorder, social anxiety disorder, generalized anxiety disorder, post-traumatic stress disorder, premenstrual dysphoric disorder, fibromyalgia, chronic musculoskeletal pain, seasonal affective disorder, smoking cessation, insomnia.

Antidepressants are mostly prescribed for psychiatric diseases, but many non-psychiatric conditions benefit from the use of these agents: for example, insomnia, chronic pain, smoking cessation, Parkinson's disease, and vasomotor symptoms of menopause.

Antidepressants can be divided in classes according to their pharmacological profile. We will focus on the most widely used drugs in clinical practice, providing an overview over less used molecules, that may be a starting point for further study.

R. Zanardi · F. Martini (✉) · L. Fregna (✉)
IRCCS San Raffaele Scientific Institute, Milan, Italy
e-mail: zanardi.raffaella@hsr.it; francesca.martini1@hsr.it; fregna.lorenzo@hsr.it

M. Spangaro · R. Cavallaro · C. Colombo
IRCCS San Raffaele Scientific Institute, Milan, Italy

Vita-Salute San Raffaele University, Milan, Italy
e-mail: spangaro.marco@hsr.it; cavallaro.roberto@hsr.it; colombo.cristina@hsr.it

F. Attanasio · J. Sapienza
Vita-Salute San Raffaele University, Milan, Italy
e-mail: attanasio.francesco@hsr.it; sapienza.jacopo@hsr.it

R. Cavallaro, C. Colombo (eds.), *Fundamentals of Psychiatry for Health Care Professionals*, https://doi.org/10.1007/978-3-031-07715-9_11

- Selective serotonin reuptake inhibitors (SSRIs).
- Serotonin–norepinephrine reuptake inhibitors (SNRIs).
- Tricyclics and tetracyclics (TCAs).
- Monoamine oxidase inhibitors (MAOIs).
- Other antidepressants (agents that do not fall into any of these classes and have peculiar mechanisms of action).

11.1.2 Selective Serotonin Reuptake Inhibitors (SSRIs)

Selective serotonin reuptake inhibitors (SSRIs) are nowadays the most widely prescribed antidepressants in clinical practice due to their effectiveness and few side effects compared to other classes of antidepressants.

The first SSRI fluoxetine (Prozac) was introduced in 1988, and, nowadays, other five molecules are marketed: fluvoxamine, paroxetine, sertraline, citalopram, and the most recent escitalopram.

SSRIs have received approval from the FDA as safe and effective in the treatment of a wide range of disorders: major depressive disorder, panic disorder, generalized anxiety disorder, social anxiety disorder, obsessive-compulsive disorder (OCD), post-traumatic stress disorder (PTSD), premenstrual dysphoric disorder (PMDD), and bulimia.

11.1.2.1 Pharmacokinetics

An important pharmacokinetics variability can be observed among SSRIs, concerning, in particular, serum half-lives and plasma protein-binding percentage.

SSRIs are well absorbed through the small intestine after oral administration, and they reach the peak concentration in 1–8 h.

They are metabolized by the hepatic cytochrome P450 (fluoxetine and paroxetine mainly by CYP2D6; fluvoxamine, fluoxetine, and sertraline by CYP3A4, fluvoxamine by CYP1A2) and undergo the first-pass metabolism. Paroxetine and citalopram are also partly metabolized by the kidneys.

The half-life of SSRIs is highly variable between molecules: from a few hours of fluvoxamine to several days (through active metabolites) of fluoxetine (see Table 11.1).

> Pay attention that CYP2D6 is also used to metabolize *dicoumarolic compounds*, so you should never give Fluoxetine/Paroxetine to a patient taking Coumadin, or you may increase the risk of hemorrhages. Sertraline or citalopram instead should be given to these patients.

Table 11.1 Pharmacokinetics properties of SSRIs

Drug	Time to peak plasma concentration (h)	Half-life	Half-life metabolite	Time to steady state (days)	Plasma protein binding (%)
Citalopram	4	35 h	3 h	7	80
Escitalopram	5	27–32 h		7	56
Fluoxetine	6–8	4–6 days	4–16 days	28–35	95
Fluvoxamine	3–8	15 h		5–7	80
Paroxetine	5–6	21 h		5–10	95
Sertraline	4.5–8.5	26 h	62–104 h	7	95

The administration of SSRIs with food does not change their absorption and could minimize the incidence of gastrointestinal adverse effects.

SSRIs are widely distributed throughout the body and, even though they can cross the placenta, most of them is considered safe during pregnancy.

11.1.2.2 Pharmacodynamics

SSRIs, by the inhibition of the serotonin transporter SERT located on the presynaptic cell, block the reuptake of serotonin (5-HT1A, 5-HT2C, 5-HT3C), increasing the level of neurotransmitter available in the synaptic cleft. They are called selective because they exert their function on serotonin with very little effect on other neurotransmitters: dopamine and norepinephrine.

Every SSRI has a similar pharmacologic profile, but slight differences explain their distinct clinical use, side effects, and drug interactions.

Among SSRIs is exhibited very different selectivity for the serotonin transporter, but the clinical efficacy appears to be not directly proportional to the selectivity.

Genetics of SSRIs

The serotonin transporter (SERT) is encoded on chromosome 17q11 which has two possible polymorphisms in the promoter region of the gene leading to two allelic variants (long "l" and short "s"). The long allele is associated with a threefold augmented transcription of the SERT gene compared to the short variant. Consequently, subjects with l/l genotype have a better response to SSRIs than subjects with l/s genotype, which has a better response than s/s genotype.

11.1.2.3 Therapeutic Indications and Clinical Use

SSRIs are the first-line treatment for several mental health disorders due to their effectiveness, few side effects, ease of dosing, and low toxicity in overdose.

They are greatly preferred over the other classes of antidepressants in children or adolescents, and they are the first choice for late-onset depression, due to their superior tolerability, minimal anticholinergic effects, and comparatively more benign safety profile.

Starting, maintenance, and high doses of SSRIs are detailed in Table 11.2.

Table 11.2 Dosage of SSRIs

SSRI	Starting (mg)	Maintenance (mg)	High dosage (mg)
Paroxetine	10	20–60	>60
Fluoxetine	10	20–60	>80
Sertraline	25	50–200	>300
Citalopram	10	20–40	>60
Escitalopram	5	10–30	>30
Fluvoxamine	50	50–100	>300

Clinical indications

- Major depressive disorder.
- Obsessive-compulsive disorder (first-line treatment: fluvoxamine).
- Generalized anxiety disorder.
- Panic disorder (first-line treatment: paroxetine).
- Social anxiety disorder.
- Premenstrual dysphoric disorder.
- Post-traumatic stress disorder (first line: paroxetine).
- Bulimia (first line: fluoxetine).
- Premature ejaculation (paroxetine and sertraline).

Other off-label uses include but are not limited to binge eating disorder, body dysmorphic disorder, fibromyalgia, premature ejaculation, paraphilias, autism, Raynaud phenomenon, and vasomotor symptoms associated with menopause.

Fluoxetine

A starting dose of 10 mg per day is usually preferred to avoid early side effects, especially in patients with concomitant anxiety or panic disorder.

Dosages between 20 to 60 mg per day are commonly required as treatment of depression. Higher dosages of 60 to 80 mg per day may be necessary for the treatment of OCD.

Fluoxetine is preferably administered in the morning and should be taken with food to avoid, respectively, the risk of insomnia or gastrointestinal side effects.

Fluoxetine's most common early side effects are anxiety, agitation, and insomnia; those effects are usually temporary and may improve with a dose reduction or combination with benzodiazepines or beta-adrenergic blockers.

Sertraline

It is recommended a starting dose of 25 mg per day in order to avoid early side effects (anxiety, nervousness, and restlessness), especially in patients with Anxiety or Panic Disorder.

The effective dosage is from 50 to 200 mg a day.

It is useful to administer the drug during a meal to avoid gastrointestinal side effects.

The most common side effects of sertraline are nausea, xerostomia, decreased libido, and fatigue.

Paroxetine

The treatment may start at a dosage of 10 mg per day, particularly for elderly patients or patients with Anxiety disorders. The dosage for the treatment is from 20 to 60 mg a day.

Paroxetine is often chosen as the first-line treatment for depression, anxiety disorders, panic disorder, and post-traumatic stress disorder.

It is usually administrated as a single dose in the evening but can be administered twice a day.

Paroxetine is the molecule with the highest risk of discontinuation syndrome because its plasma concentration rapidly decreases after suspension. As a consequence, in case of discontinuation, the dosage should be gradually decreased.

Paroxetine appears to be often more sedating and constipating than other SSRIs because of its anticholinergic activity.

Fluvoxamine

Fluvoxamine, although also effective as an antidepressant, is FDA approved only as a treatment for OCD.

The range of effectiveness for the treatment of depression in adults is 50–200 mg per day at bedtime (higher dosages can be divided into two daily administrations). Higher dosages (200–300 mg daily) may be necessary for the treatment of OCD.

Fluvoxamine's short half-life may cause a discontinuation syndrome. In case of discontinuation, the dosage should be gradually decreased.

The most common side effects of fluvoxamine are nausea, headache, and diarrhea.

Citalopram

Citalopram is a racemate consisting of a 1:1 mixture of the R(−)- and S(+)-enantiomers; serotonin reuptake inhibitory activity of citalopram is attributable to the S-enantiomer (escitalopram).

The therapeutic range of citalopram is from 20 to a maximum of 40 mg daily. A maximum dosage of 20 mg per day is suggested for the elderly and patients with hepatic impairment. It can be taken once a day, in the morning or evening, with or without food.

Citalopram is associated with the lowest rate of activating side effects such as anxiety, agitation, and insomnia. The most common side effects are nausea (often transient) and diarrhea. It also shows a low potential for drug interactions.

Escitalopram

Escitalopram is the most selective of the SSRIs. It is the active S-(+)-enantiomer of citalopram.

The dosage recommended in clinical practice is 10 mg a day, and no additional benefits were observed with greater dosages.

The most common side effects are similar to citalopram: nausea and diarrhea. It shows a low potential for drug interactions.

11.1.2.4 Side Effects

Pharmacotherapy side effects can strongly affect patients' quality of life and therefore their compliance to the treatment.

SSRIs demonstrate superiority in terms of safety when compared to older antidepressant medications. Compared to TCAs, SSRIs do not show cardiac toxicity and orthostatic hypotension, since they are not antagonists of α-adrenergic receptors. Moreover, they have low anticholinergic activity, (except for paroxetine) and do not show typical anticholinergic side effects (dry mouth, blurry vision, constipation, drowsiness, sedation).

Nevertheless, numerous side effects can be observed in patients taking SSRIs. The main side effects of SSRIs are listed below.

- *Gastrointestinal and weight-related side effects.* They are the main SSRIs side effects. They include nausea, dyspepsia, diarrhea or constipation, loss or gain of appetite. Usually, nausea and diarrhea are transient and resolve in few weeks. In up to 25% of people, SSRIs can cause a weight gain of 10 pounds or more, with paroxetine being the most associated with weight gain. Fluoxetine usually induces anorexia and loss of weight, that decrease after a peak at about 20 weeks of treatment. To avoid dyspepsia, nausea, or vomiting, it could be useful to take SSRIs with food. SSRIs are contraindicated in patients affected by peptic ulcers and chronic gastritis.
- *Sexual dysfunction.* Inhibited orgasm or decreased libido, unlike many other SSRIs side effects, are not transient, and lasts as long as the drug is taken. Sexual dysfunction is a dose-dependent side effect, so the first-line strategy to improve SSRI-induced sexual dysfunction is decreasing the dosage of the SSRI; other options are switching or combining to another not-SSRI antidepressant.
- *Headache.* Though SSRIs are used as effective treatment options and prophylaxis for tension-related headache and migraine, SSRI-related headache has been reported in the first weeks of treatment in up to one-fourth of patients, especially with fluoxetine. It usually reduces spontaneously.
- *Anxiety, agitation, nervousness, and restlessness.* They are some of the most frequent early-side effects, especially related to fluoxetine. They usually appear at the beginning of the treatment and reduce spontaneously after a few weeks. To avoid these side effects, it may be useful to gradually increase the dose of the SSRI or to associate an anxiolytic therapy in the first weeks of treatment.
- *Insomnia and sedation.* The treatment with antidepressants is usually related to sleep improvement due to the reduction of depressive or anxiety-related symptoms. However, up to 25% of people on treatment with SSRIs show sleep disturbances, such as insomnia or sleepiness. Fluoxetine is usually related to insomnia; sertraline can cause both insomnia and drowsiness, while citalopram, paroxetine, and fluvoxamine usually induce somnolence. Molecules causing insomnia are

preferably administered in the morning, while the ones causing somnolence should be taken at bedtime. SSRI-related insomnia can be treated with hypnotics or switching to another SSRI.

- *Extrapyramidal symptoms.* About one patient on ten shows tremor during an SSRI treatment. Other extrapyramidal symptoms are extremely rare. Parkinson's disease symptoms may get worst after taking SSRIs.
- *Bleeding.* Due to serotonin depletion in platelets, SSRIs are rarely related to *easy bruising*.
- *Rash.* In up to 5% of patients taking an SSRI, a skin rash may appear. In case of allergic reactions, the SSRI has to be suspended.
- *Vivid dreams and nightmares.* Some of patients on treatment with SSRIs report vivid dreams and nightmares. This problem is usually solved switching to another molecule of the same class.

Mood Switch

Antidepressant monotherapy is related with an increased risk of mania in patients with bipolar disorder. The risk of a mood switch is far reduced by combining a mood stabilizer and choosing low dosages of antidepressant as maintenance therapies.

Suicide Risk

It is important to identify patients with suicidal risk and to monitor them during the first weeks of SSRI therapy, because the suicidal risk is higher at the early stages of relieves from depression.

Serotoninergic Syndrome

In extremely rare cases SSRIs can induce a serotoninergic syndrome, usually when combined with other antidepressants (especially MAOIs) that inhibit the reuptake of serotonin.

If plasma levels of serotonin rise to a toxic level, an excess of serotonin activity in the central nervous system and peripheral serotonin receptors may produce specific symptoms: headache, agitation, hypomania, mental confusion, hallucinations, sweating, hyperthermia, hypertension, tachycardia, nausea, diarrhea, myoclonus, hyperreflexia, tremor, and coma.

Mild cases usually resolve in a few hours by discontinuing medications causing an increase in serotonin plasma concentrations. In moderate to severe cases, it may be necessary the administration of a serotonin antagonist (e.g., cyproheptadine) and gastrointestinal decontamination with activated charcoal. Benzodiazepines can be useful to control agitation.

Discontinuation Syndrome

The abrupt suspension of drugs determining serotonin reuptake inhibition may lead to a discontinuation syndrome. It is characterized by asthenia, anxiety, agitation, gastrointestinal distress, myalgias, flu-like syndrome (without fever).

Longer durations of treatment and the use of molecules with a short half-life (paroxetine and sertraline) are associated with a higher risk of the above syndrome.

This problem can be avoided by gradually reducing the dosages of SSRI before stopping it.

11.1.2.5 Drug Interactions

SSRIs are metabolized in the liver by CYP450, and their administration can slow or block the metabolism of other drugs.

Fluoxetine and paroxetine reduce the analgesic effect of opioids such as codeine or hydrocodone (CYP2D6). Fluoxetine increases the plasmatic level of carbamazepine by slowing its metabolism. Fluoxetine, sertraline, and paroxetine increase the plasma level of tricyclic antidepressants with possible consequent toxicity. Fluvoxamine interacts with the metabolism of drugs such as theophylline, clozapine (CYP1A2), alprazolam, and clonazepam (CYP3A4).

SSRIs interact with other antidepressants because of their common mechanism of action, increasing the risk of serotoninergic syndrome.

All SSRI displace warfarin from plasma proteins causing higher hemorrhage risk in patients treated with warfarin. Sertraline and paroxetine have the highest potential risk.

Exposure to SSRIs may reduce dopamine turnover, leading to a rise in prolactin levels; they should not be administered with dopamine antagonists.

Citalopram, escitalopram, and sertraline are considered to have fewer pharmacological interactions than the other molecules.

11.1.3 Serotonin–Norepinephrine Reuptake Inhibitors (SNRI)

The serotonin–norepinephrine reuptake inhibitors (SNRIs) are a class of drugs blocking neuronal uptake transporters for both serotonin and norepinephrine.

The most commonly used SNRIs are venlafaxine and duloxetine (the only two available in Italy and examined in this chapter).

11.1.3.1 Pharmacokinetics

Venlafaxine is well absorbed by the gastrointestinal tract and reaches peak plasma concentration in less than 3 h. The degree of binding of venlafaxine to human plasma is only 25–30%. It has an extensive first-pass metabolism through the liver and, then, it is primarily eliminated by the kidneys; clearance is therefore reduced among patients with cirrhosis and severe renal disease. The half-lives of venlafaxine and its active metabolite (O-desmethylvenlafaxine) are short: respectively 4 and 10 h.

Duloxetine is well absorbed from the gastrointestinal tract and reaches peak plasma concentration within 6 h. The degree of binding of duloxetine to human plasma is about 90%. It is primarily eliminated by the kidneys after hepatic oxidation. It has an elimination half-life of 12 h, and the steady-state is reached within 3 days of oral dosing.

11.1.3.2 Pharmacodynamics

SNRIs are monoamine reuptake inhibitors; specifically, they inhibit the reuptake of serotonin and norepinephrine.

Venlafaxine at the lowest therapeutic doses inhibits the reuptake of serotonin, acting as an SSRI; its noradrenergic effects enhance progressively as the dose is increased.

Duloxetine is a more potent norepinephrine reuptake inhibitor than venlafaxine and acts concomitantly on serotonin and norepinephrine receptors at any concentration.

What distinguishes SNRIs from TCAs is selectivity. SNRIs have a relative lack of affinity for other receptors (muscarinic, histaminergic, α- and β-adrenergic receptors), presenting a more favorable tolerability profile than TCAs.

11.1.3.3 Therapeutic Indications and Clinical Use

Venlafaxine

Venlafaxine in 1993 was marketed in an immediate-release form that has to be taken 2–3 times per day. In 1997, it was introduced an extended-release form that allows a once-daily administration. The therapeutic dose is between 75 mg and 375 mg per day; the dosage range is the same for GAD as for the treatment of depression.

> The half-lives of *venlafaxine* and its active metabolite are short (4–10 h), to be kept in mind for the *discontinuation syndrome*. If discontinued, venlafaxine should be gradually reduced in 2–4 weeks.

FDA Indications: major depressive disorder, generalized anxiety disorder, social anxiety disorder, and panic disorder.

Duloxetine

The therapeutic dose of duloxetine is between 60 mg and 120 mg per day.

FDA indications: major depressive disorder, generalized anxiety disorder, diabetic peripheral neuropathy, fibromyalgia, musculoskeletal pain, osteoarthritis.

11.1.3.4 Side Effects

SNRIs may have a variety of side effects related to their mechanism of action: some due to their serotonergic activity, others due to the noradrenergic one.

Nausea, dizziness, insomnia or somnolence, headache, hypertension, sexual dysfunction, asthenia, sweating, constipation, and dry mouth are the most common side effects.

Abrupt discontinuation may produce a discontinuation syndrome. To avoid it, when discontinued, venlafaxine should be gradually reduced in 2–4 weeks.

Side effects usually appear in the first weeks of treatment. A gradual increase in dosage or the use of extended-release forms may reduce the incidence of those side effects.

11.1.3.5 Drug Interactions

No clinically significant drug interactions have been documented but SNRIs should be prescribed carefully in association with other drugs metabolized by CYP450 and with any drug that increases serotonin concentrations (especially MAOIs) to avoid the risk of serotonin syndrome (the effect of SSRIs and SNRIs on CYP-450 isoenzymes is illustrated in Table 11.3).

11.1.4 Tricyclics and Tetracyclics (TCAs)

The name tricyclic and tetracyclic derives from the number of rings present in their chemical structure, three-ring (*tri*) or four-ring (*tetra*).

Table 11.3 Inhibitory effect of SSRIs and SNRIs on CYP-450 isoenzymes

	1A/2	*2C9*	*2C19*	*2D6*	*3A4*
SSRI					
Paroxetine	+	+	+	+++	+
Fluoxetine	+	++	+/++	+++	+/++
Sertraline	–	+	–	+/++	+
Citalopram	–	–	–	+	–
Escitalopram	–	–	–	–/+	–
Fluvoxamine	+++	++	+++	+	++
SNRI					
Venlafaxine	–	–	–	+	+
Duloxetine	+	–	–	+	–

– no clinically significant inhibition, + potentially clinically significant inhibition, but only at high doses, ++ moderate clinically significant inhibition, +++ potent and highly clinically significant inhibition ad all doses

Clomipramine, Imipramine, and Amitriptyline are the only TCAs available in Italy.

TCAs are classified according to their receptor selectivity

- The highest inhibition of serotonin (5-HT) reuptake: clomipramine, imipramine.
- The highest inhibition of norepinephrine (NE) reuptake: desipramine, nortriptyline.
- Equal inhibition of both 5-HT and NE reuptake: amitriptyline, amoxapine, doxepin.

11.1.4.1 Pharmacokinetics

TCAs are rapidly and completely absorbed in the small intestine within 2–8 h. Because of their lipid solubility, these compounds concentrate in different tissues having a high volume of distribution. Since they cross the placenta, they are contraindicated during pregnancy. TCAs are extensively bound to plasma proteins (90%). The concomitant administration of acetylsalicylic acid may reduce the plasmatic protein binding of TCAs, increasing their free levels.

They are metabolized in the liver by the cytochrome P450 system (CYP2D6), with an extensive first-pass metabolism through two main metabolic pathways.

- *Demethylation* converts the tertiary amine to a secondary amine. While demethylated amines cause a stronger inhibition of NE reuptake, tertiary amines exert their effect relatively more on the 5-HT reuptake.
- *Hydroxylation* of the ring structure produces active hydroxyl-metabolites.

TCAs half-lives are approximately 24 h or longer, and, hence, the drugs can be given once a day.

11.1.4.2 Pharmacodynamics

Cyclic antidepressants act on five different neurotransmitter systems: serotonin reuptake inhibition, noradrenaline reuptake inhibition, α1 adrenergic receptor inhibition, H1 histaminergic receptor blockade, and muscarinic cholinergic receptor blockade. Each molecule differs in its affinity for each of these transporters. Therefore, secondary effects vary considerably among TCAs.

11.1.4.3 Therapeutic Indications and Clinical Use

Clinical Indications:

- Depression, second-line treatment.
- Obsessive-Compulsive Disorder (OCD), second-line treatment (clomipramine).
- Panic Disorder (imipramine).
- Generalized anxiety disorder (doxepin).
- Chronic pain and Migraine prophylaxis (amitriptyline).

Each clinical indication for TCAs is also an indication for SSRIs, which are usually preferred to TCAs in clinical practice due to their better tolerability profile. However, TCAs are a valid alternative as second-line treatment in nonresponder patients or patients who cannot tolerate SSRIs' adverse effects.

Overdosage may lead to symptoms like agitation, delirium, seizures, tendon hyperreflexia, bladder and rectal paralysis, blood pressure, temperature alterations, mydriasis, and changes in the level of consciousness until coma.

11.1.4.4 Side Effects

TCAs may have a variety of side effects related to their multiple mechanisms of action: α1 adrenergic receptor blockade may cause orthostatic hypotension and dizziness; muscarinic receptor blockade and the resulting anticholinergic action may cause dry mouth, blurred vision, urinary retention, constipation, and memory disorders; H1 histaminic receptor blockade may cause sedation and weight gain.

In case of TCA overdose, blockade of sodium channels in the heart and brain may cause cardiac arrhythmias and seizures.

- *Cardiovascular:* orthostatic hypotension, tachycardia, arrhythmia, and conduction delay. Check QTc during dosage increase.
- *Central Nervous System*: seizures—fine rapid tremor, confusion, or delirium.
- *Autonomic Nervous System*: dry mouth, constipation, blurred vision, urinary hesitancy, ocular crises in patients with narrow-angle glaucoma.
- *Other:* Increased appetite and weight gain, increases in liver enzymes, sexual dysfunction, allergic skin rash, blood dyscrasias (very rare).

> The use of TCAs is controindicated in patients affected by known cardiopathy (in particular, arrhytmias and conduction disorders), benign prostatic hyperplasia, glaucoma (narrow-angle).

11.1.4.5 Drug Interactions

TCAs are metabolized by the liver and may have significant interactions with drugs metabolized by CYP2D6, such as fluoxetine, sertraline, and paroxetine (SSRIs), phenothiazines (antipsychotics), carbamazepine (anticonvulsant), propafenone, flecainide, and quinidine (antiarrhythmics).

11.1.5 Monoamine Oxidase Inhibitors (MAOIs)

Monoamine oxidase inhibitors (MAOIs) were the first class of drugs to be approved for the treatment of depression. The antidepressant effect of the first molecule of this class was discovered by chance: *isoniazid* was originally developed as a treatment for tuberculosis.

Table 11.4 Classification of MAOIs

Nonselective		Selective IMAO-A	Selective IMAO-B
Hydrazine	Nonhydrazine		
• *Isocarboxazide* • *Isoniazide* • *Nialamide* • *Phenelzine*	• *Tranylcypromine*	• *Moclobemide* • *Pirlindole* • *Toloxatone*	• *Rasagiline* • *Selegiline*

MAOIs are effective antidepressants, but they are not commonly used in clinical practice because of the related risk of severe hypertension and the need for dietary control. MAOIs are now rarely used as a treatment for nonresponder patients.

MAOIs are a class of drugs inhibiting the activity of one or both monoamine oxidase enzymes: monoamine oxidase A (MAO-A) and monoamine oxidase B (MAO-B). They include a variety of molecules classified according to their level of selectivity for the type A or B isoenzyme (nonselective/selective for MAO-A/selective for MAO-B) and their chemical structure (hydrazine/nonhydrazine, see Table 11.4).

11.1.5.1 Pharmacodynamics

The monoamine oxidase is an enzyme on the outer membranes of mitochondria present in two different forms: MAO-A is responsible for the deamination of serotonin and catecholamine (norepinephrine, dopamine, melatonin), while MAO B is responsible for the deamination of phenylethylamine and dopamine. MAOIs act by inhibiting the activity of monoamine oxidase thus preventing the breakdown of monoamine neurotransmitters and increasing their availability.

Monoamine oxidase is located in the brain, liver, intestine, and endothelial cells. MAO-A is mostly located in the placenta and cholinergic neurons, while MAO-B is mostly located in platelets and serotoninergic neurons.

When ingested orally, MAOIs inhibit the catabolism of dietary tyramine, normally metabolized by the hepatic MAOs (primarily MAO-A). High levels of tyramine compete with tyrosine for transportation across the blood–brain barrier entering terminals of the adrenergic nerves and causing a hypertensive crisis through the release of norepinephrine.

> People taking MAOIs generally need to avoid foods and beverages containing tyramine: for example, red wine, beer, cheese, soy sauce, and salami.

11.1.5.2 Therapeutic Indications and Clinical Use

MAOIs are used to treat atypical depression, anxiety, panic and phobias, bulimia, and PTSD.

Because of potentially lethal dietary and drug interactions, IMAOs are nowadays reserved to treatment-resistant conditions as a last-line treatment.

11.1.5.3 Side Effects

They show many side effects related to their mechanism of action. The most common side effects are cardiovascular effects like dizziness, orthostatic hypotension, and peripheral edema; central nervous system effects as sleep disturbance, sedation, mood switching; general side effects as weight gain and sexual dysfunction. Rarer and more dangerous side effects are hypertensive crisis and serotonin syndrome.

11.1.5.4 Drug Interactions

MAOIs may have interactions with every drug acting on serotonin, norepinephrine, and dopamine: reuptake inhibitors (e. g. SSRIs, SNRIs, TCAs), releasers (e.g., amphetamine, ephedrine), and precursors (e.g., L-dopa, phenylalanine, tryptophan, tyrosine). They also may interfere with drugs metabolized by monoamine oxidase (e.g., phenylephrine).

> When switching from an irreversible MAOI to a different antidepressant, a *minimum washout of 2* weeks is required to allow complete recovery of MAO activity.

11.1.6 Other Antidepressants

11.1.6.1 Vortioxetine

Vortioxetine is a novel antidepressant with multiple pharmacological activities, approved by the Food and Drug Administration (FDA) in 2013 for the treatment of major depressive disorder in adults. In Italy, vortioxetine is available from May 2016.

Vortioxetine has a complex mechanism of action that includes the inhibition of serotonin reuptake through the inhibition of SERT. Moreover, it exerts a direct action at multiple 5-HT receptor subtypes: it acts as an agonist at 5-HT1A receptors, as a partial agonist at 5-HT1B receptors, and as an antagonist at 5-HT3, 5-HT1D, and 5-HT7 receptors.

Some studies have shown a positive effect on cognitive functions in elderly depression.

Nausea, constipation, and vomiting are the most common, dose-related side effects. No significant effect on body weight has been reported.

11.1.6.2 Bupropion

Bupropion is a mild dopamine and norepinephrine reuptake inhibitor and acts as an antagonist at the nicotinic acetylcholine receptors.

Bupropion is indicated for the treatment of depressive disorders (often used as augmentation therapy of other antidepressants) and smoking cessation.

It is rapidly absorbed by the gastrointestinal tract and then metabolized in the liver by CYP450. It has a half-life of 20 h and an 80–90% of protein binding percentage.

Bupropion side effects are nausea, dry mouth, excessive sweating, tinnitus, rash, insomnia, anxiety, agitation, tremor, seizure. Sexual impairment has not been observed during treatment with bupropion.

11.1.6.3 Mirtazapine

Mirtazapine is an antidepressant with a substantial anxiolytic and sedative effect.

Clinical indications of mirtazapine include major depressive disorders (often use as augmentation of other antidepressants), insomnia, and anxiety.

Mirtazapine increases the release of serotonin and norepinephrine by exerting antagonist effects on the central presynaptic alpha-2-adrenergic receptors. It also acts as a strong antagonist of histamine H1 receptors, with marked sedative and appetite-enhancing effects.

Mirtazapine is rapidly absorbed by the gastrointestinal tract, with a hepatic metabolism by CYP450. It has a half-life of 30 h and a steady state reached after 6 days of therapy. The dose range is between 15 and 60 mg per day.

The main side effects are sedation (somnolence occurs in more than 50% of patients), dry mouth, increased appetite, and weight gain.

11.1.6.4 Trazodone

Trazodone is an antidepressant characterized by high sedative qualities.

It was approved by FDA in 1981 for the treatment of major depressive disorders.

Clinical indications of trazodone include depressive disorders, insomnia, and anxiety.

The strong sedative properties, even at a lower dosage, limit its clinical use. On the other hand, the sedative proprieties of trazodone make it beneficial in patients with insomnia.

Trazodone potently antagonizes serotonin 5-HT2A and 5-HT2C receptors, while it just weakly inhibits serotonin reuptake. It has no anticholinergic effects.

Trazodone is rapidly absorbed by the gastrointestinal tract, is metabolized in the liver by CYP450, and is eliminated by kidneys. It has a half-life of 3 h, and the steady-state is reached after 36 h, with a 90% protein binding percentage. The dosage varies between 50 and 300 mg per day.

The side effects are related to its antagonism at peripheral α-adrenergic receptors: orthostatic hypotension, dizziness, dry mouth, and priapism.

11.2 Anxiolytics and Sedatives

11.2.1 Introduction

Anxiolytics are a class of drugs used to treat patients with panic disorder, generalized anxiety, and various other conditions. Sedatives (hypnotics) are a class of medications used in a variety of situations, from the treatment of insomnia to the management of epilepsy. Both classes of medications have a wide range of uses and are effective when used in the correct dosage and under the guidance of trained medical professionals. However, their use is burdened by the risk of abuse, which can lead to adverse and potentially fatal consequences.

In this subchapter, we will discuss barbiturates, benzodiazepines, "z-drugs" and antihistamines.

11.2.2 Barbiturates

Barbiturates were introduced at the beginning of the twentieth century as sedative drugs. Due to their narrow therapeutic window and high susceptibility to addiction, they have been gradually replaced by other sedative agents such as benzodiazepines, Z-drugs, and antihistamines. Nowadays, barbiturates are mostly used as anesthetics or anticonvulsants.

11.2.2.1 Pharmacokinetics

When administered orally, barbiturates are rapidly and completely absorbed by the small bowel, with a latency of action of 10 min to an hour. Barbiturates are metabolized in the liver by CYP450, have different plasma protein binding percentages, and cross the placenta.

11.2.2.2 Pharmacodynamics

Barbiturates exert their action by binding the γ-aminobutyric acid (GABA) receptor and increasing its duration of opening. On the contrary, benzodiazepines only increase the GABA receptor's frequency of opening, resulting in better safety and tolerability. See the paragraph about benzodiazepines for a focus on GABA receptors.

11.2.2.3 Therapeutic Indications and Clinical Use

Nowadays, barbiturates are only used as anesthetic agents (e.g., for electroconvulsive therapy) and anticonvulsants (in particular phenobarbital).

11.2.2.4 Side Effects

Barbiturates have a strong depressive effect on all the excitatory cells. Moreover, they have a very narrow therapeutic range, and they are associated with a high risk of dangerous overdoses. The symptoms of an acute overdose include motor and speech impairment, respiratory depression with tachycardia, hypotension, hypothermia, oliguria. They can also cause a cardiorespiratory collapse because of a depression in brainstem activity.

11.2.3 Benzodiazepines

Benzodiazepines are a class of drugs widely used in clinical practice worldwide for their anxiolytic, hypnotic, sedative, myorelaxant and anticonvulsant effects. Since their introduction in the early 1960s, they have largely replaced older drugs (such as barbiturates) used for the treatment of anxiety and insomnia, thanks to their highly safe profile and tolerability.

11.2.3.1 Pharmacokinetics

Benzodiazepines are completely absorbed after oral administration and reach peak serum levels in 30 min to 2 h. Intramuscular absorption is slower than the oral one, while the onset of action is very rapid with intravenous administration. Benzodiazepines are lipid-soluble with a binding to plasma proteins from 70 to 99%. As such, they are distributed widely in adipose tissue. They undergo hepatic metabolism, and most of them are oxidized first by cytochrome P450; these metabolites may then be hydroxylated to another active metabolite. Some benzodiazepines (e.g., lorazepam) are conjugated directly by glucuronidation, and, therefore, have a faster metabolism.

The combination of features as lipid solubility, potency, and elimination half-life predicts the onset and duration of action as well as the appropriate frequency of drug administration. Benzodiazepines with high lipophilicity and potency and a short half-life have rapid onset but a brief duration of action, thus causing a stronger withdrawal effect and stronger cravings, whereas agents with a longer half-life have a much more gradual decrease in plasma levels with a lower risk of withdrawal symptoms such as anxiety, excessive arousal, and even seizures (see Tables 11.5 and 11.6).

Pharmacokinetic properties of benzodiazepines substantially affect their clinical use: for example, the rapid onset of action could be very important in the management of a panic attack; the half-life and duration of action are important characteristics in the treatment of insomnia, shorter half-lives are better to treat initial or middle insomnia, while longer half-lives are usually chosen for early morning awakening insomnia. Half-life is also one of the main determinants of the addictiveness of each agent: a short half-life is usually linked to a higher risk of dependence.

Table 11.5 Classification of benzodiazepines by elimination rates

Classification of benzodiazepines by elimination rates		
Rapid ($t½$ < 6 h)	Intermediate ($t½$ 6–20 h)	Slow ($t½$ > 20 h)
Midazolam	Alprazolam	Clonazepam
Triazolam	Lorazepam	Diazepam
Brotizolam	Temazepam	Chlordiazepoxide
Medazepam	Bromazepam	Alazepam
	Estazolam	Clobazam
	Etizolam	Clorazepam
	Flunitrazepam	Clorazepate
	Lormetazepam	Flurazepam
	Nitrazepam	Ketazolam
	Oxazepam	Nordazepam
	Pinazepam	Prazepam
		Quazepam

Table 11.6 Pharmacokinetics properties of benzodiazepines

Agent	Dose eq. approx.[a]	Half-life[b] [h]	Duration[b]	Usual dosage [mg die]
Alprazolam	0.25	12	Short	0.5–6
Bromazepam	1.5	10–20	Medium	3–18
Brotizolam	0.25	3–6	Short	0.25–0.50
Clobazam	5	30	Long	10–30
Clonazepam	0.5	34	Long	0.5–10
Clorazepam	1	80	Long	1–6
Clorazepate	7.5	100	Long	7.5–60
Chlordiazepoxide	10	100	Long	15–100
Clotiazepam	2.5	–	–	5–30
Diazepam	5	100	Long	2–60
Estazolam	0.33	17	Short	1–2
Etizolam	–	15	Medium	0.50–2
Flunitrazepam	–	15–30	Medium	0.50–2
Flurazepam	5	100	Long	15–30
Ketazolam	5	100	Long	5–30
Lorazepam	1	15	Short	2–6
Lormetazepam	0.5	10–15	Medium	1–2
Nitrazepam	1.5	18–30	Medium	2.5–5
Oxazepam	15	8	Short	15–60
Prazepam	10	100	Long	20–40
Temazepam	5	11	Short	20–40
Triazolam	0.1	2	Short	0.125–0.25

[a]High potency <1; medium 1–10; high potency >10
[b]Short acting: half-life <25 h

11.2.3.2 Pharmacodynamics

The primary target of benzodiazepines is the GABA-benzodiazepine receptor complex. GABA is the most important inhibitory neurotransmitter in the Central Nervous System (CNS), and its receptor is a chloride channel that opens after the binding with the neurotransmitter, allowing the chloride ions to enter and hyperpolarize the neuron.

The GABA receptor complex is tetrameric, with one α, two β, and one γ subunit. All the molecules that are able to interact with the GABA receptor are subunit-specific.

The benzodiazepine receptor is contiguous to the GABA receptor and is located at the interface between α and γ subunits. Occupancy of this receptor does not affect the chloride ion channel directly but changes the conformation of the receptor, increasing its affinity for GABA. This means that GABA is still needed to activate the neuron, and benzodiazepines only exert their action at sites where a signal mediated by GABA is physiologically present. Barbiturates, on the other hand, facilitate GABA transmission by acting on the sites directly associated with the chloride ion channel, and they hold it open continuously for long periods of time. This is why respiratory failure is one of the worst fatal effects of barbiturates (as the brainstem is prevented from working properly), while benzodiazepines have optimal tolerability and safety profile.

At high doses, all benzodiazepines mediate four actions: myorelaxant, anxiolytic, hypnotic/sedative, and anticonvulsant; however, at the usual therapeutic dosage, each agent shows an affinity for a certain receptor subtype or a certain subunit, resulting in a different predominant effect, and determining its therapeutic indication and its main clinical use.

Three major subtypes of GABA receptors have been described: two of them are defined central benzodiazepine receptors (CBR) and one is called peripheral benzodiazepine receptor (PBR).

- Ω-1 receptors contain the α1 subunit and are located throughout the CNS. These receptors are responsible for the sedative, hypnotic and antianxiety effects of benzodiazepines.
- Ω-2 receptors, containing α2, α3, or α5 subunits, are located in the cortex, hippocampus, striatum, spinal cord, and on pyramidal neurons. They mediate anxiolysis, muscle relaxation, sedation, and psychomotor impairment and, partly, the anticonvulsant effect.
- Ω-3 receptors (PBRs) are found on glial and other brain cells as well as throughout the body. These receptors, which bind benzodiazepines and the endogenous inverse agonist, are mainly located in the outer mitochondrial membrane and may contribute to tolerance and withdrawal.

11.2.3.3 Therapeutic Indications and Clinical Use

Benzodiazepines are FDA approved for the treatment of Generalized Anxiety Disorder, Panic Disorder, Post-Traumatic Stress Disorder, Social Anxiety, insomnia, catatonia, acute agitation, delirium, seizure, alcohol withdrawal and addiction, neuroleptic side effects, anesthesia and conscious sedation, spasticity.

Insomnia

Benzodiazepines are widely used to treat insomnia: they increase the quality and length of sleep and reduce sleep latency and REM phases. However, benzodiazepines, in general, should always be used for a predetermined and specific duration, that should be as short as possible. They should be not used for more than 7–10 days without further investigations on the patient's sleep disturbance, because tolerance to the hypnotic effect and dependence develop after few weeks.

Initial insomnia and sporadic nocturnal awakenings should be treated with a short life (such as triazolam) or mid-life benzodiazepine (such as lormetazepam, alprazolam, or lorazepam). On the contrary, early morning awakening insomnia should be treated with longer-life benzodiazepines (such as flurazepam or delorazepam).

Anxiety and Depression

Benzodiazepines are the first-choice agents for short-term treatment of anxiety symptoms and acute treatment of panic attacks. Their rapid onset of action, their effectiveness in reducing anxiety symptoms, and their tolerability and safety make them particularly useful for the acute management of these conditions. However, benzodiazepines should be used only as a symptomatic short-term treatment because of the risk of addiction and withdrawal. Both in anxiety disorders and depression benzodiazepines are widely used in combination with antidepressants during the first weeks of treatment, as they are very effective in reducing the anxiety symptoms that often appear as early side effects of the antidepressant therapy. They are then usually reduced in few weeks, once the antidepressant starts to be effective.

Acute Agitation

Acute agitation may occur in a variety of medical and psychiatric conditions. On these occasions, it is often necessary to rapidly calm the patient to prevent dangerous behaviors for the patient himself and others, as well as to properly recognize and treat the underlying condition. Intramuscular benzodiazepines, alone or combined with an antipsychotic, are a widely used and effective strategy to achieve this goal. Lorazepam is the most used agent, among benzodiazepines, for acute agitation treatment.

Alcohol Withdrawal

Alcohol causes symptoms of toxicity by a direct effect on the benzodiazepine receptor. That is why benzodiazepines may be useful in the management of alcohol addiction in order to avoid alcohol withdrawal symptoms such as tremors and dizziness, difficulty in sleeping, nausea and vomiting, irritability, headaches, pain, anxiety, and panic. The goal of the treatment is to minimize withdrawal symptoms and then lead to a controlled tapering of the drug. Diazepam, chlordiazepoxide, and lorazepam are often used to control alcohol withdrawal. Benzodiazepines, in general, can be used with different strategies. A fixed dosage can be administered concomitantly with the discontinuation of alcohol and then slowly tapered, but, in a hospital setting, benzodiazepine dosage may also be patient-tailored, depending on the symptoms and the clinical situation.

Seizures

Benzodiazepines exert a general action of reduction of neuron firing, by eliciting the action of the most important inhibitor neurotransmitter, GABA. That makes them effective for the treatment of seizures. In chronic conditions, benzodiazepines are used as adjunctive therapy to anticonvulsant agents, and clonazepam is the most frequently used agent because of its long duration of action. Benzodiazepines are also useful for the management of acute seizures: in particular, rectal diazepam has a fast onset of action and is an accessible way of administration also during a convulsive seizure and in out-of-hospital settings. Intravenous diazepam, lorazepam, and midazolam are often and effectively used in the treatment of the status epilepticus.

Neuroleptic Side Effects

Benzodiazepines can be used to treat neuroleptic side effects: akathisia and acute dystonia in particular. Akathisia is a common and very annoying side effect of antipsychotic agents, characterized by restlessness and mental unease. Benzodiazepines may be effective in reducing or solving this condition, resulting in relief for the patient and making the treatment more tolerable.

11.2.3.4 Side Effects

The therapeutic action of benzodiazepines may also result in side effects: in particular, their sedative effect often causes daytime drowsiness and decreased concentration, which may interfere with driving and global functioning. Other side effects of benzodiazepines are mild memory impairment, anterograde amnesia, psychomotor impairment with increased risk of falling, agitation, and depression. They also often cause a worsening of Obstructive Sleep Apnea Syndrome (OSAS).

Dependence and Withdrawal

Discontinuation of treatment with benzodiazepines may lead to different clinical scenarios:

1. Relapse: the return of the same symptoms treated with benzodiazepines.
2. Rebound: the intensification of symptoms treated with benzodiazepines. It starts within a few days of drug discontinuation and lasts from days to weeks, according to the duration of the treatment.
3. Withdrawal: the onset of new autonomic symptoms that are not components of the original disorder but a consequence of tolerance and dependence. It begins within hours to days of discontinuation, and it lasts for days to weeks, depending on the drug's half-life: from few days for shorter-acting benzodiazepines to 2 weeks for longer-acting ones. The symptoms of the withdrawal syndrome include anxiety, sleep disturbances, photophobia, phonophobia, mild hypertension and tachycardia, gastrointestinal distress, sweating, tremor, headache, seizures.

In order to avoid addiction and discontinuation syndrome, benzodiazepines should be used at the lowest effective dose and for the shorter possible period. In particular, short half-life agents and drop formulations appear to be the most addictive and are more likely to cause withdrawal syndromes when discontinued.

Discontinuation of treatment with benzodiazepines needs to be considered if: symptom relief is achieved, duration of treatment is longer than 1 month, the patient is taking an excessive dosage, the patient is over 65 years old, the patient is taking concomitant CNS sedatives, the patient is affected by cognitive disorders, in case of traumatic brain injury and the patient has a history of substance or alcohol abuse, or the abuse is still active.

Different strategies should be applied to properly discontinue treatment with benzodiazepines: an individualized and gradual tapering scheme, patient education, a possible concomitant cognitive-behavioral approach, and proper management of the withdrawal symptoms.

Tapering can be started with the same medication or after a switch to an equivalent dose of a longer half-life agent, and it should consist of a 25% reduction every 2 weeks, followed by a slower reduction of 12.5% every 2 weeks. Adjunctive medications may be useful to reduce the negative effects of benzodiazepine discontinuation, such as antidepressants or anticonvulsants (e.g., gabapentin, pregabalin).

Benzodiazepine overdose is usually not life-threatening: the main symptoms include CNS depression, impaired balance, ataxia, and slurred speech. Rarely, and usually when benzodiazepines are combined with alcohol or other drugs, more severe symptoms may appear, including coma and respiratory depression.

Flumazenil is indicated for the treatment of benzodiazepine overdose. It is a benzodiazepine receptor antagonist: acting on the GABAA receptor, it counteracts the action of benzodiazepines. It has a 1-h half-life and is administered intravenously at a dosage of 0.2–0.3 mg over 15 s, followed by repeatable 0.1 mg every 60 s, until a maximum dose of 1–2 mg.

11.2.3.5 Drug Interactions

Antacids may reduce the gastrointestinal absorption of benzodiazepines. Some interactions are possible with drugs or substances metabolized by CYP450: inhibitors like erythromycin, ketoconazole, cimetidine, estrogen, disulfiram, verapamil, nefazodone, and grapefruit juice, increase benzodiazepine plasma levels, while carbamazepine and Saint John's Wort, which induce the enzyme, lower benzodiazepine plasma levels.

> Side effects of benzodiazepines, as well as the severity of an overdose (e.g., in a suicide attempt), are significantly increased by concomitant *intake of alcohol* because of their common receptor target and because of the hepatic metabolism of alcohol, which interferes with the metabolism of benzodiazepines increasing their blood levels.

11.2.4 Z-Drugs: Zolpidem, Zaleplon, and Zopiclone

Zolpidem, zaleplon, and zopiclone are usually called *z-drugs* because of their names starting with the letter "Z." They are sedative drugs that act selectively on the Ω-1 benzodiazepine receptor, resulting in better tolerability and fewer side effects compared to benzodiazepines.

11.2.4.1 Pharmacokinetics

Z-drugs are rapidly absorbed after oral administration. They have different half-lives: from 1 and 2 h respectively for zolpidem and zaleplon to 5–7 h for zopiclone. Zolpidem has a high bioavailability (about 70%), while zaleplon and zopiclone have a lower bioavailability (about 30%) due to extensive first-pass metabolism. They are oxidized to inactivate metabolites.

11.2.4.2 Pharmacodynamics

Z-drugs are selective for the benzodiazepine-1 receptor, without relevant effects on other benzodiazepine receptor subtypes at clinical dosages. Lack of effect on benzodiazepine-3 receptor may determine a lower incidence of withdrawal and rebound symptoms. *Z-drugs' activity* can be blocked by the benzodiazepine receptor's antagonist flumazenil.

11.2.4.3 Therapeutic Indications and Clinical Use

Z-drugs are used for the treatment of insomnia; they have a hypnotic and anxiolytic action, without the muscle relaxant and anticonvulsant effects typical of benzodiazepine. Z-drugs are effective hypnotics, but they do not help the regulation of the sleep–wake cycle. Zolpidem seems to improve total sleep time similarly to benzodiazepines, but without modifying sleep structure. It causes no daytime sleepiness or lowering in mental concentration. Zaleplon is often chosen for nighttime awakenings due to its short duration.

Therapeutic doses of Z-drugs are shown in Table 11.7.

11.2.4.4 Side Effects

Zolpidem, zopiclone and, zaleplon are generally better tolerated than benzodiazepines as regards daily somnolence, dullness, and lack of concentration, but they may have side effects as sedation, anterograde amnesia and higher risk of falling, in particular in the elderly. Tolerance and rebound insomnia are less likely to happen in patients in therapy with z drugs than benzodiazepines, but they can occur after chronic use.

11.2.5 Antihistamines

11.2.5.1 Pharmacokinetics

Antihistamines have good oral absorption and a plasma protein binding between 50 and 98%, with a wide distribution in the body and the CNS. They are metabolized in the liver by CYP2D6 and CYP3A4. Their half-life varies between 2 and 27 h, and they reach a peak in blood levels in 1–3 h.

11.2.5.2 Pharmacodynamics

Antihistamines are antagonists of histamine receptors. Agents blocking the action of H2 histamine receptors are very selective and are used as antacids. Molecules acting on the H1 receptor, on the other hand, are mainly used for the treatment of allergic conditions but also have a CNS sedative action.

11.2.5.3 Therapeutic Indications and Clinical Use

Antihistamines are mostly used for the treatment of nonpsychiatric conditions, in particular allergic symptoms. However, their sedative effect on the CNS may also be useful for the treatment of insomnia, while their anticholinergic effect can be beneficial for neuroleptic side effects such as Parkinsonism, acute dystonia, and akathisia.

Table 11.7 Dosage of Z-drugs

Drug	Initial dose (mg)	Maximum dose
Zolpidem	5	20
Zaleplon	5	10
Zopiclone	1	3

As sedative agents, antihistamines are not more effective than benzodiazepines, and their tolerability and safety have been less studied; therefore, benzodiazepines are usually preferred. Antihistamines are not used as long-term anxiolytic therapy.

11.2.5.4 Side Effects

Antihistamines' use is associated with side effects, such as dizziness, daily sleepiness, and hypotension. Impairment in attention and concentration can also appear and may be particularly dangerous for driving or specific jobs. Other side effects include symptoms related to their mild anticholinergic activity, like blurred vision, urinary retention, constipation, and dry mouth. The elderly are more likely to show side effects, including a psychomotor impairment that may lead to an increased risk of falling.

Antihistamines should be avoided during breastfeeding, since they are excreted with the milk, and in pregnancy because of potential damage to the fetus.

11.2.5.5 Drug Interactions

Because of their sedative effects, antihistamines should be administered very carefully in combination with other CNS depressants, including benzodiazepines, dopamine receptor antagonists, tricyclic antidepressants, and alcohol.

11.3 Mood Stabilizers

11.3.1 Introduction

Mood stabilizers are medications used to treat bipolar disorder, schizoaffective disorder, and borderline personality disorder.

These drugs help to maintain euthymia, reducing mood swings and preventing manic and depressive episodes in bipolar and schizoaffective disorder. They are also used to supplement other medications, such as antipsychotics or antidepressants, in the treatment of acute mania or depression.

Mood stabilizers are also effective in reducing borderline personality disorder symptoms, including emotion dysregulation, mood lability, and impulsivity. Medications commonly classified as mood stabilizers include salts, anticonvulsants, and antipsychotics. Antipsychotics are discussed in their specific subchapter.

11.3.2 Lithium

Lithium is the most prescribed mood stabilizer worldwide. Lithium is an anion, and it follows hydrogen and helium on the periodic table, making it the third simplest chemical element. How such a simple element can have such a great pharmacological effect is still not completely clear, although its effectiveness was first demonstrated in the 1950s and it was approved by FDA for the treatment of mania in 1974.

11.3.2.1 Pharmacokinetics

Lithium is rapidly and completely absorbed when administered orally, with serum concentrations peaking in 1–1.5 h. It has no clinically relevant protein-binding properties and no metabolites.

It is excreted almost exclusively by the kidneys, with small amounts are also eliminated through sweat and feces. A significant portion of filtered lithium is reabsorbed (mainly in the proximal tubules): this implies that other drugs acting at the level of proximal tubules may interfere with lithium reabsorption (e.g., thiazides). The elimination half-life of lithium is about 18–24 h, although it is considerably longer in the elderly because of the age-related decrease in the glomerular filtration rate. Therefore, older subjects usually require lower than usual dosages to achieve therapeutic serum concentration and reach a steady state over longer periods of time than usual.

Lithium is broadly distributed throughout the body, with different extent into tissues: thyroid and renal concentrations are greater than serum levels, differently from red blood cell and brain concentrations. Lithium enters and leaves the CNS slowly; for this reason, acute intoxication with high plasma levels are sometimes well-tolerated, and clinical manifestations of chronic intoxications often persist for long periods after plasma levels have decreased.

It is possible to dose both the lithium serum concentration and the concentration of lithium inside red blood cells (RBC). This is important because lithium persists inside red blood cells for about 7 days, while the serum concentration of lithium 48 h after the last administration is almost zero. RBC lithium concentration is a more reliable indicator of therapy adherence and is useful for the diagnosis and prognostic evaluation of lithium poisoning because toxic symptoms do not always correlate with plasma concentrations.

11.3.2.2 Pharmacodynamics

Although lithium is widely considered the first choice among mood stabilizers, it is not completely clear how it works. What does seem clear is that amount of lithium required for its clinical effect is much greater than the traces normally present in the body, thus the cause of bipolar disorder is not a lithium deficiency.

Ion Transport Theory

Neural transmembrane potential differences are maintained by sodium pumps and perturbations of this system seem to be present in patients with bipolar disorder. Some studies associated the alteration in sodium pump activity causing neurotransmitter aberrations to mood oscillations. Lithium is thought to stabilize membrane function.

There are several targets for its mechanism of action: from the microscopic intracellular signaling to the macroscopic brain structure. Lithium can modify nuclear transcription factors and influence gene expression: it generates *posttranslational modifications of G proteins*; it inhibits proteins such as PKC, MARCKS, GSK-3, IPPase, and IMPase. Lithium modulates dopamine, glutamate, and GABA neurotransmission. Some findings suggest that lithium's stabilizing effect might be related to the modulation of dopamine (associated with depression), NMDA or AMPA receptors for glutamate (antimanic effect), and GABA neurotransmission. Acting on BDNF and BCL2, the chronic use of lithium is probably associated with neuroprotection and neuroproliferation. Changes in the brain structure are associated with chronic use of lithium; in literature, changes in the anterior cingulate cortex, the ventral prefrontal cortex, the hippocampus, and the amygdala have been reported.

Lithium and GSK3

Lithium inhibits glycogen-synthase kinase3 (Gsk3) and inositol monophosphatase (IMPase). GSK3 is a protein kinase involved in many signal transduction cascades influencing protein phosphorylation, microtubular dynamics, cell proliferation and development, and neuronal apoptosis. Recent studies have shown that glycogen synthase kinase 3-β (GSK3-β) is involved in the control of cell behavior and in the mechanism of action of lithium and serotonergic antidepressants, and in humans a certain promoter variant was associated with reduced activity and better antidepressant response.

Gene–gene interactions between components of the monoaminergic signal transduction pathways and of plasticity related pathways can shape the individual antidepressant response. In particular, GSK3-β influences synaptic plasticity and cell resilience, counteracting the detrimental influence of the short form of the serotonin promoter on antidepressant response.

These new findings highlight the important role of GSK3 as a mediator of the action of lithium on mood stabilization.

11.3.2.3 Therapeutic Indications and Clinical Use

Lithium is FDA approved for the acute treatment for manic episodes and as maintenance therapy in bipolar disorder. Off-label uses of lithium are the prevention of depressive episodes in Major Depressive Disorder and the control of emotion dysregulation, mood lability, and impulsivity symptoms in borderline personality disorder.

Long-term treatment with lithium has proven to be an effective way to reduce frequency, severity, and duration of mood episodes in patients affected by bipolar disorder.

In the acute manic phase, lithium treatment aims at the remission of the episode. At the resolution of the episode, there should follow a continuation (6–9 months) and a maintenance phase (minimum 2 years) to prevent, respectively, relapses and recurrences of the illness.

It is often recommended to start the lithium treatment from the first episode of mania.

The discontinuation of successful lithium treatment, especially if rapid, is associated with a substantial increase of the recurrence risk. After the discontinuation, the

Table 11.8 Lithium prescribing and monitoring

Lithium prescribing and monitoring		
Before treatment	Monitoring	
Blood urea nitrogen	*Minimum recommendations*	
Creatinine	Lithium serum levels	6–8 weeks
Urinalysis	Creatinine	6–12 months
Urine volume 24 h	TSH	6 months
Creatinine clearance	fT4	6 months
TSH	Electrolytes	6–12 months
Free and total T4	Urine analysis	12 months
Free and total T3	*Additional recommendations*	
Complete blood count	Urine volume 24 h	6–12 months
Electrolytes	Creatinine clearance	6–12 months
Glycemia	Urine osmolarity	6–12 months
Electrocardiogram	Complete blood count	6–12 months
Blood pressure	Electrocardiogram	6–12 months
Pregnancy test (childbearing age)	Intraerythrocytic lithium	3 months

reintroduction of lithium therapy often appears to be less protective against mood episodes.

Before starting lithium treatment, it is important to consider the patient's characteristics such as age, comorbidities, and concomitant pharmacotherapies. Some medical evaluations are also necessary: renal and thyroid functions should be routinely tested before beginning treatment. A pregnancy test is recommended for women of childbearing age.

The therapeutical dosage is between 600–1200 mg/day divided into two doses approximately 12 h apart (morning–night).

Lithium serum levels, creatinine, TSH, FT4, electrolytes, and urinalysis should be checked regularly during treatment (every 6–12 months, see Table 11.8).

Lithium serum level should be checked every 3 months (blood should be collected 12 h after the last administration); the therapeutic range has been established at 0.5–0.8 mmol/L during active manic phases, and at 0.4–0.6 mmol/L as maintenance treatment.

Lithium and Neurodegeneration

MRI studies on patients affected by bipolar disorder showed a widespread disruption of brain white matter structure.

Long-term administration of lithium appeared to be associated with an increase of axial diffusivity (reflecting integrity of axons and myelin sheaths) in several white matter fiber tracts contributing to the functional integrity of the brain and involving interhemispheric, limbic, and large frontal, parietal, and fronto-occipital connections. This effect of lithium is related to its well-known GSK3 inhibition.

These findings further demonstrate the importance of long-term treatment with lithium in people suffering from bipolar disorder, shedding new light on its long-term beneficial effect.

Absolute contraindications: psoriasis, renal dysfunction, myasthenia gravis, acute myocardial infarction, diabetes insipidus, severe electrolyte imbalances.

Relative contraindications: thyroid diseases, chronic heart disease (e.g., hypertension, valvular disease), pregnancy, breastfeeding, Parkinson's disease, cerebellar disorders, ulcerative colitis, hypertension treatment.

11.3.2.4 Side Effects

The majority of patients taking lithium experience adverse effects, but less than 20% have more than minor complaints.

Following is a list of the main side effects:

- *Renal:* urinary concentrating defect, polyuria (nephrogenic diabetes insipidus), reduced glomerular filtration rate, nephrotic syndrome (absolute contraindication for administration of lithium).
- *Endocrine:* hypothyroidism.
- *Neurological:* dysphoria, slowed reaction time (rare), tremor (postural, occasional extrapyramidal), peripheral neuropathy, myasthenia gravis-like syndrome.
- *Gastrointestinal:* appetite loss, nausea, vomiting, diarrhea, altered carbohydrate metabolism, weight gain, fluid retention.
- *Dermatological:* acne, hair loss, rash. Psoriasis is an absolute contraindication.

Lithium can interfere with the synthesis of thyroid hormones because of its similarity with iodine. 7–9% of patients in treatment with lithium may thus develop *functional hypothyroidism* without thyroid damage. This condition should be treated with synthetic thyroid hormone levothyroxine; the disruption of effective treatment with lithium is not indicated in these cases.

11.3.2.5 Drug Interactions

Most nonsteroidal anti-inflammatory drugs reduce renal lithium clearance and increase the serum lithium concentration with potentially dangerous consequences. The newer COX-2 inhibitors, such as celecoxib and rofecoxib, are no exception.

Thiazide diuretics reduce renal lithium clearance and increase the plasma lithium concentration. Lithium retention may also be caused by potassium-sparing diuretics. Conversely, loop diuretics such as furosemide may increase renal lithium clearance.

Lithium is known to be a potential precipitant of serotonin syndrome in people concurrently on serotoninergic medications, such as antidepressants. Lithium cotreatment is also a risk factor for neuroleptic malignant syndrome in people treated with antipsychotics and other antidopaminergic medications.

Lithium Intoxication

Lithium has a narrow therapeutic index, and a higher level of lithium can lead to intoxication. Lithium intoxication leads to neurotoxicity that can, in extreme cases, result in death or permanent neurological damage. Clinical presentation of lithium intoxication includes gastrointestinal manifestations, dysarthria, ataxia, tremor, impaired consciousness, neuromuscular irritability (fasciculations, myoclonus), and seizures.

Lithium toxicity can occur due to excessive intake or decreased excretion. Some factors may decrease its excretion: kidney disease, dehydration, drug interaction, and a low-sodium diet.

The diagnosis is generally based on symptoms and is supported by a high serum (and/or RBC) lithium level.

Treatment of lithium intoxication consists of lithium discontinuation, gastric lavage, hydration, proper fluid, and electrolyte balance.

Lithium and Pregnancy

Lithium is probably the most effective treatment for the prevention of mood episodes, even during pregnancy and postpartum. Even if many studies throughout the years have suggested, especially in the first trimester, an increased risk of fetal abnormalities, more recent studies have estimated a lower risk than previously reported. More research is certainly needed to be conclusive.

Lithium could be suspended during pregnancy or after delivery but the higher risks of relapse, especially in the postpartum period, should be carefully considered. When lithium is discontinued, it should be restarted immediately after delivery for relapse prevention.

If lithium is administered during pregnancy, the dose can be reduced, especially in the first trimester, and the lithium blood level should be checked more frequently (up to once a week). Fetal echocardiography between weeks 16 and 20 of gestation may be considered due to the potential increased risk of cardiac malformations. Lithium dosage should be decreased or discontinued few days before delivery to reduce the risk of maternal and neonatal toxicity. Due to a high infant/maternal ratio of serum drug concentration, breastfeeding is not recommended. Different mood stabilizers, such as valproate or carbamazepine, have to be avoided during pregnancy. Lithium is, therefore, the first-choice maintenance therapy for pregnant women affected by Bipolar Disorder.

11.3.3 Valproate

Valproic acid is an anticonvulsant, first-line treatment of bipolar disorder (mania, mixed states, and as maintenance therapy), useful in treating epilepsy, in preventing migraine headaches, and in reducing borderline personality disorder symptoms, such as emotion dysregulation, mood lability, and impulsivity.

11.3.3.1 Pharmacokinetics

All formulations of valproate are rapidly absorbed, with peak plasma concentrations obtained 3–4 h after oral administration. The plasma half-life of valproate is 10–16 h; however, a shorter half-life is recorded when administered in combination with inductors of cytochrome P450. Valproate is highly protein-bound, primarily to albumin, but only the unbound portion of valproate is considered to be pharmacologically active since it crosses the blood–brain barrier. Higher serum levels than 45–50 μg/ml saturate all the binding sites with an increase of unbound valproate. This drug is metabolized primarily through hepatic glucuronidation and mitochondrial β-oxidation. Several valproate metabolites, as 2-propyl-2-pentenoic acid and 2-propyl-3-oxopentanoic acid, have potent anticonvulsant activity.

11.3.3.2 Pharmacodynamics

The anticonvulsant and mood stabilization mechanisms of action have not been specifically identified and may not necessarily be the same. The anticonvulsant effect is rapid in onset, while the mood-stabilizing one is slower and requires chronic administration. GABA deficit hypothesis, proposed by Emrich in 1980, suggested that bipolar illness was characterized by a relative deficit of GABA that could be ameliorated with valproate administration. Several hypotheses include enhancement of GABA levels via multiple actions of synthesis, degradation, and modulation of other neurotransmitters, voltage-sensitive Na channels, extra hypothalamic neuropeptides, secondary messenger systems, and neuroprotection. Preclinical studies have shown that valproate enhances GABAergic function in specific regions of the brain by means of a variety of presynaptic or postsynaptic mechanisms: the enhancement of GABA synthesis through the activation of glutamate decarboxylase, the prevention of GABA catabolism through the inhibition of GABA transaminase, the direct release of GABA in response to calcium influx, the reversal of the GABA transporter, or postsynaptic GABAA receptor activation effects.

11.3.3.3 Therapeutic Indications and Clinical Use

Valproate is currently FDA-approved as treatment for seizures, acute mania, or mixed states, as prevention of recurrent manic and depressive episodes (it has been recommended as a first-line option in many treatment guidelines), and as prevention of migraine headaches. It is also used off-label (for unapproved uses) for other conditions such as borderline personality disorder, impulse-control disorder, and agitated behavior in dementia, because of its effectiveness in reducing emotion dysregulation, mood lability, irritability, aggressiveness, and impulsivity.

It is particularly effective in the treatment of acute mania, also when associated with depressive or dysphoric symptoms (mixed states), with a rapid-cycling course of illness (four or more episodes in a 12-month period), and manic episodes complicated by substance or alcohol abuse. In patients with chronic alcohol abuse associated with hepatic dysfunction, lithium is preferred because it is not metabolized by the liver. Given its current neurological FDA indications, valproate may be particularly useful in bipolar disorders complicated by seizures, electroencephalogram (EEG) abnormalities, head trauma, and migraines.

11.3.3.4 Side Effects

Although valproate treatment is usually well-tolerated and safe, it is associated with serious adverse effects, including hepatic failure, pancreatitis, hyperammonemic encephalopathy in patients with urea cycle disorder, sedation in the elderly, and thrombocytopenia. To avoid these severe reactions, it is important to carry out baseline hepatic function tests before starting valproate treatment. However, serial laboratory monitoring does not necessarily predict severe hepatotoxicity or pancreatitis.

Minor, and more common, adverse reactions include gastrointestinal distress, tremor, sedation, benign hepatic transaminase elevation, leukopenia, thrombocytopenia, hair loss, increased appetite, and weight gain. Most of the common adverse effects are dose-related and may be prevented or minimized by dose reduction or adjunctive pharmacotherapy.

Valproate is contraindicated and should not be taken by pregnancy: it is associated with teratogenicity, with risk of neural tube defects and heart malformations. All nonpregnant women of childbearing age taking valproate products should use effective birth control.

11.3.3.5 Drug Interactions

Valproate is commonly prescribed in combination with other psychotropic agents. The combination with lithium is generally well tolerated and is often more effective than valproate alone. However, a problem may be the additive side-effect burden of the two medications: tremor, cognitive dulling, and dyspepsia. Typical antipsychotics and valproate are often prescribed together, especially in acute mania. Again, increased sedation and an increase in extrapyramidal side effects have been reported. More recently, atypical antipsychotics and valproate have been used together, with increased sedation and weight gain reported as side effects. Valproate has also been used as an adjunctive mood stabilizer and anticonvulsant in patients treated with clozapine. In fact, for patients who have had a seizure related to a high dose of clozapine, valproate is preferred over carbamazepine, burdened by an additional risk of agranulocytosis. The combination is generally well tolerated, except for sedation and weight gain.

In contrast to carbamazepine, the primary pharmacokinetic liability of valproate is related to its plasma protein binding capacity and to the inhibition of other drugs hepatic metabolism. For example, valproate increases the free concentration of diazepam by plasma protein binding site displacement and reduces the serum level of a diazepam metabolite: N-desmethyldiazepam. Close clinical monitoring of this coadministration is therefore indicated to avoid toxicity. Similar observations have been noted with phenobarbital, chlorpromazine, and aspirin. Because valproate has been shown to inhibit the secondary phase of platelet aggregation, close clinical monitoring is indicated when using the drug with other drugs that affect coagulation, such as aspirin and warfarin.

11.3.4 Carbamazepine

Carbamazepine is an anticonvulsant medication used primarily in the treatment of epilepsy and neuropathic pain. It has been recommended in many treatment guidelines as a second-line agent mood stabilizer in bipolar disorder.

11.3.4.1 Pharmacokinetics

Carbamazepine is absorbed slowly and inhomogeneously through the gastrointestinal tract. It is metabolized in the liver and then excreted by the kidneys, as only 1% is eliminated by biliary excretion. Peak plasma levels are achieved 2–8 h after a single dose and the molecule is 70–80% bound to plasma protein. The half-life after a single dose range from 18 to 54 h. However, with long-term intake, the half-life decreases to about 10–25 h due to the autoinduction of hepatic P450 enzymes, which increase the metabolism of carbamazepine itself. The degree of the drug autoinduction is dose-dependent but is generally complete after 3–5 weeks of treatment. As a consequence, tolerability improves after several weeks of treatment, and blood levels may be markedly reduced despite the use of the same daily dose.

11.3.4.2 Pharmacodynamics

The onset of anticonvulsant and antinociceptive effects is rapid and occur approximately after 24–48 h. Antimanic actions of carbamazepine usually begin within the first days of treatment, but often an optimal response in patients is achieved only after several weeks. The anticonvulsant effects of carbamazepine have been mainly associated with blockade of type 2 sodium channels, which, when inactivated, reduce the release of excitatory amino acids such as glutamate and inhibit sustained rapid neuronal firing. Carbamazepine also act as a potent agonist on adenosine A1 receptors, which mediate not only some of its anticonvulsant actions but also its sedative properties.

As well as lithium and valproate, carbamazepine blocks inositol transport at the myoinositol transporter. The antinociceptive effects of the drug have been mainly linked to effects at GABA-B receptors and are specifically inhibited by GABA-B antagonists. Interestingly, chronic (and not acute) administration of lithium, carbamazepine, and valproate determines upregulation of GABA-B receptors in the hippocampus. This action may represent a potential convergent mechanism for mood stabilization.

11.3.4.3 Therapeutic Indications and Clinical Use

FDA-approved indications for carbamazepine use are epilepsy (partial seizures and tonic–clonic seizures), trigeminal neuralgia, manic and mixed episodes of bipolar I disorder. Carbamazepine is also widely used in bipolar disorder as long-term prophylaxis: many guidelines suggest carbamazepine as first or second-line drug for mood stabilization. Moreover, it can be used as a substitute or as an adjunct to lithium in partial responders. Carbamazepine is also highly effective in a wide range of aggressive and impulse-control disorders that can occur in patients affected by personality disorders, affective disorders, and schizophrenia.

Even if carbamazepine appears to be approximately as effective as lithium in the prevention of affective episodes, some insights are needed. Lithium appears to be more effective than carbamazepine in preventing manic episodes in patients with bipolar disorder Type I with a positive family history of affective disorder and no alteration in thought content or comorbid substance abuse. Conversely, carbamazepine showed better prophylactic efficacy in patients with bipolar disorder Type II, schizoaffective disorder, dysphoric manic episodes, and substance abuse comorbidity, as well as a negative family history for affective disorder. In some cases, patients may respond to carbamazepine after failing to respond to valproate or vice versa, consistently with the notion that response to the anticonvulsants does not occur as a class effect.

11.3.4.4 Side Effects

Carbamazepine, when used in the treatment of psychiatric conditions, is usually administered at lower dosages than in the treatment for epilepsy. However, a variety of side effects can be observed in patients treated with low doses, too.

- *Hematological:* the most frequent hematological side effect is a clinically not-significant suppression of the white blood cell count due to the inhibition of the colony-stimulating factors. Potentially life-threatening blood dyscrasias such as agranulocytosis or aplastic anemia occur in approximately 1 in 100,000 patients. Patients should be warned that, if signs of potential hematological dysfunction (such as fever, sore throat, rash, petechiae, bruising, or overt bleeding) appear, they should immediately seek medical assistance.
- *Dermatological:* about 10–15%of people treated with carbamazepine develop a benign maculopapular rash typically emerging in the first 10–30 days after drug initiation. Drug cessation usually leads to the resolution of the rash. In rare instances, carbamazepine can induce a variety of dermatological syndromes, some of which are serious, including exfoliative dermatitis, erythema multiforme, Stevens–Johnson's syndrome, or toxic epidermal necrolysis.
- *Endocrine:* carbamazepine increases the 24-h excretion of urinary free cortisol and decreases circulating levels of thyroxin (T4) and triiodothyronine (T3), without altering TSH. However, thyroid replacement is rarely required.
- *Renal:* carbamazepine has an agonist-like effect on the vasopressin receptor and may cause hyponatremia.
- *Liver:* carbamazepine causes an increase in total serum cholesterol, high-density, and low-density lipoproteins, which can induce an elevation of hepatic enzymes. However, routine monitoring does not appear indicated because the appearance of hepatitis is extremely rare.

- *Neurological:* carbamazepine, like many anticonvulsant drugs, may cause dizziness, ataxia, or diplopia.
- *Weight:* long-term, high dosage treatments with carbamazepine are associated with small degrees of weight gain.

Carbamazepine should be avoided during pregnancy because of the risk of neuronal cord malformations such as spina bifida. Moreover, the drug is transferred to breast milk in a concentration that is approximately 50% of that present in maternal plasma.

Carbamazepine Overdose

Symptoms of carbamazepine overdose include drowsiness, stupor, coma, sinus tachycardia, atrioventricular conduction blocks, hypotension or hypertension, seizures, nystagmus, hyporeflexia or hyperreflexia, hypothermia, oral dyskinesias, and respiratory depression. Supportive interventions for overdose, include gastric lavage and instillation of activated charcoal until the patient is symptom-free.

11.3.4.5 Drug Interactions

Carbamazepine is a potent inductor of the cytochrome P450 (mainly CYP3A4) and increases the rate of metabolism of a great variety of drugs (warfarin, lamotrigine, phenytoin, theophylline, valproic acid), decreasing their serum concentration and often reducing their therapeutic effects. Carbamazepine also increases the metabolism of oral contraceptives often reducing their effectiveness. Through autoinduction, after 2–3 weeks, carbamazepine increases the rate of its own metabolism. Conversely, many common drugs can increase the blood concentration of carbamazepine: erythromycin verapamil, diltiazem, isoniazid, and fluoxetine.

11.3.5 Lamotrigine

Lamotrigine was developed as an antiepileptic drug, and it later showed to be useful in the treatment of a variety of neurological and psychiatric conditions. In particular, it has been found effective as maintenance therapy for bipolar disorder; it may also be somehow helpful in acute bipolar depression, whereas it is not effective in acute mania.

11.3.5.1 Pharmacokinetics

Lamotrigine is completely absorbed when administered orally. It has a 28-h half-life, and it has a hepatic metabolism. Other anticonvulsants may strongly affect the pharmacokinetic properties of lamotrigine. In particular, valproate can inhibit its hepatic metabolism, whereas carbamazepine can induce it. The half-life of lamotrigine can thus vary from 14 to 56 h when taken in combination therapies. It is important to consider cotreatments to choose a proper administration schedule for each patient (see Table 11.9).

Table 11.9 Dosage of lamotrigine as a maintenance treatment in bipolar disorder

Dosage of lamotrigine as a maintenance treatment in bipolar disorder				
	Week 1 and 2	Week 3 and 4	Week 5	Target dosage
Monotherapy	25 mg/day	50 mg/day	100 mg/day	200 mg/day
With valproate	12.5 mg/day	25 mg/day	50 mg/day	100 mg/day
With carbamazepine	50 mg/day	100 mg/day[a]	200 mg/day[a]	400 mg/day[a]

[a]Divided doses

11.3.5.2 Pharmacodynamics

It is not yet completely clear how exactly lamotrigine exerts its pharmacological action. It blocks voltage-sensitive sodium channels, affecting the realization of neurotransmitters glutamate and aspartate. It has also a mild effect on serotonin, dopamine, and norepinephrine reuptake in some animal models.

11.3.5.3 Therapeutic Indications and Clinical Use

Lamotrigine is FDA approved for the treatment of epilepsy and as maintenance treatment of bipolar disorder. Its stabilizing effect seems to be stronger at delaying depressive episodes than manic ones. Moreover, it is mildly effective for the treatment of acute bipolar depression, while it does not have any clinically significant effects in the treatment of acute mania.

Table 11.9 shows the standard dosage of lamotrigine and as maintenance treatment of bipolar disorder. The dose should be adjusted considering cotherapies, and it can be further corrected by measuring the blood levels of the drug.

11.3.5.4 Side Effects

The main concern about the treatment with lamotrigine is the risk of serious rashes, including Stevens–Johnson's syndrome. Some risk factors may be age (pediatric patients show a higher risk) and overdose, that might be caused by an excessive intake or by some interaction with other drugs.

Minor skin rashes are observed in up to 10% of patients taking lamotrigine. The risk of a serious event should lead to the discontinuation of the drug whenever a rash appears. That is why the therapy has to be started at a low dosage, and for the first month any new drug, food or cosmetics should be avoided.

Generally, lamotrigine is very well tolerated. The most common adverse effects in patient affected by bipolar disorder and taking lamotrigine are insomnia, daily sleepiness, nausea, fatigue and back pain.

Due to the risk of seizures after discontinuation of the drug, lamotrigine should be tapered slowly when therapy needs to be stopped.

Lamotrigine is not indicated during pregnancy because of its in vitro effects on folate metabolism.

11.3.5.5 Drug Interactions

Because of interference with lamotrigine hepatic metabolism, blood levels of the drug are decreased by the coadministration of agents such as carbamazepine and estrogens and increased by agents such as valproate and sertraline. That is why the administration schedule must be tailored for each patient (see Table 11.9).

11.4 Antipsychotics

11.4.1 Introduction

Until the 1950s, the management of chronic psychotic disorders was based primarily on assistance and support through institutionalization, with therapeutical attempts mainly related to the supposed effects of "shock"-related therapies based on the use of cool water, agents producing elevate fever, hypoglycemic states/"insulinic coma," and electroconvulsive therapy, with limited and not persistent results (ECT is still a valid option for resistant depression, but indication in chronic psychosis are very limited). The concept of "antipsychotic activity" arose accidentally from the observation of sedative effects after administration of chlorpromazine, originally prescribed as an adjunct to surgical anesthetics because of its body temperature lowering effect. Subsequently, chlorpromazine was successfully administrated to treat agitation in a manic patient with psychotic symptoms, surprisingly allowing the subject to resume normal life after 20 days of treatment.

This serendipitous finding represents the cornerstone of the game changing development of a pharmacotherapy for schizophrenia and led to the discovery and development of the so-called typical antipsychotics or first-generation antipsychotics (FGAs), which include phenothiazines, thioxanthenes, butyrophenones (haloperidol), and substituted benzamides. Originally, FGAs were defined as *neuroleptics* (according to the original presentation of chlorpromazine by Deniker in 1955 the term came from the Greek meaning "which takes the nerve") because their administration at high doses was associated with *neurolepsis*, characterized by behavioral indifference and psychomotor slowing.

The development and introduction of clozapine in the 1970s represented the following milestone in the pharmacotherapy of schizophrenia, being the first atypical or second-generation antipsychotic (SGA). Although clozapine was initially temporarily withdrawn from the market due to tolerability issues (agranulocytosis, see below), during the following years the development and use of SGAs spread worldwide. Indeed, compared to FGAs, SGAs were found to be characterized by similar antipsychotic efficacy but better tolerability, and nowadays prescription of atypical antipsychotics is consistently higher in clinical practice.

The discovery and introduction of antipsychotics represented a radical turning point in both the history of mental illness and society. The introduction of FGAs into clinical practice led to a sharp decline in hospital occupancy, to progressive closure of mental asylums, and to the development of social policies concerned with mental health. Antipsychotics enabled patients affected by severe mental disorders to participate again in social life, greatly reducing the pressure on the health system and the impact of mental illness on society. Moreover the study of antipsychotic action shed new light not only on the neurobiological correlates of psychosis, but also on the neural basis of physiological brain functioning.

11.4.2 First Generation (FGA)/Typical Antipsychotics

11.4.2.1 Pharmacodynamics

Antipsychotic efficacy of FGAs derives from blockage of dopaminergic D2 receptors (D2r), particularly in the mesolimbic dopamine pathway, which is primarily involved in the pathogenesis of positive symptoms. However, given the variable receptor distribution within the brain, in other cerebral areas (i.e., mesocortical and nigrostriatal pathways) D2r antagonism can determine significant side effects. The occurrence of these adverse effects is highly correlated with pharmacological load, and therefore the target dosage should always be the lowest effective.

Conventional antipsychotics possess affinities for muscarinic M1, histaminergic H1, and alfa-1 norepinephrine receptors, which can result in partially distinctive and overlapping side-effect profiles.

FGA Receptor Antagonism

- D2.
- Alpha-1.
- M1.
- H1.

11.4.2.2 Efficacy

Large multicenter trials adopting evidence-based approaches failed to demonstrate significant differences in antipsychotic efficacy between the different FGAs. Nonetheless, antipsychotic response is characterized by high interindividual variability, and some patients might respond better to one conventional antipsychotic agent than another. This great heterogeneity depends on individual genetic and metabolic features that to date are only partially defined, and that can also influence duration of treatment, dosages, and associations with additional medications such as mood stabilizers, benzodiazepines, or antidepressant drugs.

Antipsychotic monotherapy at the lowest possible effective dose is recommended by international guidelines, especially at disease onset, since the administration of high dosages and/or of antipsychotic polytherapy is not supported by scientific evidence: exceeding the dopaminergic receptor saturation threshold (in general, at the net consideration of interindividual variability, a dose of about 5 mg/day of haloperidol equivalent is already able to occupy D2 receptors to a degree related to clinical response in nonresistant patients) does not lead to further clinical benefits, but only increases the rates of side effects, especially extrapyramidal symptoms.

Haloperidol

Haloperidol is the progenitor of butyrophenone compounds. It was introduced in the 1970s but it is still one of the most prescribed conventional antipsychotic drugs because of its high antipsychotic potency and its relatively safe tolerability profile at low doses, which makes the drug easy to use. Indeed, haloperidol shows high affinity with D2r selectively but differs from other FGAs as it shows low anticholinergic and antihistaminic activity. The therapeutic range is between 2 and 6 mg, greater doses are used in the acute psychotic phase to enhance sedative properties, mostly to manage behavioral abnormalities. Rarely doses over 6 mg show higher antipsychotic effect due to D2r saturation threshold. Haloperidol is characterized by a high lipophilia thus peak plasma concentrations occur after 1.7–6 h after oral intake (bioavailability: 60–70%) and plasma protein binding is relatively high, settling around 90%. Metabolized in the liver by CYP3A4 and CYP2D6, excretion of metabolites occur preferentially through urine.

Concerning clinical indications, haloperidol is licensed for the treatment of acute and chronic schizophrenia and other psychotic disorders, delirium, psychomotor agitation within psychotic disorders or mania, mania, aggressiveness in Alzheimer or vascular dementia, Tourette syndrome or tic syndromes, and Huntington disease when other treatments fail. It is widely used currently.

Chlorpromazine

Chlorpromazine, mentioned before as the first antipsychotic discovered, is a phenothiazine compound classified as a low-potency typical antipsychotic. Compared to haloperidol, it shows lower D2r affinity but higher antihistaminic and anticholinergic activity, respectively associated with sedation and muscarinic side effects. To date, chlorpromazine is rarely prescribed as antipsychotic monotherapy and mainly used in clinical practice for its sedative properties, or as an adjunctive treatment as antipsychotic enhancer drug.

The antipsychotic effect is reached with dosages over 200–250 mg, typically not more than 600 mg (maximum licensed dose 1000 mg), lower ones (25–75 mg) are used in clinical practice to induce sedation.

Concerning pharmacokinetic properties, the absorption is rapid after oral administration and the bioavailability of the drug is around 30–50% with high plasma protein binding. It is metabolized by CYP2D6 in the liver and it is excreted in the urine and bile. The half-life is highly variable ranging between 6–30 h.

Clinical indications of the molecule are schizophrenia, paranoid delusional syndromes and substance-induced psychoses, delirium and mania. Other nonpsychiatric indications are vomiting, incoercible hiccups and treatment of intense pain, generally in combination with other analgesics. It is also approved for preanesthesia.

Zuclopenthixol

Zuclopenthixol belongs to the subgroup of thioxantens and like other FGA it presents good D2r affinity and, compared to haloperidol, also higher D1r affinity, thus showing a mixed D1/D2 antagonism. Indicated for schizophrenia treatment, it is largely used in intramuscular formulation, both long-acting (zuclopenthixol decanoate) and fast-acting vials (zuclopenthixol acetate) for the management of agitated behaviors by reason of its anticholinergic and sedative properties. In the clinical practice daily dosages range from 10 to 75 mg, associated with a strong antipsychotic activity and representing a valuable alternative to haloperidol.

Concerning pharmacokinetics, like other antipsychotics, zuclopenthixol is characterized by rapid absorption after oral administration with a time to peak concentration of 4 h, high serum proteins binding percentage and half-life is approximately 20 h. Zuclopenthixol is approved for the treatment of acute and chronic schizophrenia and other psychotic syndromes and clinical conditions characterized by agitation or psychomotor excitement, hostility and aggressiveness. Moreover, it is also approved for the management of mania and organic mental syndromes accompanied by delirium and psychomotor agitation. It is still used currently in both pharmaceutical formulations.

11.4.2.3 Other Typical Antipsychotics

Pimozide

Pimozide is a high potency FGA of the diphenylbutylpiperidine class, showing D2r blocking activity even greater than haloperidol, and is also antagonist at 5HT2r. Due to its low sedative properties at low doses, pimozide was the first antipsychotic considered as "atypical," paving the way to study and development of antipsychotic effective on both negative and positive symptoms (see atypical antipsychotics section). Rarely prescribed for treating schizophrenia because of its tolerability issues concerning QT prolongation and extrapyramidal symptoms, this typical antipsychotic drug is mainly used for motor neurological diseases such as Tourette syndrome and resistant tics.

Perphenazine

Perphenazine is a piperazinyl phenothiazine showing a medium-potency antidopaminergic (D2r) activity, approximately ten times higher than chlorpromazine. Indicated for the treatment of schizophrenia, other psychotic disorders, and manic phases of bipolar disorder. Currently it is used infrequently, also in formulations with associate low doses of the drug with low doses of amitriptyline.

Thioridazine

Thioridazine is an FGA belonging to the phenothiazine drug group. Originally indicated and widely prescribed for schizophrenia treatment, and was effective in a wide range of psychotic conditions, with a better EPS profile than haloperidol at low doses, but in the last years in several countries the compound is no more available because of its high propensity to prolong the QT interval in a dose-dependent manner, thus inducing severe cardiac arrhythmias and in particular "torsades de pointes." Thioridazine shows high affinity for cholinergic receptors, thus producing significantly less extrapyramidal side effects than most FGAs.

Promazine

Promazine is a phenothiazine compound with very low antidopaminergic potency: despite it was originally included among antipsychotics at the beginning of the neuroleptic era, basically it is never used as antipsychotic, being instead widely used in the management of psychomotor agitation.

11.4.3 Second Generation Antipsychotics (SGA)/ Atypical Antipsychotics

11.4.3.1 Pharmacodynamics

As mentioned above, the distinguishing feature of SGAs is the serotonin antagonism elicited through 5HT2Ar blocking activity, which, in addition to anti-D2r properties, gives a peculiar "atypical profile" to these pharmacological compounds.

Given that serotonin physiologically inhibits dopamine release, the blockade of 5HT2Ar is associated with higher dopaminergic activity in the brain. However, 5-HT2A and D2 receptors show differential distribution according to the different brain areas, and the effects of the double dopaminergic/serotoninergic antagonism are variable throughout the brain, depending on receptorial density and SGA's receptorial binding profile. By blocking 5HT2Ar activity, atypical antipsychotics increase dopaminergic activity in nigrostriatal, mesocortical, and tuberoinfundibular pathways, but fail to reverse antipsychotic D2 antagonism in the mesolimbic system. Therefore, compared to FGAs, use of SGA should be associated with better neurological tolerability, lower rate of hyperprolactinemia, lower drug-induced negative/cognitive symptoms, and similar efficacy on positive symptoms.

However, although 5HT2Ar/D2r antagonism activity explains a large part of the atypical profile, it is not sufficient to understand all the pharmacological properties of SGAs. As an example, risperidone partially loses its atypical properties at higher dosages and induces hyperprolactinemia, and amisulpride is considered atypical, although lacking 5HT2A affinity and inducing hypeprolactinemia. Generally speaking, a drug is clinically considered as an "atypical second generation" antipsychotic when showing a high antipsychotic efficacy (particularly on positive symptoms) and a low risk to induce extrapyramidal side effects and hyperprolactinemia, but this association has a wide variety of manifestation depending on peculiar profiles on doses. Every antipsychotic displays a unique receptor binding profile that also includes affinities for other serotoninergic and dopaminergic receptors (i.e., 5HT2C, D1, D3, and D4), as well as for other neurotransmitter pathways such as noradrenergic, cholinergic, and histaminergic systems. Taken together, all the single receptor affinities determine the overall mechanism of action of every SGAs.

11.4.3.2 Efficacy

The introduction of SGAs was seen as a revolution in the treatment of schizophrenia. The innovation of SGAs was a serotoninergic/dopaminergic antagonist activity, associated with better extrapyramidal tolerability and that was also thought to be effective in ameliorating positive symptoms resistant to conventional antipsychotics. Moreover, initially major claims were also made concerning a possible better efficacy of atypical antipsychotics on negative and cognitive symptomatology, which could have revolutionized the treatment of schizophrenia.

Nowadays, with the sole exception of clozapine in resistant schizophrenia, large-scale studies and meta-analytic evidence now indicate no superior efficacy of SGAs over FGAs in treating positive, negative, and cognitive symptoms. Nonetheless, the introduction of SGAs undoubtedly represented a step forward in psychiatric clinical practice, leading to lower rates of iatrogenic EPS, and also reducing the impact of other significant side effects such as sedation and/or hyperprolactinemia. However, SGAs introduced the new issue of metabolic tolerability, as they are more frequently associated with metabolic syndrome, obesity, type 2 diabetes, dyslipidemia and cardiovascular side effects.

High variability characterizes receptor binding profiles of SGAs, resulting in significant differences in drug-related side effects and antipsychotic efficacy. The tolerability profile of each drug, along with specific and individual features, should be taken into consideration before choosing a certain antipsychotic for administration.

SGAs are usually preferred and recommended in the early stages of treatment (i.e., First Episode Psychosis), although they have similar antipsychotic efficacy to FGAs. Indeed, at the onset of the disease, young patients are more likely to experience acute neurological side effects, which should be avoided to favor pharmacological compliance and long-term clinical stabilization.

Risperidone and Paliperidone

Risperidone is one of the most prescribed SGAs in clinical practice. It shows a high affinity for D2r and, at higher dosages, results similar to FGAs, potentially causing EPS and hyperprolactinemia. That means that is should be used at low doses to obtain the highest potential benefit. Risperidone shows a high efficacy on positive symptoms of schizophrenia, and it is also commonly used in the management of elderly patients with psychosis, agitation, and behavioral disturbances associated with dementia. The incidence rate of metabolic adverse effects is lower compared to clozapine, olanzapine, and quetiapine.

Risperidone is extensively metabolized in the liver by cytochrome CYP 2D6 to 9-OH-risperidone also named paliperidone, its active metabolite. Over the last decade, paliperidone has been separately produced and marketed in oral and intramuscular long-acting formulations. Compared to risperidone, paliperidone shows similar efficacy, but no hepatic metabolism and longer half-life, which makes it more suitable for patients with hepatic dysfunction.

Olanzapine

This SGA shows binding properties for several receptor classes and relatively lower affinity for D2r. It is thus associated with a lower rate of EPS and hyperprolactinemia, but its antihistaminic and serotonin 2C antagonist properties frequently determine metabolic side effects, being the SGAs most frequently associated with metabolic syndrome (together with clozapine).

Oral dose is rapidly absorbed and a relatively high percentage undergoes to a predominant CYP1A2 metabolism, but also CYP2D6 is involved to a lesser extent in the metabolism of the drug. Approximately, time to peak concentration is reached within 6 h with 93% of serum proteins binding potential.

Olanzapine is indicated for the treatment of acute phases and maintenance of schizophrenia, moderate to severe manic episode and prophylaxis of manic or depressive episodes in bipolar disorder.

Quetiapine

Quetiapine shows a very atypical binding profile, virtually causing no EPS and prolactin elevations, and it is thus frequently used among patients with Parkinson's disease and L-dopa-induced psychosis. Compared to risperidone and olanzapine is less effective in reducing positive symptoms, although there are psychotic patients that show a satisfactory antipsychotic response to this atypical compound. It is also prescribed for the treatment of bipolar disorder and in behavioral psychopathological symptoms in dementia. Quetiapine is associated with weight gain, as it strongly blocks histamine H1 receptors. Due to its antihistaminergic properties, quetiapine is also used as a second-line treatment for insomnia. Depending on dosages prescribed quetiapine shows different effects ranging from anxiolytic and hypno-inducing properties (25–100 mg), in particular in organic and psychogeriatric conditions, to antipsychotic ones which are reached in a range from 400 to 800 mg.

Concerning pharmacokinetics, quetiapine is well absorbed after oral administration with a time to peak concentration of 1 h and a half-life approximately of 6 h.

Aripiprazole

Aripiprazole has a peculiar pharmacodynamics profile, acting as a partial agonist at the dopamine D2r and the serotonin 5-HT1Ar, and as an antagonist at the serotonin 5-HT2Ar. Differently from other antipsychotics, aripiprazole does not reduce dopaminergic activity by blocking D2r but modulates its activity according to pathway-related dopaminergic activity. Indeed, aripiprazole reduces D2r activity if dopamine transmission is overly active, whereas it increases neurotransmission in case of dopaminergic hypofunction. Aripiprazole has the propensity to induce akathisia but is much less frequently associated with other D2r related side effects such as EPS and hyperprolactinemia. Differently from other SGAs, aripiprazole is rarely associated with metabolic side effects. As quetiapine and olanzapine, beside schizophrenia it is also indicated in case of bipolar disorder, both as treatment maintenance and in the management of manic episodes. The common dosages used by clinicians range from 2.5–5 to 30 mg, which is the maximum dose approved. Aripiprazole is well absorbed after oral intake reaching maximum plasma concentrations in nearly 3–5 h with a high percentage of molecules bound to plasma proteins (99%). Hepatic metabolism is mediated by CYP3A4 and CYP2D6.

11.4.3.3 Other Atypical Antipsychotics

Brexpiprazole

Brexpiprazole is a new SGA recently approved for the treatment of schizophrenia and as adjunctive therapy for major depressive disorder. Similarly to aripiprazole, it is a partial serotoninergic and dopaminergic agonist. Brexpiprazole shows a higher affinity on 5HT1A/2A receptors, that explains its indication also for mood disorders. Its tolerability profile is similar to aripiprazole, characterized by a low risk for EPS, hyperprolactinemia, and metabolic alterations.

Cariprazine

Cariprazine is a recent antipsychotic indicated for the treatment of schizophrenia and bipolar disorder, which shows serotoninergic and dopaminergic partial agonism resembling the pharmacodynamics of aripiprazole and brexpiprazole. Studies suggest higher efficacy of cariprazine as monotherapy for schizophrenia with prevalent negative symptoms.

Amisulpride

Amisulpride shows peculiar pharmacological properties. Despite being considered a SGA, it belongs to substituted benzamides, characterized by a nearly absent serotoninergic antagonism and high specificity for D2 and D3 receptors. Low dosages of amisulpride selectively block presynaptic receptors resulting in a higher release of dopamine, thus leading to an antidepressant effect. At high doses, amisulpride block D2r activity and determine an antipsychotic effect. Amisulpride induces significant QT prolongation (careful ECG management is required) as well as hyperprolactinemia, but lower EPS than FGAs.

Ziprasidone

Ziprasidone was the fifth atypical antipsychotic marketed. It shares the anti-5HT2Ar and anti-D2r profile of other available atypical antipsychotics. Moreover, it shows 5HT1Ar agonistic activity and weak inhibiting properties of serotonin/norepinephrine reuptake. Metabolized in the liver by CYP3A4 and CYP1A2, its half-life is approximately 6–7 h with an extensive serum proteins binding of 99%. Clinical indications are represented by schizophrenia and mania and therapeutic dosages range between 40 and 80 mg, with a maximum licensed dose of 160 mg.

Ziprasidone shows an optimal metabolic tolerability profile, as well as lower rates of hyperprolactinemia. However, ziprasidone has a significant impact on QTc, with an average prolongation of approximately 20 ms. Nowadays, it is infrequently prescribed.

Lurasidone

Lurasidone is approved for the treatment of patients with schizophrenia or bipolar depression; it is a potent antagonist of dopamine D2, serotonin 5-HT2A, and 5-HT7 receptors, and a partial agonist of 5-HT1A receptors. Lurasidone does not show any significant antihistaminic or anticholinergic activities and, thus, is not associated with metabolic alterations, sedation, or constipation, and some effects on cognition have been reported in literature. Its high affinity for D2r provides, on one side, good efficacy in treating positive symptoms, but, on the other, also a dose-related risk for akathisia, EPS, and hyperprolactinemia.

11.4.3.4 Pharmacokinetics of FGA and SGA

Antipsychotics are highly lipophilic, thus show high plasma protein binding. Peak plasma concentration generally is reached within 2–4 h except for haloperidol

Table 11.10 Pharmacokinetic of main antipsychotics

Antipsychotic	*T*max	Half-life (h)	Time to steady state (day)	Plasma protein binding (%)
Chlorpromazine	1–4	23–37	7	90
Haloperidol	1.7–6.1 h	14.5–36.7	5	90
Zuclopentixol	4 h	20	7	98
Amisulpride	1–4 h	12	6–10	17
Aripiprazole	3–5 h	75	14	99
Clozapine	1.1–3.6 h	9.1–17.4	7–10	95
Lurasidone	1.5–3 h	28.8–37.4	7	99
Olanzapine	6 h	33	7	93
Paliperidone	24 h	22	5	77
Quetiapine	1–1.5 h	6–7	3	83
Risperidone	1.6 h	3–22	5	89

which is more lipophilic. Typically, half-life ranges between 10 and 40 h (see Table 11.10).

Concerning metabolism, antipsychotics are manly oxidized (CYP 450) or conjugated with glucuronic acid in the liver with the aim to increase hydrophilic properties for renal excretion.

11.4.3.5 Therapeutic Indications and Clinical Use of FGA and SGA

Nowadays, antipsychotics are commonly used in clinical practice for several conditions:

- Schizophrenia spectrum disorders;
- Mood disorders (acute mania, augmentation strategies in unipolar or bipolar depression, bipolar disorder maintenance treatment, adjunctive treatment of major depressive disorder with psychotic symptoms);
- Delirium;
- Severe behavioral problems (also in childhood and autism spectrum disorders) or aggression;
- Gilles de la Tourette syndrome.
- Augmentation strategies in obsessive-compulsive disorder (OCD).

Therapeutic doses of main antipsychotics are shown in Table 11.11.

11.4.3.6 Side Effects of FGA and SGA

Extrapyramidal Symptoms (EPS) and Pse-udo-Parkinsonism

Antagonism of dopaminergic neurotransimission in the nigrostriatal pathway can lead to neurological motor side effects named extrapyramidal symptoms (EPS).

These symptoms, resembling those observed in Parkinson's disease (and therefore defined as "iatrogenic Parkinsonism"), include akinesia, bradykinesia, tremors,

EPS are:
- Dose-related.
- Most likely with high doses of FGA.
- Less common with SGA, particularly clozapine and quetiapine.

rigidity, and acute dystonia (as detailed in Table 11.12). Long-term antipsychotic treatment, particularly with FGAs, can also induce the development of tardive dyskinesia, which is one of the most insidious side effects of antipsychotics, highly interfering with daily functioning and quality of life and often irreversible.

Table 11.11 Daily dosage of main antipsychotics

Antipsychotic	Starting (mg)	Maintenance (mg)	High dosage (mg)
Chlorpromazine	25	300–400	600–1000
Haloperidol	1	3–6	8–10
Zuclopenthixol	10	20–40	75–100
Amisulpride	25	400–800	1200
Aripiprazole	5	15–20	30
Clozapine	25	400[a]	800[a]
Lurasidone	37	74	148
Olanzapine	5	10–15	20
Paliperidone	3	6–9	12
Quetiapine	50	400–600	800
Risperidone	0.5	3–5	6–8

[a]Mean dose, prescription should be based on clozapine's plasmatic levels (range 350–600 ng/mL)

Table 11.12 Extrapyramidal symptoms

EPS	Symptoms	Onset	Mechanism	Prevalence
Acute dystonia	Acute EPS: muscular spasm, typically of tongue, eyes, face, neck	Hours/days	Unknown	2.5–5%
Akathisia	Subacute EPS: inner tension and restlessness; inability to sit still or remain motionless	Days 1–60	Unknown	25%
Parkinsonism	Subacute EPS: tremor, hypokinesia, muscular rigidity, postural instability, amimia	Days 5–30	DA antagonism in basal ganglia	20–40%
Tardive dyskinesia	Involuntary and spontaneous muscular movements	Years	DAr supersensitivity	20–30% Annual incidence: 4–5%

The time of presentation of acute EPS is usually days to weeks after antipsychotic drugs are started or the dose is increased, whereas the onset of tardive dyskinesia may occur even after years of treatment.

Anticholinergic Agents

In the nigrostriatal pathway a reciprocal interconnection exists between dopamine and acetylcholine.

Dopamine resizes acetylcholine activity: if dopamine receptors are blocked, acetylcholine becomes overly active contributing to EPS.

Anticholinergic drugs reduce EPS decreasing acetylcholine activity.

Treatment: reduction of the antipsychotic dose, switch to an antipsychotic with lower propensity to induce EPS (lower D2r affinity) or prescription of an anticholinergic medication are therapeutic options that need to be considered. When none of these options result effective, antipsychotic switch to clozapine, whose other indication is schizophrenia intolerant to treatment, beside resistance, is often resolutive.

Acute Dystonia

Uncontrolled muscular spasms can occur within hours to days after antipsychotic treatment is started, more frequently using FGAs. Intramuscular or intravenous administration can induce dystonic contractions within minutes. Common manifestations are oculogyric crisis, spasmodic torticollis, mandibular or buccal dystonia, blepharospasm, limbs, and back pain. The two most concerning presentations are laryngospasm, rare but life-threatening, and oculogyric crisis, a highly distressing tonic and recurring deviation of the eyes.

Treatment: depending on the severity of symptoms, anticholinergic drugs can be administered intramuscularly or (with extreme caution and in the most severe cases, due to cardiological risk of injecting an anticholinergic) intravenously, as well as intravenous benzodiazepines. Subsequent switching to an antipsychotic with a low propensity to induce EPS is recommended as patients who experienced acute dystonia are more vulnerable.

Akathisia

Most antipsychotics (both FGA and SGA) can induce a subjective sense of mental unease, a inner feeling of tension, and observable motor restlessness. Typically, it is observed an inability to remain still and to maintain postions, "on-site" march, and sometimes also psychiatric symptoms such as anxiety, dysphoria, and self-aggressiveness. Akathisia is a relatively common adverse effect experienced hours to weeks after antipsychotic treatment is started.

Treatment: besides common techniques adopted to avoid EPS (antipsychotic switch or dose reduction), an effective strategy is to use benzodiazepines such as clonazepam or add propranolol. Notably, anticholinergic medications are generally unhelpful unless akathisia is associated with a iatrogenic parkinsonian syndrome.

Tardive Dyskinesia

A chronic blockade of D2 receptors in the nigrostriatal dopaminergic pathway secondary to long-term antipsychotic administration can induce a hyperkinetic movement disorder known as tardive dyskinesia (TD), a late-onset extrapyramidal syndrome. Typically, TD affects facial muscles (small muscles), causing grimacing, buccal, and tongue movements such as tongue protrusion or constant munching. Limb movements, particularly evident in distal areas, are less common. Severe orofacial dyskinesia is highly disfiguring and may greatly interfere with speech, eating, swallowing, or breathing, while truncal dystonia can be extremely distressing and interfere with gait and mobility. Respiratory dyskinesia may produce tachypnea, irregular breathing rhythms, and grunting noises that are commonly misinterpreted as primary respiratory problems.

Etiopathogenesis of TD has not been fully understood yet; however, most consistent evidence suggest that chronic antipsychotic treatment lead to a dopamine receptor supersensitivity, and to an imbalance between dopamine type 1 (D1) and type 2 (D2) receptors related activities in the basal ganglia.

The incidence of TD is about 5% for every year of conventional antipsychotics treatment maintenance, whereas SGAs are associated with a lower incidence of TD than FGAs, with an annual incidence rate approximately 25% lower and is absent or questionable with clozapine.

TD is sometimes reversible, indeed early identification and suspension of antipsychotic, lead to symptom remission in 50–90% of patients, especially among young subjects. However, it is to note that several months are usually needed for TD remission, sometimes up to 3 years after antipsychotic discontinuation.

Treatment: switch to another antipsychotic with low D2r affinity: clozapine is the first choice drug, being associated with higher rates of symptomatic resolution (the effects seems to be active due to the balance of D1 and D2 receptor blockade of this drugs, beside the low D2 activity in general). Interestingly patients with tardive dyskinesias are frequently also resistant to typical antipsychotic and for this reason may have been exposed to higher cumulative doses, with is a risk factor itself: when prescribed clozapine improves usually both resistant symptoms and tardive dyskinesia). Clozapine is also effective in treating a great number of tardive dystonias, the most resistant manifestation of long-term motor side effect of antipsychotics.

When contraindicated, quetiapine represents a possible antipsychotic alternative to clozapine. In tardive dystonia tetrabenazine, velbenazine, or deutetrabenazine may represent valid add-on treatments.

Hyperprolactinemia

Antidopaminergic properties of antipsychotics account for the disinhibition of prolactin release in the tuberoinfundibular pathway through the removal of the inhibition induced by dopamine. Increased plasma concentrations of prolactin identify a condition defined as hyperprolactinemia, which is a common side effect of antipsychotic treatments, although in clinical practice it is frequently underdiagnosed and treated only when symptomatic. Symptoms vary according to patient's sex. Typical females' symptoms are galactorrhea (breast secretions), oligomenorrhea and amenorrhea, vaginal dryness, dyspareunia, and decreased libido as well. Similarly, males can show gynecomastia (rarely galactorrhea), impotence, and decreased libido too. Hyperprolactinemia may also interfere with fertility, particularly in women.

The relationship between prolactin concentrations and occurrence of clinical symptoms of its elevation is linear, although there is high clinical variability of symptoms prevalence and severity between subjects and different antipsychotics. It is important to check for prolactin levels and related symptoms related due to different reasons including compliance, as sexual side effects are strongly related to bad compliance in (particular in young males), and health in general, as higher levels are associated with weight gain and reduced bone density (particularly in women), whereas a link with breast cancer has not yet clearly demonstrated, although suggested in some studies.

Treatment: First-line treatment is switch to other antipsychotics with a low risk of inducing hyperprolactinemia (clozapine and aripiprazole in particular). Other strategies include additional treatment with low doses of aripiprazole (partial D2 receptors agonist), or with very low-dosage of cabergoline (dopamine agonist) given weekly or biweekly. This latter strategy may, however, lead to a psychopathological worsening at higher doses or daily dosing and should be considered only if other approaches fail.

Neuroleptic Malignant Syndrome (NMS)

The neuroleptic malignant syndrome is a rare but potentially life-threatening acute disorder of thermoregulation and neuromotor control that can cause an emergency condition characterized by an extreme and generalized muscular rigidity associated with hyperpyrexia, dysautonomia, change in mental status/consciousness, renal failure and even death. Rhabdomyolysis, myoglobinuria, creatine kinase (CK) elevation and neutrophilia are usual laboratory findings.

Among patients taking antipsychotics incidence of NMS ranges from 0.02 to 3%. Again, D2 receptors antagonism is thought to be the leading pathogenetic mechanism, in particular the central dopamine receptor blockade in the hypothalamus that could lead to hyperthermia and other dysautonomic symptoms. Genetic predisposition (familiarity), high potency FGA, polypharmacy, dehydration, and male gender, are listed among the main risk factors. Although less frequently, atypical NMSs induced by low-potency antidopaminergic agents are also reported, characterized by a different clinical presentation with tachycardia, mental status changes, and diaphoresis, but not including rigidity. Early diagnosis and treatment significantly improve prognosis of NMS, that is characterized by and mortality rates ranging from 5 to 20%. Acute renal failure induced by rhabdomyolysis and myoglobinuria is considered the most serious independent predictor of mortality. Therefore, severe manifestations require intensive care unit monitoring and treatment in order to avoid permanent morbid sequelae and death. When NMS is successfully treated, symptom remission usually occurs within 10 days, although sometimes residual catatonia and motor signs are reported up to 6 month after onset, especially among patients treated with LAIs.

Treatment: removal of the causative antipsychotic agent, rehydration (cold intravenous fluids), benzodiazepines and administration of bromocriptine and/or dantrolene.

11.4.3.7 Other Side Effects

As previously discussed, D2 receptors antagonism is the main pharmacodynamic activity of antipsychotics, but both FGAs and SGAs shows several other receptor binding properties associated with both clinical effectiveness and drug tolerability. Every antipsychotic has a unique receptor binding profile, and therefore can lead to different possible side effects according to its pharmacodynamics properties.

> Haloperidol has relatively low anticholinergic or antihistaminic binding properties.

Muscarinic Cholinergic Antagonism

The ability of antipsychotics to block muscarinic cholinergic receptors can lead to several adverse effects such as constipation, cognitive blunting, dry mouth, and blurred vision. The entity of cholinergic blockade may explain the lesser propensity of some antipsychotics to induce extrapyramidal symptoms, given the "brain balance" between dopamine and acetylcholine. Specifically, fewer EPS are correlated with stronger anticholinergic properties. On the other hand, a strong cholinergic blockade can worsen the cognitive impairment of schizophrenia or precipitate a delirium or a cognitive disorder.

Treatment: dosage reduction or switch to another antipsychotic with lower anticholinergic affinity.

Histaminic Antagonism

Several FGAs and SGAs have an antihistamine activity leading to side effects such as weight gain and sedation. Sedative properties vary across antipsychotics representing both a side effect, if sedation is unwanted, and/or a therapeutic effect in case of aggressive behavior or insomnia.

Treatment: dosage reduction or switch to another antipsychotic with lower histaminic affinity.

Alpha-1-Adrenergic Antagonism

Alpha-1 adrenergic blocking activity can lead to cardiovascular side effects such as orthostatic hypotension, dizziness, tachycardia, and drowsiness.

Treatment: dosage modulation, or addition of alpha-1-agonists (midodrine) counteracting orthostatic hypotension through vasopressor and antihypotensive properties. To treat tachycardia beta-blockers are often prescribed.

ECG Alteration: QT Prolongation

Several antipsychotics, both FGAs and SGAs, are associated with prolongation of QT interval. Extreme QT prolongation is associated with increased risk for developing of polymorphic ventricular tachycardia, a peculiar life-threatening cardiac arrhythmia also defined as torsades de pointes. Pimozide, ziprasidone, and amisulpride have the highest increasing effect on QT interval, while aripiprazole and lurasidone are considered the safest. Among the others, SGAs, haloperidol and chlorpromazine have a moderate propensity to induce QT prolongation, although there is great variability among atypical antipsychotics. Electrolyte disorders (sodium, potassium, magnesium) are associated with higher risk of cardiac arrhythmia, as well as use of concomitant drugs known to prolong the QT interval such as antiarrhythmics, macrolide and fluoroquinolone antibiotics, some antifungal and antiviral drugs, and TCAs. In this context, pharmacological associations require particular attention, since concomitant treatment may also alter antipsychotic metabolism leading to higher plasmatic concentrations and thus further increasing risk of QTc prolongation.

QT interval changes according to the heart rate, shortening as heart rate increases. Therefore, proper evaluation of cardiac tolerability should include the examination of QT controlled for heart rate (QTc), in order to evaluate standardized values. Moreover, sex related differences must be take into account as well, since women physiologically exhibit longer QTc intervals.

When a patient shows a basal QT prolongation, SGAs with a low propensity to interfere with cardiac conduction should be prescribed (e.g., olanzapine, aripiprazole, and lurasidone).

Metabolic Side Effects

Antipsychotics, especially SGAs, are burdened by metabolic side effects that negatively impact long-term prognosis and life expectancy (see Table 11.13). Weight gain is indeed frequently associated with atypical treatments, typically occurring from the early phases of the treatment: the most substantial weight gain is observed among antipsychotic naive patients after the first 6 weeks of treatment, usually reaching a plateau along the course of the illness. Unfortunately, weight gain is

Table 11.13 Metabolic side effects of main atypical antipsychotics

Agent	Weight gain	Dyslipidemia	Diabetes
Amisulpride	+	–	+
Aripiprazole	+/–	–	–
Clozapine	+++	++	+++
Lurasidone	–	–	–
Olanzapine	+++	++	+++
Quetiapine	++	+	++
Paliperidone	++	+/–	++
Risperidone	++	+/–	++

– not or only very rarely present, + rarely present, ++ occasionally present, +++ often present

frequently hard to reverse, even after switching to more weight-neutral antipsychotics. The underlying mechanism is not fully understood, and the weight gain does not seem dose-dependent. Nevertheless, it is a matter of fact that clozapine and olanzapine show the greatest affinity for 5-HT2C and H1 receptors and have the greatest weight gain potential, making these receptors plausible mediators of the pathogenetic process. Altered glycemia, type II diabetes, or worsening of preexisting diabetes are other metabolic adverse effects that can occur even regardless of weight gain. Moreover, alterations of lipid homeostasis are frequently reported during antipsychotic treatment, characterized by an increase in circulating levels of low-density lipoprotein (LDL) and triglycerides, and decreased levels of high-density lipoprotein (HDL) cholesterol.

The aforementioned metabolic alterations together contribute to the establishment of a condition defined as metabolic syndrome, associated with an increased risk for cardiovascular disease and type 2 diabetes mellitus. It is to note that schizophrenia is associated with metabolic and cardiovascular risk factors (sedentary lifestyle, poor quality food intake, high rates of smoking) regardless of antipsychotic medications.

Prevention and treatment of metabolic side effects and metabolic syndrome are based on aerobic exercise and diet. Moreover, dopamine partial agonists (i.e., aripiprazole), ziprasidone, and lurasidone have minimal adverse effects on metabolic parameters and may represent potential alternative treatments in case of severe metabolic alterations.

11.4.3.8 Clozapine: The Atypical SGA

Clozapine was the first licensed and marketed SGA in the 1970s and, to date, still represents the most effective and atypical SGA in resistant schizophrenia, although its precise mechanisms of action are not fully understood yet.

Clozapine is the only antipsychotic with a demonstrated efficacy in treatment-resistant schizophrenia (TRS).

> **TRS**
> Insufficient response to sequential treatment with at least 2 antipsychotics (FGA or SGA).

Clozapine has a unique receptorial binding profile, characterized by low D2r affinity relatively high affinity for D1 receptors and higher antiserotoninergic properties, showing the highest 5HT2A/D2 ratio. Moreover, differently from other antipsychotics, clozapine is a glutamatergic agonist at NMDA receptors. This peculiar activity may underlie its superior efficacy in TRS.

> Low titration of clozapine is a good clinical practice to avoid several adverse effects.

Nevertheless, to date, clozapine is not yet considered a first-line treatment due to its peculiar tolerability profile. Indeed, besides dopaminergic, serotoninergic, and glutamatergic activity, clozapine also shows a high affinity for histaminic, cholinergic, and adrenergic receptors, leading to a wide range of side effects. Moreover, it is also associated with the risk of a rare but severe hematological complication called agranulocytosis, a severe form of granulocytopenia.

Still, due to its low propensity to induce EPS, clozapine has two additional indications. First, it represents the only indicated and effective treatment for patients with schizophrenia developing severe neurological side effects such as tardive dyskinesia. Second, it is recommended for patients with comorbid psychotic disorders and Parkinson's disease.

Side Effects of Clozapine

Agranulocytosis/Granulocytopenia

In 1975 clozapine was withdrawn because of its association with agranulocytosis, a severe side effect that occurs in about 0.5–2% of patients, with up to 3% of patients that may develop neutropenia. In 1988, the publication of Kane and colleagues concerning the higher efficacy of clozapine among TRS patients paved the way to its reintroduction in clinical practice, although strictly monitored and limited to TRS. Since then, patients must have their blood count monitored weekly for the first 18 weeks of treatment (over 80% of cases occur within this period), then every month for as long as they are treated with clozapine, although worldwide there are variations among monitoring recommendations as well as granulocytes threshold for clozapine cessation. Almost 30 years after clozapine reintroduction, to date large scale studies indicate that the risk of hematological adverse effects decreases along with the duration of the treatment and with age, with 85–90% of cases occurring within 1 year after clozapine introduction. Consistently, it is esteemed that after 1 year of treatment the risk of agranulocytosis is comparable with that of other antipsychotics. Based on this evidence, different researches proposed a revision of clozapine-related surveillance protocols (not yet done), suggesting a less strict monitoring after 1 year of treatment.

Associations between clozapine and carbamazepine are to be avoided due to increased risk of agranulocytosis.

Cardiac Side Effects

Tachycardia. It is a very common side effect, especially during the first 4 weeks of treatment. It is usually benign but sometimes may persist, requiring medication with beta-blockers.

QT prolongation. With respect to other FGAs and SGAs clozapine is relatively safe; however, clozapine has been associated with possible dose-dependent QT interval prolongation.

Myocarditis. The most feared cardiovascular side effect is clozapine-induced myocarditis, a hypersensitivity response, eosinophil-mediated, hesitating in myocardial inflammation. Despite subclinical inflammation, identified by elevation of troponin, is a relatively frequent finding, particularly in the early phases of treatment, myocarditis is an emergency that requires immediate clozapine cessation. The incidence of myocarditis is debated, ranging from 1 to 3% of clozapine users. Myocarditis is most likely to occur within the first 6–8 weeks of treatment (80–90% of cases), although cases during chronic treatment are also described. Rapid dose titration, older age, male sex, and concurrent medication with other compounds such as sodium valproate appear to be risk factors.

Myocarditis and Clozapine

Clinical features

Symptoms: fever, hypotension, fatigue, tachycardia, dyspnea, and chest pain.

Laboratory findings: C-reactive protein (CRP) and troponin elevation and eosinophilia.

Diagnostic gold standard: echocardiography and cardiac magnetic resonance.

Monitoring

Daily: pulse, blood pressure, temperature, respiratory rate, subjective reports of symptoms.

Weekly: CRP, troponin, full blood count, ECG if possible.

Discontinuation of clozapine and investigation through echocardiography or MRI are recommended if either troponin exceeds twice the upper limit of normal or CRP is more than 100 mg/L.

If raised troponin or CRP not as above: maintenance of treatment and daily CRP and troponin monitoring.

Hypersalivation (Sialorrhea)

Like other antipsychotics, clozapine shows muscarinic antagonism on M1, M2, M3, and M5 receptors but it is also a full agonist on the M4 subset, highly expressed in salivary glands. Despite pharmacological bases remain unclear, the agonist activity on M4 receptors is thought to be the mechanism underlying hypersalivation. Other plausible explanations are adrenergic alpha-2 antagonism or inhibition of the swallowing reflex. Hypersalivation occurs in almost 30% of patients treated with clozapine and negatively impacts quality of life, sometimes leading to drug discontinuation.

Gastrointestinal Hypomotility

An under-recognized and potentially life-threatening adverse effect of clozapine treatment is gastrointestinal hypomotility, which can hesitate in severe constipation, paralytic ileus, and obstruction. Older age, female sex, obesity, and high doses of clozapine are among the risk factors for gastrointestinal hypomotility.

Seizures

Clozapine lowers the seizure threshold in a dose-related manner, especially if rapid dose escalation occurs. For this reason, it is not recommended in patients suffering from epilepsy.

11.4.3.9 Antipsychotics: Pharmacological Interactions

Main drug interactions of antipsychotics are related to the cytochrome P450 (CYP) pathway. Concomitant administration of CYP inducers and inhibitors may lead to increases and decreases of antipsychotic plasma concentrations, thus altering drug clinical efficacy and tolerability. Among psychiatric drugs, SSRIs such as paroxetine and fluvoxamine strongly inhibit CYP, leading to higher antipsychotic plasma concentrations, whereas antiepileptic mood stabilizers, especially carbamazepine are known to induce CYP and to decrease FGA and SGA plasmatic levels. Smoking should be also be taken into account, as it can induce CYP dose-dependently.

Beside potential variations in plasmatic concentrations, when prescribing antipsychotics clinicians should be always aware of the individual tolerability profile of each FGA/SGA in order to avoid potential cumulative tolerability issues with other concomitant drugs (i.e., antiarrhythmics and QTc prolongation). In this view, administration of antipsychotics with sedative properties (i.e., clozapine, olanzapine, and chlorpromazine) should be carefully administered in combination with other CNS depressants.

Box 11.1: Sedation and Agitation

Different compounds have sedative properties, usually perceived as side effects (sedation) but useful in clinical practice to induce sleep or manage agitation. Antipsychotics' sedative properties are dose-dependent and proportional to H1r and 5HT2Ar antagonism.

Antipsychotics' propensity to induce sedation is:

- Very high for clozapine, chlorpromazine.
- High for quetiapine, olanzapine, asenapine.
- Moderate for haloperidol, risperidone, lurasidone, aripiprazole (high dose).
- Low for paliperidone.
- Very low/absent cariprazine, amisulpride.

In clinical practice, Promazine (FGA) is widely used for the management of aggressiveness or agitation.

Quetiapine, at very low doses, is an off-label option for the treatment of insomnia. Guidelines recommended quetiapine only in patients with specific comorbid psychiatric disorders when other strategies have failed.

Benzodiazepines (BDZ) have also a sedative effect depending on dose and way of administration. Intramuscular administration of BDZ is commonly used alone or in co-administration with antipsychotics in aggressive/agitated patients to induce sedation.

BDZ are efficacious for sleep disturbances in elderly people but increase the risk of cognitive impairment, ataxia, and motor disturbances.

Molecules with primary antihistaminic properties (hydroxyzine) are useful tools to induce sedation/drowsiness, thus overcoming insomnia. Intramuscular co-administration with antipsychotics is a common practice to obtain sedation (i.e. promethazine + haloperidol). Notably, antihistaminic compounds prolong QT interval similarly to antipsychotics.

According to guidelines, management of agitation depends on the related causes and features. If agitation is associated with delirium but not with substance withdrawal, BDZ should be avoided, preferring low dose antipsychotics (i.e. risperidone, haloperidol). In presence of substance withdrawal, oral and parenteral BDZ should be prescribed, possibly associated with antipsychotics as second line treatment. Agitation not associated with delirium but with psychosis should be treated using oral/intramuscular antipsychotics as first-line treatment, eventually associated with BDZ (oral or parenteral).

Box 11.2: Long-Acting Injectables (LAI)—Depot Antipsychotics

LAIs (depot antipsychotics) are frequently prescribed to improve adherence, especially in subjects with poor insight and pharmacological compliance, and are highly effective for maintenance treatment. There are no significant differences in efficacy between oral and depot formulations, neither in FGA nor SGA depots. Before starting treatment with LAI, when possible, the oral formulation should be administered for a brief period in order to evaluate antipsychotic response and tolerability. Haloperidol, risperidone, paliperidone, aripiprazole, and olanzapine are the LAIs most frequently used in clinical practice. Time administration interval for LAIs administration is usually 28 days, but can range from 15 days (risperidone) up to 3 months (paliperidone), according to the different antipsychotic.

Notably, there are two different long-acting formulas for paliperidone (28 days or 3 months).

Further Reading

American Psychiatric Association. Diagnostic and statistical manual of mental disorders (DSM-5). Washington, DC: APA; 2013.

Balamane M, Kolakowsky-Hayner SA. Psychopharmacology. In: Kreutzer JS, De Luca J, Caplan B, editors. Encyclopedia of clinical neuropsychology. Berlin: Springer; 2018.

Belli H, Ural C, Akbudak M. Borderline personality disorder: bipolarity, mood stabilizers and atypical antipsychotics in treatment. J Clin Med Res. 2012;4(5):301–8.

Benedetti F, Dallaspezia S, Lorenzi C, Pirovano A, Radaelli D, Locatelli C, Poletti S, Colombo C, Smeraldi E. Gene-gene interaction of glycogen synthase kinase 3-β and serotonin transporter on human antidepressant response to sleep deprivation. J Affect Disord. 2012;136(3):514–9. Epub 2011 Nov 25. https://doi.org/10.1016/j.jad.2011.10.039.
Fava M. Weight gain and antidepressants. J Clin Psychiatry. 2000;61(Suppl 11):37–41.
Fountoulakis KN, Vieta E, Sanchez-Moreno J, Kaprinis SG, Goikolea JM, Kaprinis GS. Treatment guidelines for bipolar disorder: a critical review. J Affect Disord. 2005;86(1):1–10. https://doi.org/10.1016/j.jad.2005.01.004.
Hasan A, Falkai P, Wobrock T, Lieberman J, Glenthoj B, Gattaz WF, Thibaut F, Möller HJ. World Federation of Societies of biological psychiatry (WFSBP) task force on treatment guidelines for schizophrenia. World Federation of Societies of biological psychiatry (WFSBP) guidelines for biological treatment of schizophrenia, part 1: update 2012 on the acute treatment of schizophrenia and the management of treatment resistance. World J Biol Psychiatry. 2012;13(5):318–78. https://doi.org/10.3109/15622975.2012.696143.
Huhn M, et al. Comparative efficacy and tolerability of 32 oral antipsychotics for the acute treatment of adults with multi-episode schizophrenia: a systematic review and network meta-analysis. Lancet. 2019;394(10202):939–51.
Khan SJ, Fersh ME, Ernst C, Klipstein K, Albertini ES, Lusskin SI. Bipolar disorder in pregnancy and postpartum: principles of management. Curr Psychiatry Rep. 2016;18(2):13.
Lim B, Sproule BA, Zahra Z, Sunderji N, Kennedy SH, Rizvi SJ. Understanding the effects of chronic benzodiazepine use in depression: a focus on neuropharmacology. Int Clin Psychopharmacol. 2020;35(5):243–53.
Mauri MC, Paletta S, Di Pace C, Reggiori A, Cirnigliaro G, Valli I, Altamura AC. Clinical pharmacokinetics of atypical antipsychotics: an update. Clin Pharmacokinet. 2018;57(12):1493–528. Erratum in: Clin Pharmacokinet 2019 Jun 19. https://doi.org/10.1007/s40262-018-0664-3.
Mayers AG, Baldwin DS. Antidepressants and their effect on sleep. Hum Psychopharmacol. 2005;20(8):533–59. https://doi.org/10.1002/hup.726.
Meyer JS, Quenzer LF. Psychopharmacology: drugs, the brain, and behavior. 1st ed. Sunderland, MA: Sinauer Associates; 2005. ISBN 0–87893–534-7. LCCN 2004020935.
Olfson M, King M, Schoenbaum M. Benzodiazepine use in the United States. JAMA Psychiat. 2015;72(2):136–42.
Ravindran LN, Stein MB. The pharmacologic treatment of anxiety disorders: a review of progress. J Clin Psychiatry. 2010;71(7):839–54. PMID: 20667290. https://doi.org/10.4088/JCP.10r06218blu.
Ronaldson KJ, Fitzgerald PB, Taylor AJ, Topliss DJ, McNeil JJ. A new monitoring protocol for clozapine-induced myocarditis based on an analysis of 75 cases and 94 controls. Aust N Z J Psychiatry. 2011;45(6):458–65. Epub 2011 Apr 27. https://doi.org/10.3109/00048674.2011.572852.
Rowe MK, Chuang DM. Lithium neuroprotection: molecular mechanisms and clinical implications. Expert Rev Mol Med. 2004;6(21):1–18.
Sadock BJ, Ruiz P. Sadock V. Kaplan and Sadock's comprehensive textbook of psychiatry. Philadelphia: Wolters Kluwer; 2017.
Sadock BJ, Ruiz P, Sadock V. Kaplan and Sadock's synopsis of psychiatry: behavioural sciences/clinical psychiatry. Philadelphia: Wolters Kluwer; 2017.
Soyka M. Treatment of benzodiazepine dependence. N Engl J Med. 2017;376(12):1147–57.
Stahl SM. Stahl's essential psychopharmacology. 4th ed. Cambridge: Cambridge University Press; 2013.
Stahl SM. Prescriber's Guide: Stahl's essential psychopharmacology. Cambridge: Cambridge University Press; 2020.
Stroup TS, Gray N. Management of common adverse effects of antipsychotic medications. World Psychiatry. 2018;17(3):341–56. https://doi.org/10.1002/wps.20567.
Taylor DM, Barnes TRE, Young AH. Prescribing guidelines in psychiatry. 13th ed. Chichester: Wiley Blackwell; 2018.

12 Non-Pharmacological Treatments

B. Barbini, F. Seghi, M. Bosia, L. Fregna, F. Martini, F. Attanasio, E. Manfredi, G. Vassena, C. Colombo, and R. Cavallaro

12.1 Somatic Treatments

12.1.1 Introduction

Different therapeutic approaches have been developed, both pharmacological and non-pharmacological, including chronobiologic, psychological interventions and neurostimulation techniques. Even when adequately effective, current antidepressants require several weeks to achieve at least a clinical response and, ideally, a complete remission of symptoms. The rate of response to antidepressant treatment is of 50–70% after one antidepressant monotherapy.

Unfortunately, it is estimated that up to 30% of depressed patients might fail to achieve clinical remission after multiple sequential pharmacological treatments, a condition named treatment-resistant depression (TRD). Various TRD definitions have been proposed, according to the number of pharmacological and non-pharmacological trials failed. However, it is generally accepted to have tried at least two medications at an adequate dosage for an adequate duration without any response or with rapid loss of effectiveness, especially if the drugs belong to different pharmacological classes.

B. Barbini · F. Seghi · L. Fregna (✉) · F. Martini (✉) · G. Vassena
IRCCS San Raffaele Scientific Institute, Milan, Italy
e-mail: barbini.barbara@hsr.it; seghi.federico@hsr.it; fregna.lorenzo@hsr.it; francesca.martini1@hsr.it; vassena.giulia@hsr.it

M. Bosia · C. Colombo · R. Cavallaro
IRCCS San Raffaele Scientific Institute, Milan, Italy

Vita-Salute San Raffaele University, Milan, Italy
e-mail: bosia.marta@hsr.it; colombo.cristina@hsr.it; cavallaro.roberto@hsr.it

F. Attanasio · E. Manfredi
Vita-Salute San Raffaele University, Milan, Italy
e-mail: attanasio.francesco@hsr.it; manfredi.elena@hsr.it

R. Cavallaro, C. Colombo (eds.), *Fundamentals of Psychiatry for Health Care Professionals*, https://doi.org/10.1007/978-3-031-07715-9_12

Lack of standardized definition of treatment failures and the absence of a standardized definition of the clinical entity treatment-resistant depression (TRD) have been shown to lead to confusion in clinical practice.

TRD is defined as MDD that does not respond or remit to one or more antidepressant trials of adequate dose and duration.

Box 12.1: TRD risk factors

TRD risk factors are summarized in:

- Comorbid general and psychiatric medical disorders
- Severe intensity of depressive symptoms
- Suicidal thoughts and behaviour
- Adverse life events
- Personality disorders
- Early age of onset of major depression (e.g. age < 18 years)
- Recurrent depressive episodes
- Low socioeconomic status

There are two classifications of TRD which are generally accepted, the first is Thase and Rush (1997) Stadiation with five degrees of treatment resistance:

Box 12.2: Thase and Rush (1997) Stadiation

0. Pseudo-resistance. Wrong doses and/or timing
1. AD Trial
2. AD Trials from different classes
3. AD Trials, at least 1 TCA
4. AD Trials, at least 1 IMAO and 1 augmentation
5. ECT

The second classification of TRD is Souery where the author introduced the criteria of duration of treatment not defined in Thase and Rush Stadiation where they considered the different classes of compounds.

Box 12.3: Souery (1999) Stadiation

A. Treatment-Resistant Depression

   Not responder to at least 2 AD trials

   TRD1: 12–16 weeks

   TRD2: 18–24 weeks

   TRD3: 24–32 weeks

   TRD4: 30–40 weeks

   TRD5: 40–52 weeks

B. Chronic Resistant Depression

   Period of 12 months

After defining the clinical entity of treatment-resistant depression and above all after staging according to the two classifications described above, a neurostimulation treatment such as electroconvulsive therapy, rTMS, or tDCS can be used.

12.2 Non-Invasive Brain Stimulation

12.2.1 Introduction

Researchers have been trying to identify effective methods to modulate intra-cortical and subcortical excitability for almost a century. The purpose of these procedures was to find out non-invasive tools that could act as brain stimulators aimed at specific cerebral targets in fully awake subjects without inducing generalized or partial convulsions. Different methods have been investigated, using either electrical or magnetic fields to modulate neuronal excitability and firing. Experiments on this field date back to the early twentieth century, when electrodes directly placed over the exposed grey matter of animal models were first used to deliver electrical pulses. When applied at a threshold intensity over the primary motor area, these stimuli induced a twitch in the contralateral limb after a 1–2 ms latency, a process later defined as Motor Evoked Potential (MEP). Decades later, this research was translated in humans using an electrical circuit composed of two electrodes over the scalp attached to a battery to stimulate the underlying cortical neurons. Interestingly, this circuit elicited the same peripheral muscular twitching in conscious subjects in a non-invasive fashion, laying the background for modern neurostimulation techniques, such as Transcranial Electrical Stimulation (TES). Unfortunately, TES was poorly tolerated due to discomfort and pain in the stimulation area caused by direct stimulation of the superficial muscular structures via collateral circuits. In the search for alternative techniques, Transcranial Magnetic Stimulation (TMS) emerged as a viable option with comparable results and fewer adverse reactions in 1990, with an increasing number of applications in both clinical and research settings. On the other hand, TES safety and tolerability were improved using direct, low-intensity currents, identifying a more feasible device called transcranial Direct Current Stimulation (tDCS).

12.2.2 Transcranial Direct Current Stimulation (tDCS)

12.2.2.1 Principles and Mechanisms of Action of Transcranial Direct Current Stimulation

Pathophysiology of many neuropsychiatric diseases concerns alterations of neuroplasticity and cortical excitability. Non-invasive brain stimulation (NIBS) is an effective and highly tolerated approach to act on those aspects and modify cortical activities; one of the most valuable therapeutic NIBS approach is transcranial direct current stimulation (tDCS). Since many decades it has been known that neural activity could be modulated by the direct application of current on nervous tissues, but in the late 1960s was demonstrated how that brain neural activity or cortical excitability could also be altered via transcranial application of a direct current.

Even if the discovery of tDCS dates back more than 60 years, only in the last 20 years have its effects and potential been studied. The main tDCS effect on neurons is a subthreshold modification of their resting membrane potentials (or resting voltage) in sense of depolarization or hyperpolarization; but it not only changes their membrane potential and firing rate but also reduces membrane resistance. This excitability modulation persists for more than 1 h after the end of a single stimulation of several minutes. If the "synaptic effect" of tDCS may last for hours, its "non-synaptic effect" might contribute to explain the long-lasting effects of tDCS through the hours, days, and weeks. Repeated tDCS stimulations induce neural plasticity, acting on various neurotransmissions (NMDA and GABA), modulating transmembrane ion conductance, changing function and conformation of various axonal molecules, membrane structure, cytoskeleton, and axonal transport. tDCS has a local effect, but it also acts on functional connectivity, synchronization, and oscillatory activities modulating various neuronal networks, both at cortical and subcortical level. The effect of tDCS on cortical excitability (depolarization or hyperpolarization of resting membrane potentials) depends on the relation between the current flow direction and the axonal fibre orientation. So, by modifying the electrode polarity, it is possible to determine whether the tDCS field is excitatory or inhibitory (i.e. anodal tDCS excitatory, cathodal tDCS inhibitory). If it's more understandable the relationship between stimulation polarity and type of modulation (excitatory or inhibitory), it's unclear the relationship between other stimulation parameters: strength of stimulation, duration of stimulation, session repetition timing. For example, the relationship between intensity of stimulation and the biological effect produced appeared to be non-linear: doubling intensity from 1 to 2 mA appeared to reduce the inhibition produced by cathodal tDCS on primary motor cortex (M1). The clinical effect of tDCS is also strongly affected by some technical and neuroanatomical aspects: electrode size, their shape and their placement primarily influence the diffusion of the current through the scalp and thus the induced electric field into the brain. Taking into account all the aspects seen above, many different stimulation protocols were designed and studied through the years.

tDCS is generally well tolerated and it is associated with relatively minor side effects, including tingling and/or itching sensation at the stimulation site, moderate fatigue, headache, nausea, insomnia, and induction of hypomania.

12.2.2.2 Therapeutic Indications and Clinical Use

Even if tDCS is currently not FDA approved, in the last 2 decades tDCS was not only widely used as a tool for neuroscience research but was also applied as treatment of various neurological and psychiatric disorders with different results, including pain (neuropathic pain, migraine, fibromyalgia), Parkinson's disease, stroke, aphasia, epilepsy, Alzheimer's disease, depression, schizophrenia, substance abuse, addiction, and craving.

Box 12.4: Recommendation Level

Level B (probable efficacy)

- Anodal tDCS of the left primary motor cortex (M1) (with right orbitofrontal cathode) in fibromyalgia
- Anodal tDCS of the left dorsolateral prefrontal cortex (DLPFC) (with right orbitofrontal cathode) in major depressive episode without drug resistance
- Anodal tDCS of the right DLPFC (with left DLPFC cathode) in addiction/craving

Level B (probable inefficacy)

- Anodal tDCS of the left temporal cortex (with right orbitofrontal cathode) in tinnitus
- Anodal tDCS of the left DLPFC (with right orbitofrontal cathode) in drug-resistant major depressive episode

Depression

The rationale about the efficacy of tDCS as treatment of depressive disorders is largely the same of rTMS therapy; it is based on the knowledge of interhemispheric imbalance of neuronal activity between left and right dorsolateral prefrontal cortex (DLPFC) and ventromedial prefrontal cortex areas, functional and structural abnormalities in the DLPFC, amygdala and hippocampus in depressed patients.

As for rTMS, high frequency on left DLPFC or low frequency on right DLPFC, the current tDCS antidepressant approach is to enhance neural activity in the left DLPFC or to reduce neural activity in the right DLPFC, with anodal and cathodal stimulations, respectively.

One of the most effective protocol in the treatment of major depressive episode received a Level B of probable efficacy: 10 sessions of 20 min with 2 mA anodal stimulation of the left DLPFC with a right orbitofrontal cathode.

Many studies on healthy subjects have investigated the tDCS effect in the cognitive performance modulation showing a positive effect of tDCS on working memory, verbal fluency, language processing, and more complex cognitive functions. In contrast, only few studies have investigated the impact of tDCS on cognitive functions in depressed patients: in particular tDCS appeared to be effective in reducing both depressive symptoms and cognitive impairment when compared to sham in elderly depressed patients with cerebrovascular disease (vascular depression).

Schizophrenia

Many studies agree on the relation between some of the most drug-resistant symptoms of schizophrenia, as auditory verbal hallucinations and negative symptoms (e.g. avolition, alogia, or emotional withdrawal) and specific alterations in brain activity and connectivity.

Auditory verbal hallucinations and negative symptoms were linked to a fronto-temporal connectivity alteration and hypo-activity of DLPFCs. Auditory verbal hallucinations were related to a hyperactivity in the left temporo-parietal region too.

Considering this neurophysiological background, different therapeutic protocols based on the concept of excitatory anodal stimulation versus inhibitory cathodal stimulation were hypothesized: anodal stimulation combined with a cathodal stimulation on the hyperactive left temporo-parietal region; a different montage consists in delivering a bilateral stimulation to both DLPFC regions: anodal stimulation to the left DLPFC and cathodal stimulation to the right one.

Anyway, the results obtained made no recommendation about the efficacy of tDCS to relieve schizophrenia symptoms, either positive or negative.

Substance Abuse, Addiction, and Craving

Dysfunction of inhibitory control mechanisms and reward mechanisms are the core of disturbances related to abuse, craving, and addiction to different substances such as alcohol, drugs, food, and nicotine. tDCS, acting on the DLPFCs, seems to regulate the activity of these systems in patients with various types of addiction.

Bi-hemispheric DLPFCs tDCS (anode on the right DLPFC and cathode on left DLPFC) received a Level B of probable efficacy in reducing craving in patients with various types of addiction.

A different montage with the anode over the right DLPFC and the cathode over the left supraorbital region appeared to be effective in reducing food craving since the first tDCS session; this effect seems to last for a month after a 5-day protocol. Anyway, these results are preliminary and do not make a recommendation.

12.2.3 Transcranial Magnetic Stimulation (TMS)

12.2.3.1 Mechanism of Action

Following the Maxwell–Faraday Equation, TMS relies on the fact that a time-varying magnetic field is always accompanied by a spatially varying electric field and vice versa. Hence, an electrical current flow through a wire TMS coil generates a concomitant magnetic field with a perpendicular orientation. Rapid changes in the magnetic field (as with brief single pulses) determine the induction of circular electrical currents in the underlying area, oriented on a plane parallel to the original current flow. When applying the coil tangential to the scalp at a threshold intensity, the resulting current reaches cortical neurons, lowering their firing threshold and ultimately inducing focal action potentials. The concomitant recruitment of a sufficient number of neurons beneath the stimulation area finally results in axons depolarization, consequently propagating the signal alongside all functionally connected neural networks involving both cortical and subcortical structures in the central nervous system (CNS) and peripheral nervous system (PNS).

The choice of the appropriate coil is crucial to the application of TMS. Changes in shape and positioning determine significant variations in the spatial distribution of the magnetic field, inducing different effects. Two factors should be taken into consideration: focality of action and depth of stimulation. Circular and figure-of-eight coils are the first introduced and most commonly used. The former generates a circular magnetic field encompassing larger areas with lower intensity but widespread distribution; the latter, instead, allows for a focal and more powerful stimulation by combining two circular magnetic fields (eight-shaped) at their intersection point, with a resolution of approximately 1 cm^2 for commonly used coils. In both cases, the coils can be adapted to different cerebral areas by simply moving them along the skull. However, they show significant limitations in the spatial distribution of the magnetic field. Its intensity correlates inversely with the distance from the coil, showing a rapid decrease after few centimetres, with further interpersonal variations based on the attenuation by superficial structures (musculocutaneous layers, skull, and meninges). Therefore, the direct effects of these coils are limited to cortical stimulation; subcortical structures, instead, can only be consequently involved as part of functionally integrated networks. New coils have been introduced to overcome this limitation, with the capability to reach deeper regions within the CNS directly (e.g. limbic system). The double cone coil, for instance, is composed of two circular coils oriented in a conic structure, with the apex at their intersection. The resulting magnetic fields develop perpendicularly, interacting with each other more deeply than the classic figure-of-eight coil, allowing for a stimulation in areas located below the cortex. More recently, the same principle was applied to more sophisticated coils, namely H-coils, presented as helmets composed of multiple radially oriented coil loops. The final effect derives from the contribution of each loop intersecting at deeper cerebral regions, which are stimulated directly. More than 14 H-coils exist, differing in shape and coils orientation according to the stimulation area. Even though highly effective at deeper regions, H-coils are more expensive than classical coils, with lower adaptability to the site of stimulation and reduced focality of action.

No coil is generally preferred over the others, but the choice varies according to their availability, cost-effectiveness, and the individual protocol of stimulation, which should be tailored to each patient.

12.2.3.2 Stimulation Protocols

TMS is a versatile and multifaceted technique, allowing for both diagnostic and therapeutic uses. Since the identification of depolarizing properties of single, brief magnetic stimuli, TMS has appeared as a promising tool adapted to a large variety of situations. If single pulses exert immediate effects with a fast distribution through functional networks in CNS and PNS but a rapid resolution within milliseconds ("online" effect), repetitive patterns of stimulation can elicit a neuroplastic modulation of the cerebral activity ("offline" effect) through long-term potentiation (LTP) and long-term depression (LTD) mechanisms. Online effects proved highly useful for diagnostic purposes, especially in the neurological field. The application of single pulses over target cortical regions allows for assessing the integrity of neural pathways by eliciting immediate sensorimotor reactions (e.g. motor evoked potentials). On the other hand, offline effects opened to the chance of generating long-term modulations of the neuronal activity with therapeutic potentialities. Repetitively administered stimuli create patterns of activation or inhibition of functional networks that persist beyond the stimulation session. As a result, repetitive TMS (rTMS) can induce stable modifications of the neuronal activity, potentially regulating functional disruptions responsible for both neurological and psychiatric conditions. A specific combination of parameters determines the pattern of rTMS, which, in turn, defines its clinical effect. Virtually infinite stimulation protocols can be applied, each requiring careful validation of efficacy and safety. The following parameters define a complete rTMS protocol:

- Site of stimulation
- Intensity of stimulation (SI)
- Frequency of stimulation
- Train of repetitive stimuli and Inter-Train Interval (ITI)
- Total number of pulses per session
- Number of sessions and distribution over time

The site of stimulation is the main parameter to consider when establishing the proper rTMS protocol. Data gathered from functional imaging studies are crucial in determining neurobiological targets for each neuropsychiatric disorder. Areas of stimulation can be located both in cortical and subcortical regions that pertain to various functional networks in the brain, requiring an adequate choice of the coil type according to the depth and width of the site. Within the psychiatric field, prefrontal regions are among the most investigated, especially the left and right dorsolateral prefrontal cortex (DLPFC) implicated in depression, anxiety disorders, and addiction in association with overactive limbic structures. DLPFC is a small functional region located in the frontal cortex directly accessible with focal coils, such as the figure-of-eight one. Other areas investigated include the parietal and the motor cortex, usually targeted in neurological disorders.

After the stimulation site has been identified together with the most appropriate coil type, the proper intensity of stimulation should be assessed for each patient. SI depends on multiple biological variables, such as the thickness of the superficial structures (e.g. skin, skull, and meninges), the intrinsic neuronal reactivity, and possible pharmacological agents that modulate cortical excitability. SI is defined for each protocol as a proportion of the Motor Threshold (MT). MT is the measurement of the minimum intensity of the magnetic stimulus able to elicit neuronal depolarization over the primary motor cortex after a single pulse, observable as a muscular twitching in the contralateral hand. Based on the assumption that the same intensity to depolarize neurons in the motor cortex apply to the other cortical regions (e.g. DLPFC), SI is usually calculated within a range of 80–120% of MT, according to the specific protocol.

When the appropriate SI has been calculated for each subject, rTMS delivers repetitive trains of stimuli with variable frequency. The modulating effect of TMS depends on the frequency of stimulation by the induction of different patterns of glutamatergic receptors activation/deactivation. Low frequencies (<5 Hz) reduce the functional connection between neurons at the synaptic level, ultimately leading to LTD mechanisms, while high frequencies (>5 Hz) strengthen cerebral networks through LTP mechanisms. The proper frequency relies on functional imaging studies, revealing overactive or hypoactive brain regions that require rebalancing. Low-frequency rTMS is usually administered continuously for the whole session, while high-frequency rTMS necessitates intermittent stimulation, with trains of pulses alternating with resting periods (ITI) to avoid the spread of the depolarizing signal to other regions (e.g. convulsions). More recently, a new paradigm of rTMS called Theta Burst Stimulation (TBS) has been introduced. TBS consists of bursts of extreme high-frequency stimuli (50 Hz) instead of single pulses, continuously repeated at low frequencies with inhibition properties (cTBS) or intermittently at high frequencies with stimulating properties (iTBS). If TBS proved as effective as rTMS in multiple non-inferiority trials, it dramatically reduced the duration of the session from 20–40 to 1–3 min, improving time management and feasibility of the technique.

Irrespective of the frequency applied, each TMS session is usually composed of a total number of pulses around 1500–3000. Overstimulation should be avoided to prevent (sub-)convulsive reactions, while under-stimulation could result in a lack of efficacy. The treatment of acute conditions typically requires 20–30 sessions, even though adjunctive maintenance session can be added. Historically, daily rTMS sessions have been recommended from Monday to Friday (5 days per week), leading to a global 4- to 6-week treatment. More recently, accelerated protocols (aTMS) have been proposed, consisting of multiple daily sessions with adequate resting intervals. However, further research is needed on efficacy and timing of aTMS compared to standard rTMS.

12.2.3.3 Clinical Applications

Since its introduction, TMS has been extensively investigated in a wide range of neurological and psychiatric conditions. Researchers experimented a multitude of protocols worldwide, variably combining parameters of stimulation to achieve clinical effects. Even though far from conclusive, the evidence available led to the approval of TMS devices to treat many neuropsychiatric disorders both in the USA and the EU. The FDA first approved rTMS in treatment-resistant depression (TRD) in 2008 and later approved deep TMS to treat obsessive–compulsive disorder (OCD) in 2018. More recently, deep TMS was also approved as a short-term aid in smoke cessation in 2020. In the EU, TMS was cleared for many indications, including Alzheimer's disease, autism spectrum disorders, Bipolar Disorder (BD), Major Depressive Disorder (MDD), neuropathic pain, Parkinson's disease, Post-Traumatic Stress Disorder (PTSD), negative symptoms of schizophrenia, and smoke cessation.

Depression

Depression is the first condition that achieved clearance for clinical use of TMS, with the largest body of evidence. The classical rTMS protocol for depression, both in MDD and in BD, consists of a focal, high frequency (HF) stimulation at 120% of MT over the left DLPFC, with 4-second trains and 26-second ITIs, for a total of 3000 pulses per session delivered daily for 20–30 sessions (4–6 weeks). Alternatively, a low frequency (LF) rTMS protocol over the right DLPFC was proposed, consisting of 1500 pulses administered continuously at 1 Hz for 20–30 sessions. Both protocols showed consistent and comparable antidepressant effects, as monotherapies or in combination with pharmacological treatments. However, left DLPFC HF-rTMS has the best quality evidence, leading to the highest level of recommendation. Some authors proposed LF-rTMS over the right DLPFC as the second choice thanks to its more tolerable profile, while others tried to combine HF- and LF-rTMS using bilateral stimulation, with conflicting results. Globally, bilateral stimulation did not show clear superiority over unilateral stimulation due to low-quality evidence, being supported at a lower level of recommendation. Despite the efficacy of both left and right DLPFC rTMS compared to placebo, response rates resulted in the range between 30 and 50%. With the development of innovative TMS methods, newer protocols were tested on depression to improve outcomes. Deep TMS using H-coils to stimulate left DLPFC at high frequencies systematically showed better results than placebo with high-quality evidence, leading to the FDA approval of HF-deep TMS over DLPFC with the highest level of recommendations. Based on HF- and LF-rTMS outcomes, iTBS and cTBS protocols over the left and right DLPFC were evaluated, to significantly reduce the session duration and improve time management. Given the conflicting results, evidence supporting non-inferiority compared to standard rTMS requires further support. Finally, aTMS protocols were investigated. Despite positive findings and no adjunctive safety issues, no standardized accelerated protocol has been approved to date, making impossible the generalizability of the results.

Overall, standard HF-rTMS and HF-deep TMS over the left DLPFC are currently recommended in depression, as supported by the highest quality evidence, even though other protocols could be found at least as effective in the future. Given the recent evolution, it is still unclear when to propose TMS to a patient affected by a major depressive episode. Considering the low response and remission rates, TMS is generally recommended as an add-on therapy to pharmacological treatments, especially in patients with mild to moderate treatment-resistant depression. However, no clear indications on the level of treatment resistance, clinical profiles, and preferred combinations are currently available, leaving the choice to the psychiatrist.

Schizophrenia

Both positive and negative symptoms of schizophrenia were targeted using TMS, with conflicting results. In particular, some studies investigated the effect of rTMS over the temporo-parietal cortex (TPC), including the superior temporal gyrus and the temporo-parietal junction, in the presence of auditory verbal hallucinations.

LF-rTMS on the left TPC showed some positive but not univocal findings, leading to a low level of recommendation for treatment-resistant hallucinations, especially for young patients and females. Alternative protocols using different locations, HF stimulation, or cTBS approaches showed promising results, but demand further research before approval.

Similarly to depressive disorders, negative symptoms seemed responsive to HF-rTMS protocols over the left DLPFC using a figure-of-eight coil in schizophrenia. Again, findings regarding this set of symptoms are far from being conclusive. Different factors should be considered, including the high variability of clinical profiles and the need for larger samples to draw significant conclusions. However, the use of circular coils, deep-TMS approaches, bilateral DLPFC stimulation, or the choice of alternative sites such as the vermal part of the cerebellum showed promising results.

Despite encouraging evidence, no specific protocol or target population is currently recommended in the USA and the EU for positive and negative symptoms of schizophrenia, leaving this approach to a research setting to date.

Substance Abuse and Craving

According to the dopaminergic hypothesis, craving and addiction depend on the hyperactivation of cerebral networks such as the nigrostriatal, mesolimbic, and mesocortical pathways in the presence of a substance or other substance-related cues. The whole system encompasses multiple cortical and subcortical structures such as the ventral tegmental area, the striatum, the substantia nigra, the anterior cingulate cortex, the insula, the amygdala, the lateral habenula, and the prefrontal cortex, which constitute the reward system. More in detail, the cortical structures lose their top-down regulation over the subcortical networks when these are hyperactivated, leading to the impulsive search for the addictive substance or behaviour. As a common denominator between a wide range of substances and other addictive behaviours such as gambling, rTMS has emerged as a promising tool in targeting DLPFC to regulate the cortical/subcortical activation imbalance. Different protocols, including HF-rTMS over the left DLPFC, LF-rTMS over the right DLPFC, deep TMS using H-coils over the DLPFC, iTBS, and cTBS, provided positive results in various conditions, such as alcohol craving, methamphetamine and cocaine addiction, nicotine craving, and gambling. However, the findings were highly heterogeneous and controversial, making TMS generally recommended in the EU but without specific protocol approval in all these conditions. Instead, the FDA cleared the use of HF-deep TMS for smoking cessation in 2020 (10 Hz, 120% MT, 3 s pulse trains, 15 s inter-train interval, 60 trains, 1800 pulses per session, 15 sessions).

Obsessive–Compulsive Disorder

Non-pharmacological approaches have been investigated in obsessive–compulsive disorder (OCD), including TMS, considering low response rates after multiple pharmacotherapies either alone or in combination. Different protocols were proposed using HF- and LF-TMS over various brain regions such as the prefrontal cortex, without univocal results. Focal LF-rTMS over the right DLPFC is among the most studied, with 1200–2000 pulses for 10–15 sessions, although without approved protocol schemes. This procedure is recommended with a low level of evidence due to some negative findings, requiring further research. According to functional imaging studies in OCD, LF-TMS approaches have also been tested on other areas, such as the left and right orbitofrontal cortex and the pre-supplementary motor area, with positive but limited findings that are insufficient to draw further conclusions to date.

On the other hand, HF protocols showed heterogeneous results. If focal HF-TMS over the left and the right DLPFC did not find significant improvements in OCD symptomatology, bilateral stimulation of both DLPFCs exerted promising results. More importantly, HF-deep TMS using H-coils targeted at the bihemispheric medial-PFC anterior cingulate cortex region led to more prominent positive findings, allowing for the FDA approval of this procedure in the treatment of OCD in 2018.

Post-traumatic Stress Disorder

Few studies addressed the treatment of Post-traumatic Stress Disorder (PTSD) using TMS. Most of these focused on the right DLPFC using HF (10 Hz) and LF (1 Hz) protocols with figure-of-eight coils. Even if both therapeutic schemes showed some positive findings, HF-rTMS determined a higher reduction in anxiety symptoms, even persisting 3 months after the end of the treatment. Even if recommended, HF-rTMS in PTSD is still without clear-cut indications approved by regulatory agencies. On the other hand, LF protocols (1200–1800 stimuli/session for 15–30 sessions) found improvements in PTSD and depressive symptoms, especially when combined with cognitive therapy, even if some findings showed faster relapses after rTMS cessation. Given the limited evidence on this protocol, no recommendations or approval can be made on LF-rTMS. Finally, a single study investigated the effect of deep TMS over bilateral medial-PFC using an H-coil with promising results in PTSD, even though it requires further research.

Other Disorders

Considering the versatility and the potential of the application of TMS in a multitude of neuropsychiatric disorders, we are still far from an exhaustive knowledge of the treatment protocols. Even if still not recommended, some evidence explored the efficacy of TMS on other disorders, requiring further investigation to understand its role in their treatment plan. LF- and HF-rTMS protocols over the right DLPFC achieved some positive results in anxiety disorders, especially generalized anxiety disorder and panic disorder, with a functional rebalance in the connectivity within the limbic system. Little evidence accumulated on autism spectrum disorders, especially Asperger's disorder, using deep, 5 Hz-TMS over the dorsomedial-PFC bilaterally or LF-rTMS over the left DLPFC, with positive results in social anxiety and other symptoms. HF-rTMS was also tested in mental retardation as an augmentation strategy for language training, targeting the left Broca's area, with improvements compared to language training alone. Finally, some studies investigated the effect of deep HF-TMS in ADHD but without significant results.

12.2.3.4 Safety and Tolerability of TMS

TMS is generally a low-risk, well-tolerated technique. However, TMS operators should be aware of possible risks for the patients and themselves. TMS can be performed both as an inpatients' and an outpatients' service. As a non-invasive procedure, no special precautions are required before or after other than the removal of eyeglasses, earrings, necklaces, piercings, and all metallic materials in the head. During the stimulation, the patient is fully awake without any pre-medication or anaesthesia. If no adverse event (AE) arises, the subject is free to leave without further assessments or precautions after the session.

Before being admitted to the treatment, every patient should undergo a general screening for contraindications. As TMS is based on magnetic pulses applied over the skull and extending for several centimetres, all non-paramagnetic materials around the stimulation area serve as an exclusion criterion. Operators should extensively review medical history to identify prior accidents, surgeries, leads, or implants in cerebral structures that might contain metallic components. Furthermore, medical conditions, pregnancy, known brain lesions, and medications should be carefully considered as risk factors, potentially leading to the exclusion from TMS. At the moment, non-paramagnetic metallic parts in the skull represent the only absolute contraindication, while all other conditions are relative contraindication requiring individual assessment of the risk/benefit ratio.

If the patient is eligible for the treatment, the operator should monitor for possible AEs throughout the sessions. Even though well tolerated, some subjects might experience AEs that can lead to TMS termination, especially during the initial sessions. The following are the main AEs of TMS:

- *Epileptic seizures*. Even if TMS is considered a non-invasive neurostimulation technique, the stimulation can induce a disorganized propagation of the pulse in structures other than the original target, generating partial or generalized seizures. After the first reports of this AE, concerns about the procedure led to the development of safety guidelines. According to them, stimulating protocols should not exceed predefined combinations of parameters, including frequency, pulses/train, ITI, and SI related to the MT. Despite the diffusion of TMS on larger scales, the introduction of safety limits determined a consistent drop in seizures among patients undergoing TMS. However, few cases have been reported with all stimulation patterns, even if the risk is considered very low. Current estimates of seizures are globally around 5–10/100000 sessions, with a higher risk for HF protocols and H-coils. Moreover, the combination of medications with TMS had been a major concern, given their modulating effect on the seizure threshold. However, most of the patients suffering from severe neuropsychiatric disorders were concomitantly taking psychopharmacological agents without any substantial increase in seizure incidence. Even if medications demand caution, currently, there are no recommendations against their combination with TMS. Finally, concomitant medical conditions should be evaluated as part of the risk algorithm for seizure induction. The most important disorder to rule out is prior epilepsy. Even if it does not act as an absolute contraindication, careful revision of the medical history is mandatory, requiring special precautions during the session. Other neurological diseases can lower the seizure threshold, such as stroke, multiple sclerosis, brain lesions, and neurodegenerative disorders. Medical comorbidities, including electrolyte imbalance, metabolic abnormalities, liver/kidney failure, fever, infections, and substance abuse/withdrawal (e.g. cocaine, MDMA, alcohol) may act as predisposing factors. It is advisable to correct any disorder before administering TMS.

- *Changes in the hearing threshold.* Each TMS pulse is associated with an acoustic signal estimated to reach the maximum noise exposure limits. Few reports indicated increases in the hearing threshold considering the position of the coil close to the ears and the propagation of vibrations through the skull. Such changes appeared transient in their nature, resolving approximately 1–2 h after the session, but long-term dysfunctions of the nerve cannot be excluded. Some protocols showed consistent variations in the auditory threshold, including repetitive patterns, HF stimulation, higher SI, and longer sessions. However, ear protection devices (earplugs or earmuffs) could be offered to prevent even transient effects. After each session, the operator should evaluate the occurrence of hearing loss, tinnitus, or aural fullness.
- *Dysesthesia/headache.* If TMS delivers magnetic pulses at an intensity sufficient to elicit neuronal depolarization, superficial structures in the scalp can be stimulated as well. The direct application over nerves and muscles can result in involuntary twitches in the face, discomfort, dysesthesia, or headaches in up to 20–30% of subjects, especially during the initial sessions. In most cases, these symptoms arise during the stimulation, but few patients reported the occurrence of headaches a few hours after. Even though benign, they can be considered unbearable, requiring either a readjustment of the SI or a switch to more tolerable protocols, such as LF and low-intensity rTMS. Should the symptoms persist beyond the session, paracetamol is generally recommended as a first-line agent.
- *Burns.* The generation of a magnetic pulse depends on the flow of an electric current inside the coil. Repetitive patterns of pulses can induce progressive heating of the coil attached to the scalp, which can cause burns. Nowadays, all TMS devices incorporate air- or liquid-based cooling systems with an automatic block of the stimulation should the coil reach damaging temperatures. After the optimization of the technique, burns are no longer considered TMS side effects.

Besides risks for the patient, operators should be aware of recommended precautions for themselves. Electromagnetism can interfere with electronic or other magnetic devices, so it is advisable to keep mobile phones and credit cards away from the TMS equipment during the session. Considering long-term, daily exposures to rapidly time-varying magnetic fields, operators might exceed the “Exposure Limit Values” set as safety issues. A distance of at least 40 cm should be kept from the coil during active stimulation to avoid this inconvenience. Last, noise pollution can affect operators as well due to chronic exposure. The use of earplugs or earmuffs is recommended to prevent hearing alterations.

12.2.4 Electroconvulsive Therapy (ECT)

12.2.4.1 Definition

Electroconvulsive therapy (ECT), commonly known as electroshock therapy, is a therapeutic technique based on inducing convulsions in the patients after passing an electric current through the brain. The therapy was developed and introduced in the 1930s by the Italian neurologists Ugo Cerletti and Lucio Bini.

History

- First chemically induced convulsions were used as a treatment for schizophrenia in the 1930s.
- Ugo Cerletti developed an experimental model for epilepsy and produced the first ECT device to induce convulsions in animals. The first treatment in humans was carried out in 1938. The patient underwent 11 treatments without any adverse events.
- In 1960 the introduction of new anaesthesia techniques included neuromuscular blockade.
- In the late 1970s square wave machines were developed.
- The Royal College of Psychiatrists declared ECT effective in "depressed patients".
- Late 1970s and 1980s: seven controlled trials carried out in Britain, despite uncertainties.
- To date, up to 50,000–100,000 people every year undergo ECT in the USA.

The approach was based on the Nobel Prize winner Julius Wagner-Jauregg's research on the use of malaria-induced convulsions to treat certain nervous and mental disorders (such as paralytic dementia caused by syphilis) as well as on the theories developed by J. Meduna, according to whom schizophrenia and epilepsy were antagonistic disorders. These theories led M. Sakel to develop insulin coma therapy in psychiatry in 1933.

Cerletti used electroconvulsive therapy for the first time in April 1938, in collaboration with Lucio Bini, on a patient suffering from schizophrenia with delusions, hallucinations, and confusion; a series of therapeutic electroshocks allowed the patient to return to a normal mental state. In the following years, Cerletti and his collaborators regularly carried out therapeutic electroshocks, both on animals and on neuropsychiatric patients, establishing the reliability of the therapy and its safety and usefulness in clinical practice, especially for the treatment of acute schizophrenia, manic-depressive psychosis, and the most severe cases of depression.

Initially, the therapy was carried out on conscious patients, without the use of anaesthesia or muscle relaxants. Patients lost consciousness during the session and suffered violent uncontrolled muscle contractions, which could sometimes cause bone fractures (especially of the vertebrae) and muscle strains. Since 1960, electroconvulsive therapy has been carried out under general anaesthesia.

12.2.4.2 ECT: Mechanism of Action

Even if the exact mechanism of action is yet to be fully elucidated, there is clinical evidence of antidepressant, antipsychotic, anti-catatonic, and anti-convulsing effects.

There are significant changes in neural activity in front-limbic brain regions.

The neurotransmission is involved in the mechanism of action in terms of an enhanced serotonergic, noradrenergic, and dopaminergic functions. It was also found an increase of cortical GABA concentrations and evidence of increased seizure threshold after ECT treatment.

Dopamine system: Several evidence show that ECT activates the dopamine system.

HPA axis function: Studies show that ECT can activate the HPA axis, as indicated by the increases of cortisol, ACTH, and arginine vasopressin after ECT sessions.

It is important to take into consideration that there is an inducing effect on neuroplasticity in terms of dose-dependent hippocampus dendritic arborization and excitatory synapses in the amygdala.

12.2.4.3 Clinical Indications for ECT

- Major depressive episode with psychotic symptoms.
- Bipolar disorder, depressive episode.
- Bipolar disorder, manic episode.
- Schizoaffective disorder.
- Schizophrenia—acute onset, with confusion.
- Catatonia.
- Parkinson's disease (bradykinesia, tremors, rigidity, gait disturbances, postural instability).

As we have seen in the previous section, the mechanism of action of ECT is not selective, so it is not aimed at treating a single syndrome, although patients with melancholic depression seem to benefit more than those with other pathologies.

Depression with both unipolar and bipolar psychotic manifestations respond to treatment after 6–8 sessions twice a week.

Manic syndromes require daily treatment until the episode is resolved, while schizophrenics benefit in a very limited way.

So we might consider that in major depression the effect of ECT is syndromic in the sense that all components of the syndrome improve in a parallel way, whereas in the treatment of schizophrenia only some symptoms respond while Schneider's first-rank symptoms do not change.

Predictors of response to ECT treatment in depressed patients are the presence of delusional ideation and psychomotor retardation. About 90% of depressed patients with psychotic manifestations respond to this therapy.

Finally, a further clarification is that candidates for ECT are those who need a faster response to treatment than conventional treatment, such as catatonic patients or those with significant suicidal ideation or those who do not respond to medication.

12.2.4.4 ECT and Pharmacological Treatments

It is advisable to discontinue psychotropic drugs if not strictly necessary, especially anticonvulsants and benzodiazepines, as they inhibit convulsions, but also tricyclics/MAOs because they make the course of therapy less predictable and finally lithium, cause it can be associated to an increase in organic mental syndrome and may prolong neuromuscular blockade. On the contrary, neuroleptics can be maintained.

12.2.4.5 Contraindications

- Uncontrolled hypertension.
- Myocardial ischaemia.
- Valvular stenosis.
- Aortic aneurysm.
- Pheochromocytoma.
- Thrombophlebitis in patients not on anticoagulant therapy.
- Airway infection.
- Upper airway obstruction due to arthritis, dental abscesses, laryngeal tumours, myopathies, myasthenia gravis, muscular dystrophy.
- Recent strokes.
- Brain tumours with increased endocranial pressure.
- Acute angle glaucoma.
- Hb value < 10 g/dL.
- Hepatic or renal failure.

12.2.4.6 Adverse Events with ECT

Mortality: 2:100,000 treatments.

Seizures: prolonged over 3 min evoke confusion and memory impairment.

Other: headache, muscle pain, nausea, drowsiness, weakness, anorexia, amenorrhoea.

12.2.4.7 Cognitive Effects

Studies in the 1970s using a sine-wave machine identified side effects inherent in memory processes, both retrograde, for events very close to the treatment, and anterograde amnesia, up to 6 months post-treatment.

Seventy-five per cent of patients undergoing electroconvulsive therapy acknowledged memory loss as the most serious side effect of the treatment. Thirty per cent said their memory had not returned to normal.

These data must be interpreted with caution as most published studies consider the possibility that the self-assessment of a symptom by individuals with depressive illness is inevitably influenced by the affective state or cognitive distortions typical of these patients.

12.3 Chronobiological Treaments

The use of chronotherapies, a group of non-pharmacological therapeutic approaches to mood disorder treatment, is rooted in the hypothesis of chronobiology aetio-pathogenesis of psychiatric disorders (mainly mood disorders) which spans across several biomedical disciplines—from molecular biology to internal medicine and clinical psychology.

12.3.1 Introduction

Life on Earth is marked by rhythmic occurrence of many phenomena such as the succession of periods of light and darkness, the alternation of seasons, the repetition of changes in environmental temperature, rain, flowering and fruit production, lunar phases and tides. All organisms must swing in accordance with external environment in order to adapt and survive, and their nervous systems have developed so. "Whoever wishes to investigate medicine properly, should proceed thus: in the first place to consider the seasons of the year, and what effects each of them produces for they are not at all alike, but differ much from themselves in regard to their changes" is a quote by Hippocrates, the founder of occidental medicine, who already in fifth century BC recognized the influence of environmental cycles on the human organism. In the case of humans, rhythmicity is an integral part of the functioning of our organism: bodily functions such as heart rate, respiratory rate, hormone secretions, and sleep-wake cycle exhibit a periodicity that is influenced by exposure to environmental stimuli. This rhythm that regulates our daily living, is determined by a biological clock within our brain and is independent of sleep and wake.

Box 12.5: Chronobiology Terminology

Circadian: a cycle that lasts 24 h (there is one cyclic event per day), e.g. sleep-wake cycle

Ultradian: a cycle that lasts less than 24 h (there are many cycles in a single day), e.g. respiratory cycle

Infradian: a cycle that lasts more than 24 h (there is less than 1 cycle per day), e.g. menstrual cycle

12.3.2 Biological Clock

The endogenous rhythms are closely linked to the rhythmicity of the surrounding environment through the so-called *Zeitgeber* (from the German "time givers") like the light-dark cycle, variations in ambient temperature or noise; these are all signals that inform the person's biological clock of what is happening in the surrounding environment, leading them to adapt accordingly.

The human biological clock resides in the suprachiasmatic nucleus (SCN), which is made of 20,000 cells packed in less than 0.3 mmc in the hypothalamus. It produces a signal that paces the organism at a 24-h periodicity. Environmental stimuli, ambient light in particular, act on biological clock synchronizing it with external world every morning at dawn. In fact there is direct connection from retina to SCN to neurons and their genetic material, the DNA. There are genes that are activated by sunlight in the morning, with a cascade of genetic and metabolic events, and feedback inhibition of other genes, so that a complete cycle takes exactly 24 h to complete. From the SCN, there is a polysynaptic connection to the pineal gland: it goes from the SCN to the paraventricular nucleus (PVN) in the hypothalamus, then to medio-lateral cell in the spinal cord to the superior cervical ganglion (SCG) and, finally, to the pineal gland so that the biological clock synchronized melatonin secretion with photoperiod (that is, the season).

According to preference for rest and awakening earlier in the evening or later in the night, people can be grossly divided in two chronotypes called "larks" and "owls", respectively, which rely on a complex origin: genetic vulnerability, environment, gender, and age all are contributing factors. The magnitude of such circadian variation is genetically determined by the allelic variations that a subject carries, therefore some characteristics can be inherited or be altered by epigenetic factors. It is everyday life experience that children tend to wake up and go to sleep early, adolescents progressively delay their phases, and sleep-wake cycle gradually shows a phase advance with age.

Not only light is a biological clock determinant, but also the so-called "non-photic stimuli," for example life events and habitual behaviours, which have a deep psychological significance, can act as internal circadian regulators or disruptors. This is why a regularity of lifestyle and sleep hygiene are strongly recommended at the basis of every therapeutic approach.

Examples of the close link between environmental periodicity and pathophysiology are:

- Body temperature: normally, the body temperature is lowest in the middle of the night and gradually rises during the day, with the maximum temperature about dinner time (7 p.m.).

- Hormonal secretion: melatonin and cortisol are the principal hormones affected by circadian rhythms. Melatonin production is affected not only by light intensity but also by the photoperiod, i.e. the hours of light to which we are exposed: in winter, with 7 h of light, melatonin production starts at around 7 p.m., while in summer, with 12 h of light, production begins at 10 p.m. Melatonin not only regulates the sleep-wake cycle, but via cytokines and second messengers regulates various body systems (e.g. neurotransmitter and receptor turnover, synaptic dynamics, immune system, blood pressure, metabolic organs). Typically, cortisol secretion shows a peak a couple of hours before awakening, with the goal of preparing the body to start a new day and lowers in the evening preparing the body to melatonin actions, but depressed patients have been demonstrated to have a compromised morning secretion of cortisol and a HPA hyperactivation.
- Sleep-wake cycle: according to the two-process model of sleep postulated by Borbély, somnolence and need for sleep depend both on tiredness and "sleep debt" that we accumulate for the mere sake of being awake (the homeostatic process), and biological clock that regulates day and night rhythm and sleep architecture (the circadian process). A commonly experienced phenomenon in flight travellers is induction of depression when travelling from east to west (phase delay) and induction of manic symptoms when travelling from west to east (phase advance).
- Cognitive functions: attention, concentration, memory, thought, and executive speed (e.g. in videogames, using a language or formal logic thought process) follow the same circadian rhythm that regulates every other body function, with maximum cognitive performance in late morning to early afternoon.

The circadian clock affects multiple systems and pathways that are thought to underline mood disorders. There are reciprocal interaction among circadian rhythms, immune response, and mood regulation.

Examples of the close link between biological rhythms and psychiatric illness are:

- Major depressive disorder: It typically recurs with seasonal pattern, mainly in spring and autumn. Successful antidepressant treatment is associated with a normalization of the melatonin production in depressed patients. During a depressive episode, there is drop in daily body temperature with respect to healthy controls and patients themselves during euthymia.
- Bipolar disorder: Depressive episodes typically occur in winter, while manic episode in late spring-summer. Bipolar patients have a different pattern of melatonin secretion, and it is particularly sensitive to environmental light with respect to healthy controls: if awakened during night sleep, bipolar patients fail to secrete melatonin again so that they cannot fall asleep again, triggering manic episodes.
- Seasonal affective disorder: Patients are unable to adjust the amount of melatonin to the photoperiod. An observational study on the seasonal presentation of certain phenomena showed that Google searches for "depression", "anxiety",

and "insomnia" showed a rhythmic pattern with a marked increase in searches in spring and autumn.

- Mood fluctuations: Depressed patients present a marked worsening of symptoms in the early morning, with a progressive improvement later in the day.
- Gonadal hormones modulate neurotransmitter mechanisms: The lowering of ovarian hormones in premenstrual phase, in postpartum and menopause jeopardizes the functioning of biological clock, predisposing ground for development of cyclic mood disorder manifestations (e.g. MDE, manic episode, PMDD).
- Sleep architecture in mood disorder patients: Depressed patients have a markedly reduced REM sleep latency, a shorter first REM phase with an increment in duration and intensity of REM sleep in the following sleep cycles, frequent awakenings, and terminal insomnia. On the other hand, manic episodes are characterized by markedly reduced need of sleep and in some cases complete insomnia. Moreover, patients exhibit persistent sleep architecture alteration even during euthymia, and so do first-degree relatives, so that these hallmarks have been proposed as *endophenotypes* of depression.
- Suicide: Disrupted sleep and nocturnal wakefulness are evidence-based risk factors for suicidal thoughts and behaviours.
- Anxiety and OCD: Alertness and reactivity to threatens, altered emotional processing and response, repetitive worries, and intrusive thoughts correlate bi-directionally with sleep and circadian cycle alterations (e.g. delayed sleep phase advance, childhood insomnia).
- Schizophrenia: Frequently, circadian cycle alteration is prodromal to SKZ onset and symptom severity has been associated with circadian rhythm alteration itself; data showed that auditory hallucination manifest a circadian pattern, with higher occurrence in the evening (6–9 p.m.). At molecular level, SKZ patients have distorted melatonin secretion curves (e.g. blunted peak, reduced rhythm amplitude, phase advance, altered response to darkness, and sleep habits) and at a genetic level, post-mortem studies showed disrupted rhythmicity in cortical neuron gene expression.
- Genetic variants in clock genes, such as SNPs, have been associated with a variety of psychiatric and neurodevelopmental disorders (e.g. MDD, BD, schizophrenia, ASD): A hypothesis is that such disorders reflect a defect in synchronization between ambient and internal rhythms, so that the body—both as cellular and as an organism—fails to adapt to environmental periodic changes.

12.3.3 Chronotherapies

In recent decades, based on developing knowledge in chronobiology system and multi-factorial aetiopathogenesis of mood disorders, specific therapeutics have been developed, aimed at treating mood disorders through the manipulation of the biological clock. The main therapies are sleep deprivation, light therapy, and dark therapy.

The therapeutic effect of chronobiological techniques in mood disorders can be seen at a neurobiological level. LT and SD act on the biological clock, with an increase in the production of serotonin and activation of numerous areas involved in the emotional circuitry that is compromised at various degrees in mood disorder patients. Interestingly, these neurobiological changes in functional connectivity and neurotransmitter communication are the same as in successful antidepressant treatment and somatic therapies (e.g. TMS, ECT), furtherly supporting the exquisite role of circadian rhythm disruption in mood disorders pathophysiology. The effect of chronobiology on functional brain activity is also correlated with the genetic profile. For example, subjects with the *s/s* alleles in the SERT gene have a reduced function of the transporter and a reduced response to both pharmacological and chronobiological treatments.

12.3.3.1 Light Therapy

Light therapy (LT) has a complex mechanism of action, not fully understood yet: based on the hypothesis of Seasonal Affective Disorder (SAD) aetiopathogenesis, it has been supposed that LT functions by resynchronizing the biological clock's phase delay in wintertime, or by augmentation of serotonin dynamics. However, the mainstay of its therapeutic effect is restoration of melatonin evening peak in depressed subjects.

LT has been developed as an antidepressant treatment for SAD; in successive studies, it has been demonstrated that LT combined with antidepressant medications hastens antidepressant effects in non-seasonal depression as well, with an effect comparable to that of antidepressant medications.

Given the extreme sensitivity of the biological clock to light stimuli and the endogenous alteration in overall circadian rhythm, it has been studied when LT exerted the most beneficial effects: the antidepressant effect is greater if the light is administered 2 h earlier than the usual wake-up time, a time best individuated through the Morningness-Eveningness Questionnaire (MEQ), thus generating a phase advance.

As not all morning times are alike, so not all "lights" are therapeutics. A proper LT treatment sees the utilization in Bright Light sources (typically 10,000 Lux) with an UV-ray filter. The patient sits in front of the LT source, at about 50–100 cm from the lamp, with open eyes but not directly staring at it (for example they can read a book or a magazine), for about 30 min. After LT session, patients should not go back to bed despite any trouble in night sleep or physical tiredness or activating effects of LT on neurotransmitters and mood symptoms would vanish.

12.3.3.2 Sleep Deprivation

In 1966, Schulte firstly noted a possible correlation between reduction of sleep hours and improvement of depressive symptoms. In 1971, Pflug conducted the first therapeutic trial on Sleep Deprivation (SD) as antidepressant treatment: clinical improvement was transitory and patients relapsed the day after. In 1983 and 1987,

Gillin and Wiegand proposed that sleep disorders could be of pivotal importance in depression physiopathology, with special regard to BD.

General indications for SD treatment are:

- Supporting diagnosis and prognosis (e.g. BD versus MDD).
- Alternative or potentiation to classical pharmacological treatments.
- Treatment-resistant depression.
- Rapid cycle patients.

SD is a non-pharmacological intervention useful in bipolar depression and other depressive episodes that has a rapid and striking effect in reducing depressive core symptoms, suicidal ideation included. It has the advantages of being quick, non-invasive, costless, well tolerated by most of patients. It was observed that the anti-depressant effect of SD differs between the different categories of subjects suffering from depression.

Better effects have been observed in endogenous primary depression compared to reactive and/or secondary depression, and in the treatment of bipolar disorder (mean response rate 50–75%) compared to major depressive disorder (mean response rate 33–66%). SD is used in the treatment of bipolar depression, as an alternative or enhancement of antidepressant drugs. In fact, SD has been applied in drug-resistant depressed patients with good results: patients resistant to treatment with SSRIs had a positive response rate in 50% of cases, while patients resistant also to TCAs responded in 40% of cases.

Box 12.6: Clinical Indications for SD Treatment

Unipolar depression (MDD)
Bipolar depression
Schizoaffective depression
Reactive depression
Secondary depression (e.g. Schizophrenia, Parkinson's disease)
Depression in elderly and children
Depression associated with pregnancy and postpartum
Premenstrual dysphoric disorder

Box 12.7: Clinical Predictors of Response to SD

Circadian mood oscillations
Pronounced cardiac frequency circadian variation
Increased motor activity
Clinical diagnosis (e.g. BD)

The development of treatment protocols was subjected to several considerations:

- One night of SD is effective in more than 60% of patients, but with a relapse rate is about 80–95%.
- The antidepressant action of SD is higher and more stable if SD is continued for 36 consecutive hours (Total Sleep Deprivation, TSD).
- LT exposure in the middle of the awake night and in the morning after recovery sleep potentiates the chronobiological effects of SD.
- Concomitant medication utilization potentiates and favours the maintenance of well-being after SD.
- In BD patients, the use of lithium salts in combination with SD significantly improves clinical outcomes, compared with patients not taking lithium.

Box 12.8: SD Protocols

- Total SD
  - For three times in 1 week (see Fig. 12.1)
  - For two times in 1 week
  - Twice a week for 3 weeks
  - For two times in 1 week followed by partial SD
- Partial SD
  - Once a week for 3 weeks
  - For three times in 1 week
  - Twice a week for 2 weeks
  - Thrice a week for 2 weeks
  - For five times at 5 days intervals
  - For six times at 4 days interval

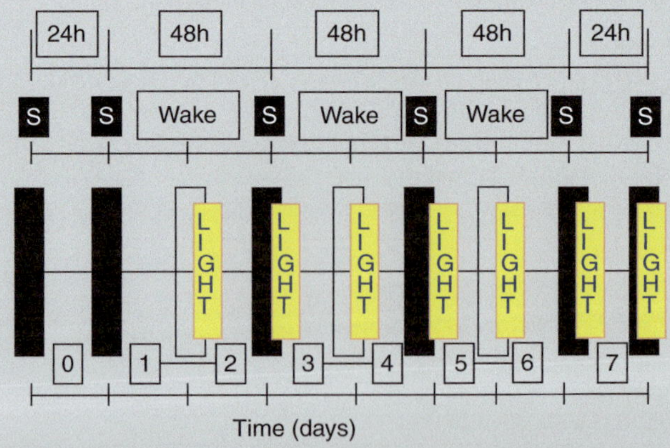

Fig. 12.1 Total sleep deprivation protocol: a robust treatment scheme consists of three forced vigils lasting 36 h, alternating with nights of sleep recovery. LT is administered during the night of deprivation at 3 a.m. and on the awakening following the night of recovery

Since sleep-wake rhythms are manipulated in bipolar patients and sleep deprivation is a trigger for the manic phase, the risk of shift into the manic phase has been worried and eventually observed. Numerous studies showed a switch rate of 6% after 3 cycles of TSD in BD type I non-rapid-cyclers, definitely lower than drug-induced mania (15–29%), while rate peaked 78% in rapid-cycle patients. Restoration of nocturnal sleep with intravenous benzodiazepines results in a disappearance of manic symptoms in 30% of patients and recover into euthymia within 3–5 days.

12.3.3.3 Dark Therapy

Environmental, psychological, and pharmacological factors that induce spontaneous SD can give rise to mania, with inverse correlation between the duration of sleep and the intensity of the manic symptoms at the beginning. As a matter of fact, SD is used in the animal model of mania (appearance of manifest behavioural changes in the rat, e.g. hypervigilance, hyperactivity, speed, irritability): the replicability of inducing manic behaviours both in lower mammals and in humans with manipulations of the sleep-wake rhythm suggests the presence of a common and phylogenetically preserved biological substrate that might confer an evolutionary advantage.

In the 1990s Wehr and Wirz-Justice reported, separately, their successful experience in treating rapid-cycling bipolar manic patients with exposure to a prolonged ambient darkness and bedtime, the basis of Dark Therapy (DT) protocols, which might be added with mood stabilizing and antipsychotic medications.

When it is impossible to reserve a quiet and isolated room for a manic patient, virtual DT comes in help: it is carried out with the use of specific glasses with lenses that filter blue light waves (those light waves that mostly activate the SCN) in order to reduce the exposure to any awakening stimuli.

12.4 Cognitive Remediation

12.4.1 Introduction

Cognition is one of the most relevant determinants of quality of life and daily functioning. Cognitive deficits may be observed in several psychiatric disorders such as schizophrenia, mood disorders, attention deficit/hyperactivity disorder (ADHD). Since neurocognitive impairment is associated with poor long-term global outcomes, including poorer quality of life and impairment in personal, occupational, and social functioning, early interventions targeting cognition are of crucial importance.

While antipsychotic treatments are usually highly effective in improving positive symptoms, they show scarce or null effects on cognitive deficits. To date, rehabilitation interventions are the best available and effective tools to treat cognitive impairment. Cognitive remediation (CR)—a set of interventions based on behavioural training whose goal is to enhance neurocognitive abilities in a generalized and durable manner—has thus become an essential component of the treatment for major psychiatric disorders.

12.4.2 Definition

CR has been defined as "an intervention targeting cognitive deficit (attention, memory, executive function, social cognition, or metacognition) using scientific principles of learning with the ultimate goal of improving functional outcomes". Such therapy aims to ameliorate the individual's everyday functioning, such as scholastic, professional, or social functioning, through the improvement of cognitive performance.

Originally, CR was designed and developed to target subjects with brain lesions and subsequently applied to other disorders that involve cognitive deterioration. Since then, interest in the effectiveness of CR has increasing grown, leading to the development of various types of CR therapies.

CR includes a range of approaches varying in the structure, intensity, and duration of treatment. However, all the approaches are typically based on the same main principles: Training, Strategy Monitoring, and Generalization.

Training consists of repetitive practice through the use of paper and pencil or computerized cognitive training exercises of increasing difficulty as patients progress through the programme. The aim of the training is to promote neuroplasticity and improve information processing.

Strategy monitoring aims to promote a metacognitive comprehension of the problem-solving strategies used during the training. It is usually facilitated by therapists which assist the patients in developing awareness of the approaches they use, generating new strategies, and acquiring flexibility in shifting strategies depending on the specific situation.

Generalization refers to the process through which patients translate the abilities and strategies developed during the treatment into everyday life. Generalization is usually facilitated by therapists through activities that allow patients to identify areas of their daily life where they could apply the strategies acquired during the therapy.

CR strategies can be divided into two methods: "compensatory" and "restorative". The "compensatory" approach aims to bypass specific cognitive deficits, through the use of the patient's residual cognitive abilities.

Conversely, the "restorative" model aims at restoring cognitive function through repetitive practice based on brain plasticity and neurogenesis. Restorative strategies involve two different approaches: bottom-up or top-down. Bottom-up approaches start targeting basic cognitive domains, such as attention, and advance to more complex cognitive functions, such as problem-solving. On the other hand, top-down approaches aim at improving specific neurocognitive functions through use of more complex skills. Therefore, while some restorative approaches involve the use of drill and practice exercises to enhance neuronal plasticity and restore cognitive functions, others involve the generation of new strategies and encourage the generalization in everyday life through the execution of activities that require their use.

CR techniques are highly variable, depending on the therapist's and the patient's characteristics, the therapy goals, and the programme format. While early CR interventions consisted of paper and pencil exercises, nowadays, computer-assisted cognitive training programmes are the most widely utilized. Computer-Assisted Cognitive Rehabilitation (CACR) represents a homogeneous subgroup of CR interventions, usually consisting of weekly sessions of drill and practice comprising both domain-specific exercises (such as verbal memory and fluency, attention, working memory psychomotor speed and coordination, executive functions) and non-specific exercises that require the simultaneous use of different cognitive functions. CogPack, CogRehab, and Circuits are among the most used types of computer-based CR programmes.

12.4.3 CR for Schizophrenia

Over the past decades, CR has been widely used in the treatment of schizophrenia, where cognitive dysfunction is prominent. Cognitive impairment represents a hallmark of schizophrenia: it occurs in about 75% of patients, and it is one of the main predictors of long-term outcomes, affecting quality of life, social relationships, independent living, and occupational functioning. Cognitive impairment appears prior to the onset of the disorder, predicts functional outcome even better than positive and negative symptoms and also limits recovery even when other support has been provided. Deficits are widespread and can be observed in attention, processing speed, memory, executive function, language, and social cognitive function.

Cognitive deficits progressively became the main target of rehabilitative interventions to achieve functional recovery.

In recent decades, a wide range of cognitive remediation therapy (CRT) programmes have been developed and their effectiveness on cognitive functions has been widely demonstrated. CR therapy (CRT) produces small to moderate effects on cognitive outcomes. CRT has been associated with improvement both in global cognition and in specific neuropsychological performances, as well as in psychiatric symptoms, especially negative ones. Finally, CRT has shown to produce improvements in general functioning, especially if the intervention is a part of a broader psychosocial rehabilitation involving the learning of other communication, social, and self-control skills. Indeed, literature suggests that for CR to improve everyday functioning, the subject needs to be in an environment where active skill acquisition or development can take place. Moreover, while CRT effects on cognitive abilities seem to persist even after 5 years, a durable functional improvement requires an integrated rehabilitative approach.

Despite its documented efficacy, CRT is associated with highly heterogeneous outcomes, with some patients normalizing their performance to healthy control levels and others showing no improvement. Several patient's or treatment's variables could influence response or resistance to CRT. Overall, CR seems to be more effective in subjects with younger age, shorter duration of illness, greater pre-treatment cognitive reserve, and fewer disorganized symptoms.

12.4.4 CR for Other Disorders

More recently, CR use has been broadened to other psychiatric disorders and the training has been successfully applied in the treatment of other psychiatric disorders, such as mood disorders, attention deficit/hyperactivity disorder, and anorexia nervosa.

12.4.4.1 Mood Disorders

Major depressive disorder (MDD) is associated with an impairment in neuropsychological functions such as attention, processing speed, learning and memory, and executive function. Cognitive decline is known to predict relapse of depressive episodes, treatment resistance, and functional impairment, besides being associated to compromised quality of life. Moreover, cognitive functioning may continue to deteriorate, not only during the depressive episode, but also in euthymic states. Although literature is still scarce, the efficacy of CR in improving depressive symptom and psychosocial function suggests that CR could represent a promising treatment option for MDD.

Bipolar disorder has been associated with neurocognitive impairment as well, particularly in attention, executive functions, working memory, and social cognition domains. Such cognitive dysfunction affects quality of life, psychosocial functioning, and productivity. Only few studies investigated the effect of CR on neurocognitive and psychosocial function in bipolar disorder and the results are inconclusive.

12.4.4.2 Attention Deficit/Hyperactivity Disorder

Although ADHD is typically characterized by attention deficits, impairments in working memory, processing speed, and executive functions have also been observed. CR studies conducted on ADHD reported improvements in overall symptoms and neuropsychological performance, especially in working memory and executive functions.

Autism Spectrum Disorder

Autism spectrum disorder (ASD) is characterized by impairment in social cognition and restricted and repetitive behaviours and interests. Moreover, subjects affected by ASD are known to experience deficits in cognitive domains such as memory and executive functions. These deficits may affect daily life and psychosocial functioning. Overall, studies investigating CR efficacy on ASD suggest CR interventions could be effective in improving social cognition and neurocognitive performance.

Substance Use Disorder

Deficits in cognitive domains such as attention, working memory, and executive functions have also been observed in subjects with substance use disorders (SUD) and may contribute to poor treatment adherence and outcomes. Literature on CR efficacy on SUD showed improvements in neurocognitive domains, such as working memory, learning, processing speed, and executive functions, but also in psychosocial outcomes such as depressive symptom, craving, and self-regulation. Overall, CR seems to be a promising approach in treating SUDs.

12.4.4.3 Anorexia Nervosa

Anorexia nervosa is not characterized by global cognitive dysfunction, but cognitive inflexibility and processing bias toward detail or local information have been reported and correlated with disease intensification and delayed recovery. CR has been used in subjects with anorexia nervosa, in addition to the usual treatment, with the aim of enhancing cognitive flexibility and central integration ability. Although studies showed mixed results, they suggest that CRT has potential as a supplementary treatment for anorexia nervosa.

Borderline Personality Disorder

Borderline personality disorder (BPD) is associated with cognitive deficits which may affect the outcome of this disorder. Few available data suggest the feasibility and potential effectiveness of CR on specific cognitive domains and psychosocial functioning measures.

CRT should not be a stand-alone therapy, but part of an integrated rehabilitation programme. In this regard, most evidence suggests that an integrated rehabilitative approach seems to be more effective, in terms of both generalizability of results to daily functioning and also durability.

Regardless of the type of mental illness, cognitive impairment affects patients' quality of life, decreases vocational and social functioning and compromises the effectiveness of therapy. Cognitive rehabilitation is as an effective strategy for addressing these cognitive impairments, and has been linked to improvements in social adjustment, vocational functioning, and other functional domains.

12.5 Psychotherapeutic Approaches

An increasing number of studies have highlighted the effectiveness of combined pharmacological and psychotherapeutic interventions for the treatment of different clinical conditions such as mood disorders, anxiety disorders, eating disorders, and personality disorders. The definition of psychotherapy varies according to the theoretical model and the intervention model. In the following paragraph, we briefly describe the two main approaches in psychotherapy: the cognitive-behavioural approach and the psychodynamic approach.

12.5.1 Cognitive-Behavioural Therapies

Cognitive-behavioural therapies represent a class of pragmatic approaches for understanding and dealing effectively with a wide range of psychiatric disorders and problems. Cognitive Behavioural Therapy (CBT) began during the 1970s when a large percentage of behaviour therapists shifted their focus to internal cognitive processes. This shift implied a central role for cognitive processes in the mediation of behaviour and therefore legitimized cognition as a viable target for clinical intervention. Specifically, cognitive-behavioural theories assume that cognition mediates emotional and behavioural responses and thereby it may influence the acquisition and maintenance of psychopathology.

Cognitive-behavioural therapies can be represented along a continuum in terms of how much cognition is included in the formulation: on the one end we find behaviour therapies that focus on behaviour and environmental determinants in terms of elementary learning theory, and at the other end of the continuum we find therapies that formulate therapy only in cognitive terms, leaving no space for behavioural intervention. Most cognitive-behavioural approaches fall somewhere in between since they emphasize the behavioural and cognitive interventions to differing extents.

Although there is much diversity among these treatments, they are all problem focused, goal directed, future oriented, time limited, and empirically based. Moreover, cognitive-behavioural therapies rely on psychoeducation whose goal is to improve quality of life through an adequate knowledge of the disorder, the capacity to recognize early symptoms of relapse, and a strong compliance with the pharmacological treatment.

12.5.1.1 Principles of CBT

Dobson (2001) has identified some basic principles that cut across the variety of treatment approaches in Cognitive Behavioural Therapy (CBT).

1. Cognitive activity affects behaviour and emotions.
2. Cognitive activity may be monitored and altered.
3. Desired emotional and behaviour change may be obtained through cognitive change.

12.5.1.2 Techniques of CBT: Integrating Cognitive and Behavioural Techniques

In order to give an overview of the features shared by cognitive-behavioural therapies, we briefly describe some basic methods and techniques that are commonly used to help patients dealing with their symptoms and problems.

Within the cognitive-behavioural framework, maladaptive thinking is both a symptom and a critical maintenance factor. Therefore, patients can overcome their problems by identifying and modifying their negative thoughts. Patients learn to recognize their automatic thoughts and they are encouraged to view them as hypothesis rather than manifest facts. The therapist induces *cognitive restructuring* by asking leading questions that guide the patient to question and alter his faulty cognition.

Another cognitive and practical technique is *problem-solving*: a self-directed process by which a person attempts to identify and implement effective solutions for a specific problem faced in everyday life. Problem-solving has been applied to a wide range of situations commonly encountered in psychiatric practice: difficulties associated with anxiety, mood, stress, substance abuse.

Moving to more behavioural techniques the use of *activity schedules* serves to counteract the patient's loss of motivation and inactivity. Since inactivity is associated with negative emotional states, the therapist and the patient may use a schedule in order to plan activities in advance.

Exposure techniques are used to treat fear, anxiety, and other intense emotional reactions. Because of the centrality of avoidance mechanisms in anxiety disorders, exposure techniques represent a major component of CBT for these clinical conditions. Exposure methods are usually graded: the patient begins to face situations or stimuli that trigger lower anxiety's levels and then is progressively confronted with stimuli that may produce higher levels of anxiety. Exposure to these feared or avoided situations allows the patient to gather data that are inconsistent with the beliefs that produce his anxiety. Exposure can be implemented in vivo or in imaginal mode.

12.5.1.3 Applications of CBT to Specific Disorders

CBT has received a considerable attention in recent years because of its strong evidence base. There is an extensive literature supporting the efficacy of cognitive-behavioural therapy for many psychiatric disorders. Cognitive and behavioural therapies were first applied to mood and anxiety disorders and then they were extended to eating disorders, somatoform disorders, substance abuse, personality disorders and, recently, even to schizophrenia in conjunction with medication. Therefore, the efficacy of CBT has been studied taking into consideration for a wide variety of clinical conditions.

Mood Disorders

There is a growing evidence supporting the efficacy of CBT in combination with pharmacotherapy in the treatment of mood disorders.

Depression

CBT of depression focuses on the cognitive restructuring of the major cognitive patterns that induce the patient to see himself, his future and his experiences in an idiosyncratic manner. At the same time, it involves behavioural strategies, such as weekly activity schedule, that are used not only to change the behaviour but also to elicit cognitions associated with specific behaviours. Different studies have demonstrated the efficacy of CBT in reducing the symptoms of depression and patients maintain their treatment gains at 3 and 6 months follow-up.

Bipolar Disorder

CBT in conjunction with mood stabilizers and regular psychiatric follow-up reduces relapse rates in the short term and improves symptoms and social functioning in the long term.

Anxiety Disorders

CBT is considered the gold standard psychotherapy for anxiety disorders. Recent meta-analyses have demonstrated that CBT is efficacious in treating anxiety disorders and there is some evidence suggesting its long-term efficacy. CBT is associated with greater improvement on measures of anxiety, panic, fear, avoidance, depression, and quality of life. Moreover, CBT reduces the risk of relapse relative to pharmacotherapy alone.

Panic Disorder
Several studies showed the important therapeutic role of cognitive techniques through which the patient identifies maladaptive beliefs with regard to bodily sensations and then modifies them with more adaptive beliefs.

Social Phobia
Treatments for social phobia include exposure, social skills training, and cognitive restructuring. Therefore, on the one hand, cognitive-behavioural therapists employ exposure methods to habituate anxiety and, thereby, enable the patient to function in the presence of other people. On the other hand, cognitive interventions target the negative beliefs about the self, helping the patient construct a more accurate image about himself as a social actor.

Specific Phobia
Exposure-based treatments represent the treatment of choice for specific phobia; however, adding cognitive restructuring appears to produce better outcomes than exposure alone.

Generalized Anxiety Disorder
CBT therapists help patient to recognize anxious thoughts, seeking helpful alternatives and taking action to test these alternatives. At the same time, interventions include psychoeducation, applied relaxation, and imaginal and in vivo exposure.

CBT has been found to produce better outcomes for patients with GAD than psychodynamic therapy.

Obsessive-Compulsive Disorder
Most behavioural and cognitive-behavioural treatments for OCD induce change via exposure and ritual prevention. Exposure and response prevention (ERP) is considered to be the first-line psychotherapy for the disorder. ERP aims to break this cycle of symptoms by eliminating rituals and avoidance, thereby teaching patients how to tolerate distress without engaging in counterproductive behaviours and providing "corrective information" that challenges people's existing fear response. More cognitive approaches to OCD help the patient to identify, evaluate and alter problematic beliefs.

Post-Traumatic Stress Disorder (PTSD)
Trauma-focused cognitive-behavioural methods, such as stimulus confrontation and cognitive restructuring, have proved to be efficacious and are often the treatments of choice for individuals with PTSD.

However, EMDR, an evidence-based treatment, has become one of the most used treatment for PTSD and its effectiveness has undergone the scrutiny of several meta-analyses. EMDR consists of a standard protocol which includes eight phases and bilateral stimulation (usually horizontal saccadic eye movements) to desensitize the discomfort caused by traumatic memories and the aim of the therapy is to achieve their reprocessing and integration within the patient's standard biographical memories.

Eating Disorders

Eating disorders provide one of the strongest indications for CBT since their core psychopathology, the over-evaluation of shape and weight, is cognitive in nature.

CBT is the treatment of choice for bulimia nervosa and for binge-eating disorder. Regarding to anorexia nervosa CBT has made great advances over the last 10 years appearing to be a viable and promising treatment for patients with AN, with about 40% of adults and almost 60% of adolescents reaching and maintaining a normal weight range. The increase in weight is accompanied by a decrease in eating disorder psychopathology and over half of adult and about 80% of adolescent patients reach and maintain minimal residual psychopathology.

Personality Disorders

There is a strong support for the use of cognitive-behavioural psychotherapy for personality disorders in terms of efficacy and effectiveness. The intervention typically includes clinical assessment, cognitive conceptualization, technical interventions, and building and using the therapeutic relationship. Among the empirically investigated evidence-based therapies we can mention Dialectical Behaviour Therapy (DBT), Schema Therapy (ST), Acceptance and Commitment Therapy (ACT), and Rational-Emotive Behaviour Therapy (REBT). The majority of the studies focused on Borderline Personality Disorder (BPD) and DBT has been studied the most and is currently considered the most effective treatment for BPD. DBT conceptualizes BPD with the biosocial theory which suggests that BPD is primarily a dysfunction of the emotional regulation system and BPD arises from a transaction between biological vulnerability and invalidating environments. Since the DBT model assumes that individuals with BPD lack key interpersonal, self-regulation, and distress tolerance skills, the therapy is designed to facilitate the learning of these skills through individual therapy and group therapy.

12.5.2 Psychodynamic Psychotherapy

Psychodynamic therapy derives from psychoanalysis whose origins date back to Sigmund Freud in the late 1890s and the early 1900s. However, today psychodynamic therapy differs dramatically from the original approach.

Gabbard, in order to describe contemporary long-term psychodynamic therapy, proposed the following definition which is partly based on Fonagy's conceptualization: a set of psychotherapeutic treatments, some specifically tailored to disorders and others more general that are based on a thoroughgoing understanding of human subjectivity and how it interacts with the individual's relationship with both the external and internal environments.

In order to distinguish psychodynamic therapy from CBT, seven distinctive features of technique have been identified by Blagys and Hilsenroth (2000):

1. Focus on affect and expression of emotion.
2. Exploration of attempts to avoid aspects of experience.
3. Identification of recurring themes and patterns.
4. Discussion of past experience.
5. Focus on therapeutic relationship.
6. Exploration of wishes, dreams, and fantasies.

Another distinctive feature of psychodynamic psychotherapy is that it focuses on the unique characteristics of the individual, therefore the strategies and the techniques are tailored to the patient.

The duration of the therapy represents a feature that has always distinguished psychodynamic psychotherapy from other form of psychotherapy such as the cognitive-behavioural psychotherapy. However, this aspect has changed over time and time-limited psychodynamic psychotherapies have arisen. Therefore, psychodynamic psychotherapy can be carried out both as a:

- Short-term (time-limited) treatment: the number of sessions is predetermined (STPP).
- Long-term open-ended treatment: the end of therapy is designed naturalistically.

More specifically, short-term treatments (STPP) usually last between 7 and 24 sessions, whereas Gabbard (2004) uses the term long-term psychodynamic psychotherapy (LTPP) when the treatment is longer than 24 sessions or 6 months.

12.5.2.1 Interventions

With regard to the interventions used, psychodynamic therapy operates on a continuum from the expressive or interpretative pole, on the one hand, to supportive and emphatic pole, on the other hand.

Thus, the therapist may be more expressive or exploratory at some time while shifting to a more supportive style at another, depending on the patient's needs. The more severely disturbed a patient is or the more acute his or her problem is, the more supportive and the less expressive interventions are required and vice versa.

On the expressive pole, the type of interventions that characterize psychodynamic psychotherapy consist in statements made by the therapist to explain the patient's feelings, thoughts, behaviours, and symptoms linking them to unconscious fantasies, meanings, or childhood origins. On the supportive pole of the continuum, therapist's interventions may include specific advice to patients on how they should live their lives and how they should behave in certain situations of their life.

12.5.2.2 Applications of Psychodynamic Psychotherapy to Specific Disorders

Research on the outcomes of psychodynamic treatments has been relatively scarce for years. However, in the last years, there is a growing body of efficacy research. We have now empirical evidence that supports the efficacy of psychodynamic psychotherapy. In addition, he pointed out that patients maintain therapeutic gains and often continue to improve after treatment has ended when the research involves follow-up measures.

Mood Disorders

The efficacy of short-term psychodynamic psychotherapy (STPP) for depression is debated. However, in the last years, a number of large-scale and high-quality studies have been conducted.

Depression

Psychodynamic approaches to the treatment of depression focus on the patient's internal world, emphasizing "how (unconscious) motivational factors lead the patient to (mis)perceive and (mis)interpret external reality and experiences and to create, unwillingly, problems that maintain depressive symptoms, particularly in interpersonal relationships". Recently, there has been an increasing number of studies supporting the efficacy of STPP for depression. STPP results in symptom reduction and function improvement during treatment and the gains are either maintained or further improved at follow-up.

Anxiety Disorders

Panic Disorder

Panic-focused psychodynamic psychotherapy consists in 24 sessions, twice a week. It uses substantially different techniques from CBT since it does not involve homework or exposure protocol. The therapy explores the personal meanings of panic symptoms in the light of common psychodynamic conflicts in panic disorders such as separation and autonomy.

Social Phobia

The study of Knijnik and colleagues (2004) aimed to assess the effectiveness of psychodynamic group therapy (PGT) in patients with generalized social phobia. PGT resulted superior to a placebo control group as shown by the significant changes in the anxiety's scales used.

Generalized Anxiety Disorder

In a randomized controlled trial (Leichsenring et al. 2009) short-term psychodynamic psychotherapy (30-session treatment), which focuses on the therapeutic alliance as a corrective emotional experience that allows the patient to approach the feared situations, was associated with significant improvements in measures of anxiety.

Post-Traumatic Stress Disorder

Psychodynamic therapy delves into the construed meanings of the traumatic event, the individual's response to it, and the behaviours that developed after it, in the hope of helping patients develop insights into the factors that activate traumatic re-experiencing and gain mastery over their internal experiences through more effective coping.

Empirical and clinical evidence suggests that psychodynamic approaches may result in improved self-esteem, increased ability to resolve reactions to trauma through improved reflective functioning, increased reliance on mature defences and decreased reliance on immature defences and improved social functioning. In addition, psychodynamic psychotherapy results in further improvement after treatment ends.

Eating Disorders

Psychodynamic therapists focus on the meaning of the symptom in terms of the patient's history and of their experience with their family and also on the effects of the symptom and its influence upon the patient's current relationship. Psychodynamic psychotherapy produces significant improvements in symptomatology; however, once treatment is over, some improvements, such as weight gain, are nor mantained.

Personality Disorders

Dynamic therapist working with patients with personality disorders utilizes models of personality pathology that emphasize developmental distortions and anomalies in cognitive and affective processes in the understanding of self and others. These underdeveloped or distorted internalized representations are the object of assessment and change.

Transference-Focused Psychotherapy (TFP), a form of LTPP, has proved to be an efficacious treatment for BPD through a significant diminution of the symptom criteria and significant changes in the levels of impulsivity, irritability, and verbal and physical aggression.

A noteworthy psychoanalytically oriented and fully manualized treatment for BPD is the *Mentalization-Based Treatment* (MBT). MBT was founded on the specific theoretical basis that vulnerability to frequent loss of mentalizing is the underlying pathology that gives rise to these characteristic symptoms. Therefore, the treatment focuses on increasing mentalization that entails making sense of the actions of oneself and others on the basis of intentional mental states, such as desires, feelings, and beliefs.

Overall, psychodynamically orientated psychotherapy reduces personality pathology, reduces symptoms and improves social functioning in patients presenting with a mixture of PD clusters A, B, C, and NOS (Not Otherwise Specified).

Further Reading

Ballard ED, Vande Voort JL, Bernert RA, Luckenbaugh DA, Richards EM, Niciu MJ, Furey ML, Duncan WC Jr, Zarate CA Jr. Nocturnal wakefulness is associated with next-day suicidal ideation in major depressive disorder and bipolar disorder (Erratum in: J Clin Psychiatry 2016 Dec;77(12):e1655). J Clin Psychiatry. 2016;77(6):825–31. https://doi.org/10.4088/JCP.15m09943.

Bateman A, Fonagy P. Psychotherapy for borderline personality disorder: mentalization-based treatment. Oxford: Oxford University Press; 2004.

Benedetti F, Colombo C. The use of SD in mood disorders. Neuropsychobiology. 2009;64(3):141–51.

Benedetti F, Smeraldi E. Neuroimaging and genetics of SD. Curr Pharm Des. 2009;15(22):2637–49.

Benedetti F, Barbini B, Colombo C, Smeraldi E. Chronotherapeutics in a psychiatric ward. Sleep Med Rev. 2007;11(6):509–22. https://doi.org/10.1016/j.smrv.2007.06.004.

Biagianti B, Castellaro GA, Brambilla P. Predictors of response to cognitive remediation in patients with major psychotic disorders: a narrative review. J Affect Disord. 2021;15(281):264–70. https://doi.org/10.1016/j.jad.2020.12.011.

Blagys MD, Hilsenroth MJ. Distinctive features of short-term psychodynamic-interpersonal psychotherapy: a review of the comparative psychotherapy process literature. Clin Psychol Sci Pract. 2000;7(2):167–88.

Buonocore M, Spangaro M, Bechi M, Baraldi MA, Cocchi F, Guglielmino C, Bianchi L, Mastromatteo A, Bosia M, Cavallaro R. Integrated cognitive remediation and standard rehabilitation therapy in patients of schizophrenia: persistence after 5years. Schizophr Res. 2018;192:335–9. https://doi.org/10.1016/j.schres.2017.05.022.

Charrier A, Olliac B, Roubertoux P, Tordjman S. Clock genes and altered sleep-wake rhythms: their role in the development of psychiatric disorders. Int J Mol Sci. 2017;18(5):938. https://doi.org/10.3390/ijms18050938.

Dallaspezia S, Benedetti F. Chronobiology of bipolar disorder: therapeutic implication. Curr Psychiatry Rep. 2015;17(8):606. https://doi.org/10.1007/s11920-015-0606-9.

Fosse R, Read J. Electroconvulsive treatment: hypotheses about mechanisms of action. Front Psych. 2013;27(4):94. https://doi.org/10.3389/fpsyt.2013.00094.

Gabbard GO. Long-term psychodynamic psychotherapy. A basic text. Washington, DC: American Psychiatric Publishing; 2004.

Gabbard GO. Major modalities: psychoanalytic/psychodynamic. In: Gabbard GO, Beck JS, Holmes J, editors. Oxford textbook of psychotherapy. Oxford: Oxford University Press; 2007. p. 3–13.

Grant P, Young PR, DeRubeis R. Cognitive and behavioral therapies. In: Gabbard GO, Beck JS, Holmes J, editors. Oxford textbook of psychotherapy. Oxford textbook, vol. 2007; 2007. p. 15–25.

Kim EJ, Bahk YC, Oh H, Lee WH, Lee JS, Choi KH. Current status of cognitive remediation for psychiatric disorders: a review. Front Psych. 2018;1(9):461. https://doi.org/10.3389/fpsyt.2018.00461.

Klomjai W, Katz R, Lackmy-Vallée A. Basic principles of transcranial magnetic stimulation (TMS) and repetitive TMS (rTMS). Ann Phys Rehabil Med. 2015;58(4):208–13. https://doi.org/10.1016/j.rehab.2015.05.005.

Lefaucheur JP, Antal A, Ayache SS, Benninger DH, Brunelin J, Cogiamanian F, Cotelli M, De Ridder D, Ferrucci R, Langguth B, Marangolo P, Mylius V, Nitsche MA, Padberg F, Palm U, Poulet E, Priori A, Rossi S, Schecklmann M, Vanneste S, Ziemann U, Garcia-Larrea L, Paulus

W. Evidence-based guidelines on the therapeutic use of transcranial direct current stimulation (tDCS). Clin Neurophysiol. 2017;128(1):56–92. https://doi.org/10.1016/j.clinph.2016.10.087.

Lefaucheur JP, Aleman A, Baeken C, Benninger DH, Brunelin J, Di Lazzaro V, Filipović SR, Grefkes C, Hasan A, Hummel FC, Jääskeläinen SK, Langguth B, Leocani L, Londero A, Nardone R, Nguyen JP, Nyffeler T, Oliveira-Maia AJ, Oliviero A, Padberg F, Palm U, Paulus W, Poulet E, Quartarone A, Rachid F, Rektorová I, Rossi S, Sahlsten H, Schecklmann M, Szekely D, Ziemann U. Evidence-based guidelines on the therapeutic use of repetitive transcranial magnetic stimulation (rTMS): an update (2014–2018) [Erratum in: Clin Neurophysiol. 2020 May;131(5):1168-1169]. Clin Neurophysiol. 2020;131(2):474–528. https://doi.org/10.1016/j.clinph.2019.11.002.

Linehan MM. Cognitive-behavioral treatment of borderline personality disorder. New York: Guilford Press; 1993.

Rasmussen K. The practice of electroconvulsive therapy: recommendations for treatment, training, and privileging (second edition). J ECT. 2002;18(1):58–9. https://doi.org/10.1097/00124509-200203000-00015.

Rossi S, Antal A, Bestmann S, Bikson M, Brewer C, Brockmöller J, Carpenter LL, Cincotta M, Chen R, Daskalakis JD, Di Lazzaro V, Fox MD, George MS, Gilbert D, Kimiskidis VK, Koch G, Ilmoniemi RJ, Lefaucheur JP, Leocani L, Lisanby SH, Miniussi C, Padberg F, Pascual-Leone A, Paulus W, Peterchev AV, Quartarone A, Rotenberg A, Rothwell J, Rossini PM, Santarnecchi E, Shafi MM, Siebner HR, Ugawa Y, Wassermann EM, Zangen A, Ziemann U, Hallett M. Safety and recommendations for TMS use in healthy subjects and patient populations, with updates on training, ethical and regulatory issues: Expert Guidelines. Clin Neurophysiol. 2021;132(1):269–306. https://doi.org/10.1016/j.clinph.2020.10.003.

Segalàs C, Labad J, Salvat-Pujol N, Real E, Alonso P, Bertolín S, Jiménez-Murcia S, Soriano-Mas C, Monasterio C, Menchón JM, Soria V. Sleep disturbances in obsessive-compulsive disorder: influence of depression symptoms and trait anxiety. BMC Psychiatry. 2021;21(1):42. https://doi.org/10.1186/s12888-021-03038-z.

Shapiro F. Efficacy of the eye movement desensitisation procedure in the treatment of traumatic memories. J Trauma Stress. 1989;2:199–223.

Souery D, Amsterdam J, de Montigny C, Lecrubier Y, Montgomery S, Lipp O, Racagni G, Zohar J, Mendlewicz J. Treatment resistant depression: methodological overview and operational criteria. Eur Neuropsychopharmacol. 1999;9(1–2):83–91. https://doi.org/10.1016/s0924-977x(98)00004-2.

Thase ME, Rush AJ. When at first you don't succeed: sequential strategies for antidepressant nonresponders. Clin Psychiatry. 1997;58(Suppl 13):23–9.

Walker WH 2nd, Walton JC, DeVries AC, Nelson RJ. Circadian rhythm disruption and mental health. Transl Psychiatry. 2020;10(1):28. https://doi.org/10.1038/s41398-020-0694-0.

Wehr TA, et al. A circadian signal of change of season in patients with seasonal affective disorder. Arch Gen Psychiatry. 2001;58:1108–14.

Wirz-Justice A. The implications of chronobiology for psychiatry. Psychiatr Times. 2011;28(10):10.

Psychiatric Emergencies 13

M. Locatelli, E. Manfredi, L. Fregna, F. Martini,
D. Pratesi, G. Perrozzi, and C. Colombo

13.1 Introduction

A psychiatric emergency is an acute onset of severe psychopathological symptoms, such as thought, affect and motor disturbances, requiring rapid diagnostic and therapeutic interventions in order to minimize risks for the patient, and bystanders too.

Up to 50% of patients presenting with psychiatric emergencies have a coexisting medical disease. Thus, medical screening with physical examination, neurological examination and laboratory tests is mandatory in order to identify any physical conditions, which would be the cause or a precipitating factor to the psychiatric emergency itself.

About 6–25% of all emergency department (ED) access are due to psychiatric emergencies, with or without concomitant physical conditions.

M. Locatelli · L. Fregna (✉) · F. Martini (✉)
IRCCS San Raffaele Scientific Institute, Milan, Italy
e-mail: locatelli.marco@hsr.it; fregna.lorenzo@hsr.it; francesca.martini1@hsr.it

E. Manfredi · D. Pratesi · G. Perrozzi
Vita-Salute San Raffaele University, Milan, Italy
e-mail: manfredi.elena@hsr.it; pratesi.donata@hsr.it; perrozzi.giulia@hsr.it

C. Colombo
IRCCS San Raffaele Scientific Institute, Milan, Italy

Vita-Salute San Raffaele University, Milan, Italy
e-mail: colombo.cristina@hsr.it

R. Cavallaro, C. Colombo (eds.), *Fundamentals of Psychiatry for Health Care Professionals*, https://doi.org/10.1007/978-3-031-07715-9_13

13.2 Assessment

13.2.1 Psychiatric and Medical History

The clinical interview with complete clinical history collection is fundamental and it could be integrated indirectly by interviewing also the caregivers or bystanders present at the onset of the current acute symptoms. In the ED, a specific requirement for the psychiatric consultant is to gather information thoroughly and concisely. The main aims of emergency evaluation of a psychiatric patient would be assessing the need of hospitalization and relieving the acute distress. In a second moment, when the emergency is passed, there will be time to investigate furtherly in the ward or in the outpatient clinic setting.

13.2.2 Physical Examination

Priority is to ensure stabilization of vital signs (airways, breathing and circulation). A complete physical examination, including neurologic examination, should be performed on all patients.

Physical symptoms may indicate an organic cause, including infections, endocrine abnormalities, autoimmune dysfunction, central nervous system disease, substance intoxications or withdrawals. Prompt assessment and correction of physical disorders generally relieves psychiatric symptoms. Here we describe a few and frequent examples:

- Hypoxia and hypotension can manifest with agitation, disorientation, and numbness.
- Hypertension or hypertensive crisis can present with psychomotor agitation or anxiety.
- Tachycardia can be associated with agitation or acute anxiety, but also could be a symptom of an underlying medical disorder like infectious state with fever, intoxication, withdrawal, and cardiac disease.

The neurologic examination should involve a general assessment of orientation (in space, time and autobiographical clues), memory, cranial nerves, language, motor system and reflexes, and cerebellar function. If any focal neurologic deficit emerges, the patient must undergo further tests to examine acute neurologic causes of their presentation (for example, stroke, encephalitis/meningitis, nerve palsy, faints, peripheral motor disorders, neurodegenerative disorders…).

Eventually, atypical features of psychiatric illness (for example, visual hallucinations) should suggest an organic disorder aside from a primary psychiatric illness.

Red flags for a probable organic origin of the acute psychiatric condition:

- Acute onset.
- First episode.

- Elderly.
- Physical comorbidities.
- Substance abuse.
- Non-auditory hallucinations.
- Neurological signs.
- Disorientation, loss of memory, difficulty concentrating, apraxia.

13.2.3 Routine Laboratory Testing

In order to exclude medical conditions, it is helpful during the evaluation of a patient with a psychiatric emergency to perform lab tests that include full blood count, metabolic panel, toxicology screening, thyroid panel, liver and kidney function tests, glucose tests and C-reactive protein.

13.2.4 Neuroimaging

Whenever a patient with acute psychiatric complaints shows also neurologic symptoms, and/or has a positive history for head trauma, it is recommended to perform a brain imaging to rule out neurological acute disorders. The fastest imaging technique is computed tomography (CT scan), but several hospitals also perform magnetic resonance imaging (MRI) in acute settings.

13.2.5 Additional Testing

Additional testing in the ED should include EKG in patients with heart concerns (for example arrhythmia or ischemia). A spinal puncture, to rule out meningitis or encephalitis, if the patients present with neurological focal signs and fever, leucocytosis, delirium, or altered consciousness. An EEG should be requested for patients with personal history positive for seizure disorder.

13.3 Classification of Emergencies

Conditions requiring a specialistic psychiatric evaluation usually are alternations in mood, thought and behaviour. Such disturbances from normal functioning (referred to a person's previous state or regarding the general population "normal" activity) can take place alone or with a various degree of physical illness. There are in fact many medical disorders and diseases that are accompanied by anxiety, psychomotor agitation, low mood, hallucinations, etc. For example, an acute anxiety crisis can represent a pure anxiety disorder, but also a heart attack patient is likely to feel anxious; low mood can be a pivotal symptom in depressive disorders, or it can be a collateral feature of endocrine disorders or neurological diseases.

Conversely, psychological disorders can manifest with a preeminent physical complaint, and only later in time the patient will "get in touch" with their mood and thoughts. For example, syncope can be a manifestation of a pure cardiologic or neurologic disorder, or it can be a conversive disorder symptom. Low appetite can be due to cancer, or to depressive disorder. Lastly, there are substance intoxication and withdrawal which manifest with life-threatening alterations in consciousness and vital signs, being a medical emergency, but substance abuse disorders are a big issue in psychiatric care settings.

13.3.1 Anxiety Disorders

Among anxiety disorders, panic disorder is the most frequently encountered in emergency care settings, requiring a prompt and clear differential diagnosis with other serious physical pathologies.

13.3.1.1 Assessment

The patient refers to the emergency department worried that the symptoms presented (palpitations, pain in the chest or arm) indicate a serious disease, statically a heart attack or a stroke are the mostly reported. When a panic attack comes, a vicious circle is created in which anxiety causes hyperventilation that produces alarming somatic symptoms, which in turn sustain the fear of tremendous physical illnesses, making the subject overwhelmed and worsening hyperventilation, and so on.

Having ruled out any physical condition, panic attack is a diagnosis of exclusion.

An attack lasts generally 10–30 min, so the healthcare professional is often confronted with "post panic attack anxiety" characterized by a state of serious concern and profound asthenia. An acute anxiety crisis can be quite similar to a panic attack, but it usually lasts longer and symptoms mitigate more softly. In both cases, if post-acute mental and physical distress are particularly high, pharmacological intervention can be offered.

Later, it is extremely important not to dismiss the patient's symptoms but calmly explain the patient the psychogenic origin of the distress and to provide them with information regarding panic and anxiety disorders, potential treatment and prognosis (highlighting that is highly favourable in most cases) and addressing them to speciality setting.

13.3.1.2 Differential Diagnosis

- Myocardial infarction.
- Pulmonary embolism.
- Acute surgical disease.
- Hyperthyroidism.
- Pheochromocytoma.

- Hypoglycaemia.
- Diabetic ketoacidosis.
- Cushing's syndrome.
- Hypocalcaemia.
- Encephalopathies.
- Asthmatic disorders.

13.3.1.3 Management and Treatment

Immediate treatment for a panic attack relies on short-peak and half-life benzodiazepines (e.g. alprazolam) by mouth (both drops and tablets are available). If post-acute anxiety is particularly high, immediate pharmacological intervention mid-long half-life benzodiazepines (e.g. delorazepam) may be necessary, by mouth or in some cases it is preferable to administer them parenterally which is more effective.

13.3.2 Mood Disorders

13.3.2.1 Depression

Depressed patients can present to the emergency department or be referred by the GP or a relative in diverse acute situations: suicidal thoughts or attempts, delusionary thoughts or disorganized behaviour, alarming changes in nutritional status up to cachexia. Usually, these are specifiers of severity of the depressive episode and hospitalization is required.

Assessment

The presence of thoughts about death is quite common in depressed patients: it is fundamental to assess suicidal ideation during the clinical interview, to investigate the issue with the patient, and evaluate the presence of other risk factors, in order to establish the best preventive strategy.

Delusions and hallucinations in depressive episodes are usually mood-congruent, but they can be incongruent and associated with dysphoric and irritable mood, which make the patient at greater risk of impulsive and potentially self-damaging behaviour. Disorganized behaviour in depressed patients can be another psychotic-spectrum symptom or a manifestation of pseudodementia, together with confusion, difficulties in memory and disorientation.

Another pivotal area that is affected is neurovegetative system, so that a depressed patient typically loses appetite and the sense of thirst. Another case is that they starve due to lack of driving force to cook and eat at minimum suffice or due to delusions or hallucinations. Depression frequently is accompanied by drastic weight loss and dehydration: fluid and electrolyte imbalances can figure a medical emergency, especially in elderly patients or in those with physical comorbidities. In all these cases, a complete lab testing is helpful to assess the severity of physical compromission and necessity of supplementation.

Management and Treatment

As stated above, depressive episode emergencies typically require hospitalization to address the acute distress, monitor and optimize the pharmacotherapy and evaluate further medical interventions as needed.

In case of suicidal ideation, pharmacological interventions include lithium salts as first choice, or other anti-impulsive agents if impulse dyscontrol is extended to many areas, such as substances or behaviours, and lithium is contraindicated: the most frequently administered are valproate, aripiprazole and carbamazepine.

Treatment of psychotic features requires antipsychotic agents, being the fastest and more effective haloperidol, but it easily causes extrapyramidal adverse effects. For this reason, second generation antipsychotics are usually preferred.

13.3.2.2 Mania

Manic patients are often agitated and not cooperative as they frequently have poor insight.

Assessment

Typical symptoms are elated mood with or without irritability, racing thoughts and speech, insomnia, hyperactivity and higher risk of aggressive behaviour; psychotic symptoms such as delusions of grandeur or persecution can also be present.

Differential Diagnosis

Clinical cases resembling a manic state can be mainly due to stimulants intoxication, alcohol withdrawal, steroids-induced mania and neurodegenerative disorders. It is fundamental to review the psychiatric and general medical history of the patient, to evaluate the presence of previous mood episodes, the consumption of drugs and substances and the presence of other medical comorbidities.

Management and Treatment

Managing mania can be challenging, because of the lack of insight of the patient and the high levels of agitation and risk of aggressive behaviour. Pharmacological interventions include sedative and hypnotic, antipsychotic and mood stabilizing medications and, in some cases, compulsory hospitalization and treatment may be necessary.

13.3.3 Acute Psychosis

An acute psychotic episode is a sudden manifestation of perceptual alterations (typically auditory hallucinations), thought alterations (e.g. persecutory delusions) and insomnia, so that agitation, behavioural abnormalities and the risk of aggressive behaviour put the patient in need of medical treatment, whether they are aware or not of their disorder. Patients experiencing an acute psychotic episode may present in the clinics of the general practitioner, in an emergency department or in the psychiatric facilities. The serious distortions of the judgment of reality, the alterations in the state of consciousness and the emotional involvement compromise the reliability of the information provided by these patients. Therefore, the collection of anamnestic data needs the intervention of additional figures (e.g. family, friends, neighbours, colleagues).

13.3.3.1 Assessment

It is important to identify possible triggers and precipitating factors, in particular external stressors, the use of recreational drugs, concomitant medical conditions and adherence to psychotropic medication already in use. Moreover, additional information need to be gathered: previous psychopathological episodes, premorbid personality, social and occupational adaptation, prodromal symptoms, modalities of onset and the evolution of psychotic manifestations, the presence of psychiatric or neurological diseases in the family.

13.3.3.2 Differential Diagnosis

All these information will lead the clinician to orientate between two big and discrete chapters:

- Primary Psychotic Disorders: schizophrenia, schizoaffective disturbance, brief psychotic disorder, major depression with psychotic features, mania with psychotic features.
- Secondary Psychotic Disorders: psychotic disorder derived from a general medical condition and substance-induced psychotic disorder.

13.3.3.3 Management and Treatment

The alternative that arises is that between hospitalization, either voluntary or forced, and outpatients' clinics. The severity of the productive symptomatology is not always a sufficient reason to decide on hospitalization. On the contrary, the presence of psychomotor arrest or severe agitation, especially if accompanied by aggressive behaviours, marked confusion or the suspicion of an underlying medical condition should lead to hospitalization. Similarly, an unfavourable familial and social environment with misunderstandings, conflicts, lack of an adequate caregiver represent further elements in favour of hospitalization. In some cases, hospitalization is conducted without the patient's consent, if clinical conditions require so.

Pharmacological treatments include sedative, hypnotics and antipsychotics. In case of acute agitation, intravenous benzodiazepines (e.g. delorazepam) and antipsychotics—either via oral (e.g. haloperidol, clotiapine) or intramuscular (e.g. aripiprazole, clotiapine, olanzapine) administration. When a psychotic disorder is likely to be secondary, it is fundamental to seek and treat the underlying condition.

13.3.4 Psychiatric Emergencies Due to Acute Intoxications or Withdrawal

Urgent clinical conditions (intoxication and withdrawal) can be associated with illicit drugs (opioids, cocaine, stimulants), alcohol or medications (benzodiazepines, antidepressants, antipsychotics, mood stabilizers).

These topics are discussed in the specific chapters.

The aim of emergency intervention is to quickly stabilize the patient's clinical condition and determine the most appropriate treatment. It is a priority to identify the substance, the route of administration, the quantity, the time passed since the intake, and the time of symptom onset.

The purpose of acute drug therapy is to treat the state of intoxication or withdrawal as well as any psychiatric comorbidity.

13.3.4.1 Assessment

Whenever a patient comes declaring a positive history for drug abuse or the clinical suspicion is high, a toxicological screen and blood alcohol concentration test must be prescribed. Not all hospitals are equipped for dosing medications, in case the best option is to ask the consultation of a clinical pharmacologist or the local poison and drug information service.

13.3.4.2 Management and Treatment

It is fundamental to stabilize vital signs, assure adequate hydration and nutritional support. Withdrawal can be managed with intravenous benzodiazepines (e.g. lorazepam, delorazepam), if agitation or hallucinations are present haloperidol can be added.

13.3.5 Personality Disorders

Personality disorders are "persistent patterns of internal and behavioural experiences that differ or deviate from the expected social and cultural norms causing disruption and distress leading to difficulties in daily functioning" (DSM 5, APA). They typically have onset in adolescence or early adulthood and model the overall functioning of the person throughout all their life. Personality disorders are relatively common: general population prevalence is about 10%, but rates peak up to 25% of primary care patients and 50% of psychiatric outpatients. These patients frequently seek healthcare services intervention and recurrently refer to emergency departments.

The specific disorders are described in the Personality Disorders chapter.

13.3.5.1 Assessment

In acute settings, the most frequent personality disorders encountered are cluster B personality disorder (borderline, narcissistic, histrionic, and antisocial) due to abrupt emotional dysregulation episodes and tendency to hetero- and self-aggressive behaviour. The risk of violent acts is greater in case of comorbid substance abuse, history of childhood abuse, personal and family history of violence.

The principal and most dangerous reason for emergency department referral is an episode of self-harm or suicidal behaviour. Such behaviours can be sustained by categorical suicidal ideation or have demonstrative purposes towards family or friends.

The clinical interview in acute settings should be focused on clarifying the presence of stressors or triggers for the present distress, assessing the presence of mood and thought disturbances, assessing suicidal ideation and verifying the presence of protective factors against future suicidal conducts, running laboratory and toxicological screening exams to rule out substance intoxication.

13.3.5.2 Management and Treatment

Based on the reason of referral, healthcare workers should assess vital signs, treat eventual intoxication, and administer drugs to reduce anxiety or angst: for this purpose, benzodiazepines are first choice drugs, orally or intravenously (e.g. delorazepam, lorazepam).

In case of main mood disturbances, persistent suicidal ideation and the absence of a supportive socio-familial network, hospitalization should be proposed to the patient.

13.3.6 Psychomotor Agitation

When a patient suffering from a physical or mental pathology changes their level of consciousness, have gross distortion of the judgment of reality, agitation, disorganized gestures up to aggressiveness towards objects or people, so that they express the imminent risk of disorganized or violent behaviour, family members, friends, neighbours and other specialty colleagues frequently ask for a psychiatric consultation.

13.3.6.1 Differential Diagnosis

- Substance intoxication: one of the most frequent is alcohol intoxication: the typical evolution starts from disinhibition and euphoria; then the person totally loses judgment capacity, with impulsive and hetero- and self-aggressive acts, road accidents, various injuries and trauma; then, the subject presents mood depression, and impulsive suicidal behaviour is not rare.
- Substance withdrawal: especially from alcohol. At the first stages of delirium tremens, there are severe alteration of consciousness or cognitive abilities.
- Delirium induced by substances: alcohol, anxiolytics, corticosteroids, atropine.
- Delirium due to medical condition: metabolic disorders, post-ictal states, head trauma, focal lesions.
- Mental retardation.
- Dementia: Alzheimer's or vascular type.
- Psychotic disorders.
- Mood disorders: in particular manic or mixed-mood episodes.
- Personality disorders.

13.3.6.2 Risk Factors for Violent Behaviour in Agitated Patients

- Young age.
- Male sex.
- Low intellectual, cultural, socio-economic level.

- Deviant family environment.
- Massive stressful events (childhood abuse).
- Resistance or poor adherence to drug treatment.
- A previous history of violent behaviours in the medical history (best probability indicator).

13.3.6.3 Violence Risk Assessment

The aim of psychiatric emergency consultation is to assess the patient and offer adequate treatment to reduce agitation. The patient might have been conducted in the ED by a family member, or by public force: if so, the patient might be even more threatened and agitated.

The assessment concerns the risk of short-term violence based on the available information. International guidelines on this regard help the clinical team reduce the risk of violence towards healthcare professionals and bystanders.

First, a quiet and comforting milieu should be assured, and the clinician should approach the patient calmly and preventing further escalation in psychomotor agitation. There should not be objects that can be used as a means of aggression, the clinician should keep an adequate distance from the patient avoiding to exhibit a threatening appearance. Moreover, the clinician should stand near a safe exit door, it should be easy to call for rapid intervention, if accompanying people increase the agitation they should be kept away from the room.

Verbal and motor signs that may indicate the possibility of violent behaviours are loud, threatening, or provocative speech, motor hyperactivity, tension; mydriasis; violent impulses or acts against objects.

The patient's requests should be discussed directly, paying attention to them in order to try to establish a collaborative relationship.

13.3.6.4 Management and Treatment

Psychopharmacological therapy can be usefully administered to sedate the agitated patient. A quick sedating therapy can be proposed, while the routine exams are running.

Once the urgency passes, the specific therapy of the condition can be evaluated and discussed with the patient.

- Agitation whose cause is unclear: benzodiazepines and antipsychotics.
- Acute alcohol or substance intoxication: antipsychotics.
- Delirium due to medical condition: therapy for the medical condition.
- Alcohol withdrawal: benzodiazepine.
- Psychotic disorder, manic phase: antipsychotics.

13.3.7 Suicide

Suicide is the result of a complex interaction of psychological, biological and social factors. The subject loses the habitual points of reference, feels anguished, frustrated, expresses feelings of hopelessness and helplessness. The will to die might be ambivalent: rather they would like to live but their anguish is more unbearable than ever. Up to 10% of suicides are committed by people with previous psychiatric hospitalization, but the greatest risk is conferred by a psychiatry disorder diagnosis independent of previous hospitalizations.

The psychiatric pathologies most frequently associated with suicide are:

- Major mood disorders: major depressions and bipolar disorders.
- Alcohol abuse.
- Schizophrenia.
- Borderline personality disorders.
- Antisocial personality disorders.

Around 1 million people commit suicide worldwide every year. It is a rare event in children under the age of 12 and becomes more common after puberty, reaching its peak after the age of 65.

Intense depressive experiences with suicidal ideation can be triggered by bereavement, particularly the loss of a partner or a very close person; this ideation can be more intense if the loss is accompanied by conditions of social isolation or dependence on institutions. Other common triggering events can be the breakdown of interpersonal relationships, separations or divorces, emigration. A typically younger age phenomenon is the so-called "Werther effect" (from the novel "The sorrows of young Werther" by W. Goethe): disclosure of suicide through mass media increases the suicide rate for the immediately following period.

13.3.7.1 Risk Factors for Suicidal Behaviour

- Male.
- Age 45–64.
- Anniversaries of particular significance.
- Unemployment or financial difficulties.
- Mourning, especially for spouse's loss.
- Recent separation or divorce.
- Recent arrest or legal troubles.
- Family or personal history of suicide, previous suicide attempts, drawing up detailed suicide plans, taking steps to implement the plan.
- Family history of mental disorder.
- Depressive episode, especially at major depression onset or in bipolar disorder.
- Significant motor agitation, restlessness and anxiety with severe insomnia.
- Marked feelings of guilt, inadequacy and despair; perception of being a burden to others; self-denigration; nihilistic delusion.
- Delusional ideation with somatic content (fear of being suffering from a serious or lethal disease) or of ruin.
- Personality disorders, especially borderline or antisocial.
- A chronic, painful, disabling physical disease, especially in previously healthy patients.

- Alcohol or drug abuse especially if recent use has increased.
- Use of drugs that can contribute to suicidal behaviours (e.g. abruptly stopping paroxetine and some other antidepressants can cause increased depression and anxiety, which in turn increase the risk of suicidal behaviours).

13.3.7.2 Assessment

Suicide is frequently the terminal act of a history of suicidal ideation, often communicated to others but not adequately considered. Patient management should aim at reducing social risk factors for suicide.

The physician in suspicion of suicide risk should listen carefully and refrain from any judgment; they should reconstruct the patient's history and verify the social and family support available to the patient. It is important to discuss the suicidal ideation with the patient without fear that this could increase the risk: conversely, it could be of help to explore hopelessness, anhedonia, insomnia, anxiety and psychomotor agitation, and the patient could feel relief seeing that their profound distress can be talked about and understood by the clinician. It often happens that the patient at risk of suicide presents with the following features: pervasive sadness, depressed mood, affirmations like "I wish I was dead", "I can't do anything", "I can't go on like this anymore", "I'm a loser", "Others will be better off without me". Anyone who threatens to harm or kill themselves, or looks for means (e.g. firearms, drugs), or talks about death, which is unusual for such a person, should be considered at high risk for suicide. Furthermore, a high risk of suicide is associated with feelings of despair, uncontrollable anger and impulsivity, acting recklessly or risky, seeking revenge, feeling trapped and without a way out. The risk is also associated with alcohol and drugs consumption; separation from friendships, family, and social contacts; besides anxiety, agitation and sleep disturbances are always identifiable in the presence of suicide risk. The individual at risk often reports marked changes in mood, lacks reasons to live, and cannot identify the meaning of life.

The clinician should focus on the patient's psychic status and physical examination (especially if there has been a suicide attempt). The most adopted suicide methods in Europe are hanging, defenestration, gunshot and drowning. In the United States, the use of firearms is more common, given the ease of obtaining guns and rifles in stores. Inhalation of vehicle exhaust gases appears to be significantly increasing among suicidal methods. Among the para-suicidal methods, on the other hand, drug overdose (especially benzodiazepines) and, to a much lesser extent, cutting (e.g. at the height of the wrists) are common self-harm methods.

In the case of suicidal behaviours, the psychiatrist might ask "Why now?" as to explore the latest vicissitudes that led to the act. Suicidal behaviour is typically one of the most frequent motivation for hospitalizing a patient: it permits the psychiatrist to observe the patient in a protective environment preventing further acts, to collect a complete history of the patient, to choose the best pharmacological and multidisciplinary treatment, to address any drug side effect, and to take time to help the person rebuild social relationships, or create new ones (e.g. getting in touch with family, friends, partners or children).

Further Reading

American College of Emergency Physicians Clinical Policies Subcommittee (Writing Committee) on the Adult Psychiatric Patient, Nazarian DJ, Broder JS, MEW T, Wilson MP, Zun LS, Brown MD. Clinical policy: critical issues in the diagnosis and management of the adult psychiatric patient in the emergency department. Ann Emerg Med. 2017;69(4):480–98. https://doi.org/10.1016/j.annemergmed.2017.01.036.

Baracaia S, McNulty D, Baldwin S, Mytton J, Evison F, Raine R, Giacco D, Hutchings A, Barratt H. Mental health in hospital emergency departments: cross-sectional analysis of attendances in England 2013/2014. Emerg Med J. 2020;37(12):744–51. https://doi.org/10.1136/emermed-2019-209105.

Barsky AJ, Orav EJ, Bates DW. Somatization increases medical utilization and costs independent of psychiatric and medical comorbidity. Arch Gen Psychiatry. 2005;62(8):903–10. https://doi.org/10.1001/archpsyc.62.8.903.

European Drug Report. Trends and developments: European Monitoring Centre for Drugs and Drug Addiction. Lisboa: EMCDDA; 2015. p. 2015.

Fleury MJ, Fortin M, Rochette L, Grenier G, Huỳnh C, Pelletier É, Vasiliadis HM. Assessing quality indicators related to mental health emergency room utilization. BMC Emerg Med. 2019;19(1):8. https://doi.org/10.1186/s12873-019-0223-8.

Nentwich LM, Wittmann CW. Emergency department evaluation of the adult psychiatric patient. Emerg Med Clin N Am. 2020;38(2):419–35. https://doi.org/10.1016/j.emc.2020.02.001.

Parmar P, Goolsby CA, Udompanyanan K, Matesick LD, Burgamy KP, Mower WR. Value of mandatory screening studies in emergency department patients cleared for psychiatric admission. West J Emerg Med. 2012;13(5):388–93. https://doi.org/10.5811/westjem.2012.1.6754.

Pluym ID, Holliman K, Afshar Y, Lee CC, Richards MC, Han CS, Krakow D, Rao R. Emergency department use among postpartum women with mental health disorders. Am J Obstet Gynecol MFM. 2021;3(1):100269. https://doi.org/10.1016/j.ajogmf.2020.100269.

Robinson RL, Grabner M, Palli SR, Faries D, Stephenson JJ. Covariates of depression and high utilizers of healthcare: impact on resource use and costs. J Psychosom Res. 2016;85:35–43. https://doi.org/10.1016/j.jpsychores.2016.04.002.

Sood TR, Mcstay CM. Evaluation of the psychiatric patient. Emerg Med Clin N Am. 2009;27(4):669–83. https://doi.org/10.1016/j.emc.2009.07.005.

Sweeny A, Keijzers G, O'Dwyer J, Stapelberg NC, Crilly J. Patients with mental health conditions in the emergency department: why so long a wait? Emerg Med Australas. 2020;32(6):986–95. https://doi.org/10.1111/1742-6723.13543.

Tucci VT, Moukaddam N, Alam A, Rachal J. Emergency department medical clearance of patients with psychiatric or behavioral emergencies, part 1. Psychiatr Clin N Am. 2017;40(3):411–23. https://doi.org/10.1016/j.psc.2017.04.001.

Urbanoski K, Cheng J, Rehm J, Kurdyak P. Frequent use of emergency departments for mental and substance use disorders. Emerg Med J. 2018;35(4):220–5. https://doi.org/10.1136/emermed-2015-205554.

Wheat S, Dschida D, Talen MR. Psychiatric emergencies. Prim Care. 2016;43(2):341–54. https://doi.org/10.1016/j.pop.2016.01.009.

Psychiatric Legislation and Forensic Psychiatry

14

G. Travaini, R. Zanardi, L. Fregna, F. Martini, D. Pratesi, A. Sarzetto, G. Perrozzi, and C. Colombo

14.1 Introduction

The aim of this part of volume is to provide some medical-legal and psychiatric forensic notions that could be useful in the future medical profession. The normative references are mainly to the Italian legislation even if some general concepts are common to the most modern legislations, for example the issue of the consent to a medical act or criminal responsibility.

It is easy to understand the reasons of the strong link between law and psychiatry. It is psychiatric knowledge that can help the legal operator to understand if someone has functioning cognitive abilities and is able to understand the value and the effect of their action. There are many contexts in which it is necessary to perform such evaluation, including civil law and criminal law. Of extreme importance is the issue of the evaluation of accountability of subjects who engaged in criminal behaviour, such that based on the psychiatric evaluation, in the eye of the law, the actor is a criminal or simply a subject with illness. This is a complex assessment that could have important consequences on a person's freedom. But also, to understand the ability of a person to carry out important acts such as making a will or deciding to

G. Travaini · D. Pratesi · A. Sarzetto · G. Perrozzi
Vita-Salute San Raffaele University, Milan, Italy
e-mail: travaini.guido@hsr.it; pratesi.donata@hsr.it; sarzetto.alessandro@hsr.it; perrozzi.giulia@hsr.it

R. Zanardi · L. Fregna (✉) · F. Martini (✉)
IRCCS San Raffaele Scientific Institute, Milan, Italy
e-mail: zanardi.raffaella@hsr.it; fregna.lorenzo@hsr.it; francesca.martini1@hsr.it

C. Colombo
Vita-Salute San Raffaele University, Milan, Italy

IRCCS San Raffaele Scientific Institute, Milan, Italy
e-mail: colombo.cristina@hsr.it

R. Cavallaro, C. Colombo (eds.), *Fundamentals of Psychiatry for Health Care Professionals*, https://doi.org/10.1007/978-3-031-07715-9_14

give the consent to a medical act. Finally, some information about professional secrecy will be provided. Any violation of this rule can have civil, criminal and ethical consequences.

14.2 Brief Overview About Accountability

In this chapter we will discuss basic principles involving the evaluation of accountability of subjects who engaged in criminal behaviour, referring to Italian law. Italian juridical terms are substituted by Anglo-American ones in order to allow a smoother reading. However, we acknowledge that Anglo-American terms do not always have the same meaning. Therefore, we report below a list of the juridical terms coupled with the Anglo-American translations that we have adopted.

Imputabilità → Accountability.
Infermità mentale → Insanity.
Capacità di intendere e di volere → Capacity.
Capacità processuale → capacity to stand a trial or fitness to plead.
Perito → expert.

Many psychiatric conditions can inhibit mental capacity, intellectual abilities and judgment, in different ways; therefore, in every case, it is crucial to determine whether a criminal act was committed voluntarily or whether a mental disorder reduced or removed the patient's mental capacity to act freely.

Hence, in the following paragraphs we will try to answer these questions

1. How can we establish if a subject who committed a crime was affected by a mental condition that prevented him to act freely?
2. If that's the case, is he/she punishable?

Historically and across different cultures, people have had to interact with the problem of criminal responsibly of mentally insane subjects.

Hebrew Scriptures, in the sixth century B.C.E., report that in some offences fault could not be imposed. This happened in those offences committed by children that were considered unable to understand moral consequences of their behaviour. Intellectually disabled and insane people were also considered as children, making them not punishable.

Later, in the ancient Rome there were two particular categories of subjects that were not punishable: "*furius*" and "*fatui*". "*Furius*" referred to mad people, whereas "*fatui*" were subjects with intellectual disabilities. However, since madness can be intermittent, a "*furius*" could have committed the crime in a moment of mental lucidity. Latin law was quite modern in this point and stated that if the crime was committed in a "*lucidum intervallum*" the punishment was not avoidable. Afterwards, in the sixth century, Code of Justinian implemented categories that were excluded from criminal punishment, which became: "*dementia*", "*insania*", "*fatuitas*", "*mania*" and "*amentia*".

Subsequently, in the twelfth century, in England, Lords of state began granting pardons to individuals who were convicted of a crime and obviously mad. Instead, they ordered the accused to commitment and treatment in a mental institution. Mental institutions, at the time, lacked adequate facilities and specific treatments were not available.

Later, in the nineteenth century we can find similar concepts in the Napoleonic Code. A particular article stated that if the subject at the moment of the action was affected by dementia or obliged by an exterior force, the crime could not be considered as such.

Coming to Italian history, Code Zanardelli in 1889 reported similar notions. Article 46 explained that it is not possible to punish someone who committed a crime when he/she was mentally insane. Moreover, this mental insanity had to be so intense that the subject was unable to act freely or even not conscious of his actions in that specific moment. In code Zanardelli we find an intermediate category between accountability and non-accountability: partial insanity. This concept refers to cases when insanity reduces, without completely abolishing, one's ability to act freely.

Currently, in Anglo-American society accountability in people with psychiatric conditions is regulated by "insanity defence"; this is a legal construct that permits defendants with mental illness to be excluded from legal responsibility of a criminal act, if determined conditions occur.

In Italy, law discusses this subject in its penal code, which is in force from 1930. In particular article 85 introduces the concept of "capacity". Capacity, in brief, is the determination of one's ability to act following his will and to understand the consequences of his behaviour. If, during a trial, capacity is found to be abolished, it is not possible to punish the subject who committed the crime.

In Italian law, capacity is formally divided in two concepts:

- Capacity to understand the value of the action (or omission), both legally and morally. Of note, ignorance does not affect one's capacity to understand. Consequently, it is not relevant if the subject ignores juridical consequences of his behaviour. Similarly, subjects who do not share the ethical and emotional value of a rule, are also punishable.
- Capacity to want or to decide, which refers to capability of self-determination in the moment of committing a crime. This reflects the fact that in certain cases, a mental condition can diminish, or abolish, one's ability to act freely. Furthermore, capacity to want requires cognitive and affective integrity, two elements often greatly affected by psychiatric conditions, reducing the subject's ability to choose his behaviour.

Aside from psychiatric disorders, there are other conditions in which accountability is excluded, for example the subject's age. Accountability is excluded a priori for subjects below 14 years old (art. 97 c.p.). Between 14 and 18 (art. 97 c.p.) accountability has to be considered for every single case. Instead, once the subject is 18 (art. 98 c.p.), accountability is presumed to be present. This concept is accepted legally but controversial from a moral and medical point of view. In other words, there is a very slight physical and mental difference between subjects of 17 and 18 years old, but their punishment for the same crime can be enormously different.

Other cases when accountability is excluded are:

- When incapacity is induced by others (art. 86 c.p.)
- When the subject is deaf-mute (art. 96 c.p.)
- In case of accidental drug or alcohol intoxication (art. 91, 93 c.p.)

Of note, all these conditions share the fact the individual had no choice; therefore free will is compromised. This is either because capacity is compromised by an uncontrollable external factor, or because of an invalidating previous condition which the subject did not choose (i.e. being deaf-mute).

Nevertheless, the most common cause of exclusion of capacity is mental insanity (art.88 c.p.), characterized by a psychiatric/mental disorder, able to narrow one's ability to control, inhibit and choose his behaviour, compromising his intellectual abilities. However, the ways psychiatric diseases affect one's mind are extremely heterogeneous; this makes necessary the existence of a third, intermediate condition between capacity and incapacity, to include the cases when there are some residual intellectual abilities. This is called partial insanity and, when present, can mitigate the sentence (art. 89 c.p.).

It is important to clarify also the role of emotional and passion states; these conditions, like psychiatric disorders, can modify transiently one's psychological balance. However, they are not pathological, hence they are not considered when doing capacity evaluation (art. 90 c.p.).

The final decision regarding one's capacity is done by the judge. However he/she has not the legal and substantial ability to formulate a psychiatric diagnosis, therefore, the judge asks an expert to perform the evaluation. Since in the majority of the cases the evaluation is related to a psychiatric condition, the expert is a medical doctor or a psychologist; the evaluation is a complex process, involving both law and medicine, which have different languages, theories and needs. The final report is a scientific document and, as such, it uses a nosography shared by the scientific community, coming from international psychiatric manual/classifications. The most commonly used are DSM and ICD (Diagnostic and Statistical Manual of Mental Disorders, International Statistical Classification of Diseases and Related Health Problems). The use of a comprehensible and shared language is also important since psychiatry is a heterogeneous science, with different possible interpretations of the same condition. Consequently, the use of common diagnostic criteria permits to have more objective evaluations and also allows the judge to control reliability and soundness of the evaluation.

During this process, the expert collects data about the subject (medical anamnesis, family, job and affective relationships) and can have access to useful judicial acts. Psycho-diagnostic tests can also be performed. Once all the information is gathered, the expert has the role of understanding and communicating:

- if the subject, at the moment of the crime, was affected by a psychiatric illness;
- if there was a causal association between the psychiatric condition and the crime.

Once the evaluation is completed, it is assessed by the judge to decide on the defendant's accountability. As mentioned before, psychiatry is widely heterogeneous, and evaluation varies depending on different factors: the expert formation, his subjective view, the experience he has in a determined psychiatric condition. However, an expert can be hired by the defendant attorney as well, so that the judge has to take in consideration also a second opinion.

To obviate the problem mentioned above regarding different interpretations in psychiatry, a new and more objective method of judging was sought and this need has been partially fulfilled by neurosciences. Consequently, the use of neuroscientific findings progressively gained importance in the evaluation of insanity. Various neuroimaging techniques were discovered and permitted to have a precise insight of our brain structures: computed tomography (CT), magnetic resonance (MR), positron emission tomography (PET) and single photon emission tomography (SPECT). Moreover, new sequencing techniques widened our genome understanding. In particular, it was noted that precise gene variants have an influence on our behaviour. Some studies even explored the neural basis of free will, and some scientists postulated the non-existence of free will. These theories raised scepticism and critics and currently free will is still a highly controversial subject in neuroscience.

The free will concept is strictly bound to criminal responsibility, especially in the context of mental disorders. For example in a patient affected by schizophrenia free will is probably compromised when auditory imperative hallucinations order him to do something. However, we have seen that free will is a highly debated concept and its evaluation is not contemplated in most countries' standards for insanity defence. Hence forensic psychiatry used neurosciences more in relation to a second concept, which is decision-making, whose neural mechanisms are currently more studied and understood. Decision-making can be conceptualized as the ability to select one course of action amongst several possible options. Neuroscientific literature concerning decision-making process has been widely used in evaluation of criminal responsibility and had a huge impact on judgment of mentally ill defendants.

As an example we report the famous case of S.A. who in 2009 segregated the sister and obliged her to take a huge dosage of benzodiazepines. The amount of the sedative drug was so high that she died, and S.A. burnt her corpse. While she was suspected of murder, she also attempted to kill her mother suffocating her with a belt. Police intervened quickly, saved the mother and arrested S.A. During the trial, traditional psychiatric evaluations found a dissociative disorder; however what was crucial in establishing her capacity, was the neuroscientific evaluation that revealed the presence of alterations of grey matter and of three genetic alleles associated with aggressive/impulsive behaviour. Based on these findings, partial insanity was then recognized and the punishment was reduced to 20 years of imprisonment.

However, neuroscience potential was perhaps overestimated. In fact, neuroscience is contributing enormously to the understating of our brain processes, but at the moment we are far from a full comprehension of the neural basis of human behaviour. The way we choose our behaviour is extremely complex and involves the interaction of genetic, neurological, neurodevelopmental and environmental factors. Therefore, the mere consideration of determined alleles of structural alterations appears insufficient to decide if someone committed a crime voluntarily or not. This brought "Corte suprema", the maximum decisional court, in 2019, to highlight the importance of relating neuroscientific analysis to other environmental considerations. In this judgment, there was also a warning to find a better harmonization with traditional methods of investigating capacity.

Finally, for sake of completeness, we report two concepts, which are strictly connected to accountability: "capacity to stand a trial" and "capacity to testify". These share with accountability the fact that they require an evaluation of one's intellectual/psychic abilities; however, in these cases, the evaluation considers one's capability of absolving different tasks, which are part of the trial.

"Capacity to stand a trial" is the ability to participate consciously to the trial. In order to do so, a defendant should be able to understand his rights and faculties. Someone may think that a subject, who is judged mentally insane at the time of the crime, is automatically incapable of standing a trial; however, Italian law rejects this assumption, and if the defendant is thought to be incapable, the judge is obliged to request another evaluation. In fact, "capacity" and "capacity to stand a trial" refer to two completely different moments. The former regards only the moment when the subject committed the crime; the latter regards the timespan of the whole trial. Between these moments, one's mental state may have changed completely. Of note, also the abilities to understand the value of his behaviour (and to act freely) are different from those required to participate to a trial. Furthermore, if someone is judged not to have capacity to stand a trial this may not be a definitive decision. In fact, because of the non-chronic nature of some psychiatric condition, incapacity to stand a trial can be reversible. In other words, if the judge thinks that the defendant can recover from his mental condition, the trial is postponed and capacity to stand a trial is periodically re-evaluated. Conversely, if the mental condition, according to the current scientific knowledge, is irreversible, incapacity to stand a trial is also irreversible. This is the only case in which the trial is definitively closed.

The final concept we want to discuss is "Capacity to testify". It refers to one's eligibility for witnessing. A "capacity to testify" evaluation is requested by the judge, in order to establish if a witness is able to testify reliably. There are two principal reasons for requesting this kind of evaluation. The first one is the presence of a mental condition that could make the witness unreliable. The second is the case of an underage witness. Hence, depending on the reasons why the judge requested the evaluation, the expert has to make different kinds of considerations. Concerning psychiatric/mental conditions, the expert will explore if a causal relation exists between the disease and subject's unreliability. Conversely, in case of underage witness, the expert attention will be more focused on other aspects: one's suggestibility and one's capacity to differentiate between reality and fantasy. It is really important to remember the complexity of this type of evaluation which must be carried out with high level of competence and prudence.

14.3 Brief Overview of Informed Consent, Obligatory Treatment and State of Necessity

14.3.1 Informed Consent

Consent, in law, is a voluntary agreement with an action proposed by another person. In medicine, this action is represented by a treatment or a procedure that a physician considers for the patient's diagnosis or treatment.

Before performing the treatment/procedure, the doctor must obtain the patient's consent. Since most patients do not have a sufficient understanding of their medical condition and treatment options, the doctor should give first accurate explanations so that the patient is fully informed.

Informed consent in medicine has gained more and more importance in the last decades. In fact, according to the ancient paternalistic conception, the patient had to completely rely on his doctor, who acted with his conscience and knowledge to protect his patient's health. Then, patient–doctor relationship modified, becoming progressively more balanced. Today, it is stressed the importance of the subject to freely make decisions about his own health. The juridical explanation relies on article 32 of Italian constitution, which states that no one can be obliged to any medical procedure, however, with the exception of mandatory commitments.

In the current doctor–patient relation, it is fundamental to obtain informed consent before every medical procedure. Otherwise, in case of absence of consent or whether the doctor did not give enough explanations, the physician could face serious legal consequences. Firstly, in case of damage he has to compensate the patient. Secondly, sometimes the matter has also penal relevance: for Italian criminal law, a doctor can be accused of personal injury, voluntary or culpable.

Coming back to obtainment of informed consent, there are fundamental requirements for its acquisition. Indeed, the patient must:

- Have sufficient mental capacity, which is the competence to understand and to decide voluntarily.
- Be in possession of all essential information: diagnosis, nature of the proposed treatment, potential risks and the probability of success, medically recognized alternative measures, consequences of patient's decisions.
- Have the ability to act, which is acquired at 18 years old. It is the capability of the subject to acquire and exercise subjective rights and to assume obligations.

Characteristics and requirements of informed consent:

1. Personal (non-delegable and unavailable).
2. Knowledgeable and informed.
3. Current (therefore also revocable).
4. Manifest, clear and unequivocal.
5. Freely given.

6. Not imposed and free from mistake, violence and malice.
7. Complete.
8. Free of charge (not provided for payment or as consideration for a favour or advantage).
9. Receptive (takes effect when the recipient-therapist becomes aware).
10. Requested (the professional must request consent).

Furthermore, understanding and giving consent is a complex action, where a series of abilities are required: competence to understand and to decide, voluntary decision-making, comprehension of terms of decision in favour of a plan, authorization of the plan. In psychiatry there are several conditions that might affect these abilities. For example, in schizophrenia patients experience with the time a progressive deterioration of their intellectual abilities making them possibly unable to understand given information. Bipolar patients during a euphoric phase can underestimate the risks of a medical procedure. Delusional patients may refuse a treatment if they believe the existence of a conspiracy to harm them. Moreover, there are neurological conditions that hamper comprehension, such as cognitive disorders or mental retardation.

What happens if the subject is not able to give consent? There are some protective measures depending on the degree of incapacity. These are, in the Italian law, three legal institutions of protection and support for those who are unable to carry out actions.

- Interdiction: It is adopted in case of habitual and definitive condition of incapacity. It is established by a judge following a careful evaluation of clinical documentation and with possible use of a consultant. In this case a guardian ("tutore") is appointed.
- Incapacitation: It is used in less severe clinical conditions. It is established by the judge. The subject has the possibility to carry out actions with limited legal effects. It can be applied to regular alcohol and drug users to protect the assets. A guardian ("curatore") is appointed.
- Support administration: A support administrator is appointed by a judge, when the person who, as a result of an illness or a physical or mental impairment, is unable, even partially or temporary, to look after his own interests.

When these legal institutions are adopted, the guardian/administrator becomes legally responsible for signing informed consent. He becomes the principal interlocutor of the doctor regarding patient's health, even if it is often helpful to involve the patient in the decisional process.

Finally informed consent is not specific for medical doctors, as in recent years this procedure was extended to other health professionals like nurses and physiotherapists. However, in the particular field of psychiatry various professionals usually work in a team; in this case the consent acquired by one is extended to the other members.

Psychiatric intervention themselves are medical treatments; hence they also require informed consent. In this case the psychiatrist has the duty to explain the treatment with particular attention to the patient psychic abilities. If during the medical interview, the psychiatrist understands that the patient is not able to give his consent, the treatment has to be postponed unless one of the previously mentioned measures is already active.

Informed consent in psychiatry assumes peculiar characteristics. Informed consent is based on the fundamental and unavoidable assumption that the patient has understood the nature of the illness from which he is suffering, the possible treatments, and the advantages, disadvantages and risks of each of them. It is assumed that to do so, the subject:

- has an awareness of the disease and its treatment;
- is able to make a free choice between the various options.

In the case of mental illness, both of these assumptions can sometimes be compromised. It is possible to encounter situations in which, in addition to the unwillingness of the subject to be treated, there is a real risk that the patient may engage in conduct detrimental to himself and others. Not perceiving himself as sick and in need of care, there could be a refusal of therapeutic options. In this case, not only the treatment but the entire therapeutic/rehabilitative pathway is questioned. The refusal in a patient in a manic state, for example, is to be considered an act of freedom that the patient makes? Is it a refusal that can be considered valid, since it is consciously expressed by the patient? These questions already contain an answer: In these cases, the patient's freedom of choice is not influenced by external events, but is compromised by the pathology itself. The same observation can be made in the case of a patient who presents paranoid delusions and identifies the medication as the poison provided by the persecutors. Another example is anorexia. These patients often, obsessed by thinness, not perceiving the severity of this, and not aware of the seriousness of their physical health condition, categorically refuse to eat, implement extreme calorie restriction, and sometimes end up dying of malnutrition. The application of Compulsory Health Treatment (we will refer to the involuntary commitment or compulsory health treatment with the Italian acronym TSO, which stands for *Trattamento Sanitario Obbligatorio*) in anorexia nervosa is an issue that has been debated for several years. Some authors have argued that subjecting the patient to forced nutrition may compromise the therapeutic relationship and have an uncertain impact on the long-term course and outcome of anorexia nervosa, while others have argued that coercive treatment such as anorexia nervosa should be considered as compassionate treatment and is therefore justified, even though it may only provide short-term benefits. Nonetheless, even though awareness is not always present in psychiatric patients, this should not prevent health professionals from proceeding with information that takes into account the patient's state of illness and leads to the transformation of passive adherence into possible consent to treatment.

But what happens when the patient needs treatment and does not give consent?

In the psychiatric field in Italy, this question was introduced with the law of December 23, 1978, n. 833, art. 34 with the very significant title "Voluntary and compulsory health checks and treatments for mental illness", which we will discuss later. Although psychiatric patients can have recourse to compulsory health treatment, based on art. 34 and 35 of n.833/78, health professionals must seek the patient's consent to treatment from the very beginning, to fall, in all cases where it is possible, within the general rule.

The committee emphasized the gradual and changing nature of the capacity or incapacity to make decisions even in the case of psychosis, since between absolute incapacity, typical of dementia, and "normality" there are a series of intermediate degrees where cognitive deficits and affective alterations can determine reductions, but not the absence of capacity. This, of course, cannot legitimize the renunciation of information to the psychiatric patient, but it does imply a greater caution in evaluating case by case the appropriate measure for the individual patient concerning his or her situation.

In general, from a clinical point of view, the first element to be taken into consideration is the presence/absence or partial presence of awareness of the illness. What establishes the validity of consent, is not so much the diagnostic framework itself, as the current clinical condition, derived by the psychopathological state, the developmental stage of the disorder and the mode of response to the therapeutic context.

The Oviedo Convention (1997) establishes the necessity of the consent of a "representative" in case the patient is unable from expressing himself. "When, according to law, an adult, because of a mental handicap, illness or similar reason, cannot give consent to an intervention, the intervention may not be carried out without the authorization of his representative, an authority or a person or body designated by law" (Article 6).

The Medical Code of Ethics of 2014 (Article 33) provides that the doctor must inform the patient, taking into account the patient's ability to understand, but without omitting information.

Deontological recommendations that must be kept in mind to operate correctly in the psychiatric field (but the rule is generalizable to all medical action, generic or specialist) can still be derived from the other articles of the Medical Code of Ethics. Articles 3, 4, 5 and 17, for example, remind everyone—albeit indirectly—of a basic attitude for the process of knowledge and intervention: neutrality and respect for the fundamental rights of the person. Recommendations that we can translate into the following obligations: maintenance of emotional distance, abstention from manipulative techniques and intrusive practices, careful avoidance of any passage to the act. Article 18 recommends competence, utmost care and commitment. The professional style, in other words, must be made of expertise, diligence, ethics and adherence to precise theoretical and methodological references. Articles 15 and 50, finally, prohibit the doctor to participate or collaborate or implement treatments that "affect the physical or mental resistance of the patient" or that are "cruel, inhuman or degrading".

Compulsory Health Treatment for the mentally ill must therefore be considered an exceptional eventuality, a derogation expressly authorized by law to the principle of necessity.

For the sake of simplicity, it seems appropriate to start with a simple definition of compulsory medical treatment: it is an administrative procedure that allows hospitalizing a patient for medical treatment against their will but in their own interest.

14.3.2 Obligatory Treatment

Law 180/78 mainly covers compulsory medical treatments of a psychiatric nature; still, it includes mention of several other cases in which it may be required, such as Infectious Diseases.

Art.33 of law 833/1978 was meant to guarantee the principles of the Italian Constitution concerning the freedom and inviolable rights of all citizens. It states that "all tests and treatments are voluntary" and that "health authorities may order involuntary tests and treatments in respect of the dignity of a person and of his civil and political rights as guaranteed by the Constitution". Therefore, the compulsory psychiatric treatments are to be considered as an exception, authorized by law, and to be employed as an extreme solution to be chosen only once all other means of obtaining consent have been endeavoured. An additional guarantee of the right of freedom of the subject is imposed by the Legislator in stressing that compulsory health treatment, although it is a coercive measure, in no way impairs the "civil and political rights guaranteed by the Constitution"; such rights concern also to "the free choice of a physician and place of treatment" (art.33, 2) and the patient's right "to communicate with whomever he may wish" (art.33, 7).

Article 34 of law 833/78 dictates that treatment "should take place in a hospital regime only in those cases where psychic conditions are such as to require urgent therapeutic intervention if treatment is not consented to by the patient, and where circumstances do not allow the adopting of timely and adequate measures in an outpatient regime" (art.34, 4). It is to be noted that not all psychiatric facilities present in the area are accredited to perform this function.

The territorial outpatient departments are designated to provide not only for treatment, but also for diagnosis and prevention of psychiatric illness by using a network of facilities capable of guaranteeing an adequate therapeutic continuum.

Article 35 of law 833/78 establishes the elaborate bureaucratic process underlying compulsory psychiatric treatment. Following a motivated proposal from a physician (generally from outside the structure where it has to be carried out), sustained by a second physician (working in a public structure), the Mayor, as local highest health authority, will authorize patient placement and notify the Tutelary Judge within 48 h.

The compulsory psychiatric treatment can be carried out without resorting to hospitalization: at the Mental Health Centre, the clinic, the patient's home, the emergency room.

If hospitalization is required, it can only be performed at the Psychiatric Diagnosis and Care Service of the Health Authority. In other words, not all psychiatric facilities present in the area are accredited to perform this function.

If the motion is to be applied to foreign citizens or stateless subjects, the Prefect will notify the Home Ministry and the Consulate concerned.

The order can only be issued if three conditions are present at the same time:

- Necessity and urgency that cannot be delayed.
- The subject refuses the intervention of the health workers.
- It is not possible to take appropriate extra-hospital measures.

The duration of compulsory psychiatric treatment cannot exceed 7 days, at the end of which the physician in charge may request an extension, if the conditions of the patient do not allow his voluntary placement or outpatient treatment. The Mayor should be notified of the interruption of compulsory psychiatric treatment by the physician in charge of the psychiatric services; the above will subsequently notify the Tutelary Judge of the same within 48 h.

Art. 35 also provides for the immediate interruption of compulsory psychiatric treatment if the correct procedures have not been performed, foreseeing the offence of omission in public proceedings (art. 328 of the penal code). Whosoever (either the person who has undergone it or any other person involved) may appeal against the motion approved by the Tutelary Judge. These dispositions are aimed at providing a further guarantee of the citizen's rights in the face of possible errors or abuse of the Health Authorities. The Court Magistrate will hear both parties and will decide on the request for suspension of the compulsory psychiatric treatment within 10 days.

One of the more controversial features about the application concerns the legality of using methods of physical or pharmacological containment on the subject undergoing placement.

If the restraining measures are improperly operated or are not connected in the very short term with a therapeutic programme, they can represent an unacceptable limitation of personal rights and an unnecessary application of violence that could expose the physician to the crime of:

- Private violence (art. 610 c.p.), which sanctions the behaviour of "anyone who compels with violence or threat others to do, tolerate or omit something" and which provides for the sanction of imprisonment of up to 4 years. This rule protects the free will of each individual, obviously also the freedom to accept or not health care.
- The violation of art. 613 c.p., state of incapacity procured through violence, which establishes behaviour as a crime of "Anyone, by hypnotic suggestion or by administering alcohol or drugs, or any other means, places a person, without consent, in a state of inability to understand and want". The punishment is imprisonment for up to 1 year.

- Abduction (art. 630 c.p.), which provides the punishment of the conduct of "Anyone deprives someone of personal freedom and is punished with imprisonment from 6 months to 8 years".

14.3.3 State of Necessity

There may be circumstances in which the patient has to be restraint before fulfilling the procedures for compulsory medical treatment. In such situations, the conduct of the doctor will be sustained by art. 54 of the Criminal Code, concerning acting in case of the state of necessity.

The State of necessity (art. 54 Penal Code.): "It is not punishable if committed for having been forced by the need to save himself or others from the current danger of serious harm to the person; such danger is not voluntarily caused by him, nor otherwise avoidable, given that the fact is proportionate to the danger".

Hence, the state of necessity is a hypothesis of force majeure stating that it is not possible to defend oneself without harming the rights of others. First of all, it is possible to act in a state of necessity only to avoid serious harm to the person. Such harm, in order to be free from responsibility, must be ongoing and not voluntarily provoked by the agent.

Finally, the danger must be inevitable; it is not possible, unlike in the case of self-defence, to invoke the state of necessity when danger could be avoided by escaping.

In the psychiatric field, the state of necessity justifies physical or pharmacological retention against the consent of the patient. The retention must be kept under the direct supervision of the doctor, explained with proper motivations, circumscribed over time and recorded in the patient's medical record. Since this is a borderline measure and in an area where there are still disputes fuelled by antipsychiatry pressures, it is good practice for the Primary to request a specific report on the reasons that led to the measure. The collection of such documents is a tool for monitoring incidents of violence and acts of detention to guarantee patients and psychiatrists. If the clinical picture requires an extension of physical retention, the use of TSO becomes inevitable.

14.4 Brief Overview of Professional Secrecy

In the course of their work, health care professionals often learn information about patients. Generally, this information must be treated in confidence and must not be disclosed. The presence of professional secrecy allows the patient to freely confide information about his or her state of health without fear of it being revealed to others. Professional secrecy has always been linked to the profession of the doctor, so much so that it is listed in the principles of the Hippocratic Oath:

> …And whatsoever I shall see or hear in the course of my profession, as well as outside my profession in my intercourse with men, if it be what should not be published abroad, I will never divulge, holding such things to be holy secrets.

More generally, the issue of data privacy is common in most countries. In Europe, the reference law is the Privacy Regulation 2016/679, more commonly referred to as GDPR (General Data Protection Regulation) where at article 9 we find the part dedicated to the correct way to manage health data. In addition, with regard to Italy, there has been a precise provision on the subject of the Guarantor of Privacy: the provision No. 55/2019 to which we refer for further information.

Professional secrecy is particularly important in psychiatry, since mental health is a particularly sensitive topic for most patients; in fact, even nowadays, the fact of undergoing a mental health treatment (either psychological or psychiatric) can be a reason for stigmatization by the context in which they live. Therefore, the patient can easily present resistance to open communication; this resistance can certainly be alleviated when the patient is reassured about the presence of professional secrecy. Therefore, during the first interview, the psychiatrist must assure his patients, for both legal and ethical reasons, that he will not reveal anything he learns in the course of therapy. By law, confidentiality must be maintained, both while the patient is alive and after his or her death. Disclosure of information protected by professional secrecy is an extremely serious matter and falls under the heading of so-called malpractice, i.e. a case in which a patient is harmed by poor management of medical procedures.

The consequences of this violation can be very serious and in order to better understand them, it is useful to know more about the type of responsibility linked to professional secrecy. At the regulatory level, it is mainly divided into two types: judicial and deontological. The first responsibility derives from the violation of rules of the penal and civil code and the doctor/health professional can therefore incur in a trial and a sentence; from the deontological point of view instead, the violation is examined by the professional order that can decide which measures to adopt which can vary considerably from economic fines to the expulsion.

There are, however, exceptions, when it is not punishable to reveal the secret. These are valid only from the criminal point of view, while from the deontological point of view the disclosure of the secret is never allowed. It is important to consider then, that in some cases the health professional is in a controversial situation because he is legally authorized (or even obliged) to reveal the secret, and at the same moment he is danger of incurring in a punishment from his order. It is advisable in these cases to act with consciousness and following one's ethics.

In any case, disclosure of professional secrecy is not punishable by the law when:

- It is required by law: including reports, lawsuit of public officials, mandatory certifications, reports, certificates, etc.
- It is authorized by the interested party.
- The disclosure is requested by the legal representatives of the minor or the incapacitated person in the interest of the latter. Regarding the information requested

by the parents of the minor and the guardian, the doctor is not obliged to disclose the secret if he does not consider it appropriate. This principle is particularly valid in all cases in which parents request information on the sexual life of their son/daughter who have reached the age of 14 (in the case of a daughter between 14 and 18 years of age, on her possible state of pregnancy or if they express a request for its interruption).

Amongst the cases foreseen by the law, we recall, once again, the state of necessity. In fact, a medical doctor may recognize in the words or in attitudes of his patient, a danger for the patient himself or of third parties. The danger must be actual and the damage highly probable or certain. In psychiatry this always regards patients who express intentions of harming their selves or others. For example, a psychotic patient, with delusions of persecution or jealousy, may show reliable projects of destructiveness towards certain people. In this case, while the T.S.O. is pending or if the patient is untraceable, it is necessary to inform the possible victims and the authority of public safety. In these cases, the disclosure of professional secrecy is not only unpunishable, but in case the patient or other persons should be harmed, the doctor can be prosecuted.

It is different in the case where one learns of a crime already committed. Obviously, in this case, the state of necessity does not apply. However, it is the duty of a physician working in the public facility to report a crime if he becomes aware of it during his profession. This occurs for a particular type of offences called ex officio prosecutions. These terms mean that in order for a criminal proceeding to be initiated, it is not necessary for the victim to make a complaint. For example, if during an interview a minor patient reveals to his doctor that he has been the victim of sexual violence, the doctor is obliged to report the fact to the competent authorities. In addition to the non-punishability for disclosure, the doctor is required by law to break the bond of professional secrecy and in case of inaction, he may incur in the crime of omission of report.

Another case in which the revelation of the secret is allowed is the request of a judge. In fact, the penal code states that in the case where there is a legitimate order from the public authority, the disclosure of the secret is not punishable. However, again, the Penal Code specifies that the practitioners of health professions cannot be obliged to testify about what they learned during their profession.

14.5 Conclusion

At the end of this chapter, it is evident the importance of interconnections between psychiatry and law. It is important to know and understand these judicial connections for many reasons: for the protection of the patients, to interact well with the legal bodies and to reduce wrongful actions in your profession. These are necessary simplified information but, in the end of the chapter, you can find further readings to develop the issue.

Further Reading

Cary P. A brief history of the concept of free will: issues that are and are not germane to legal reasoning. Behav Sci Law. 2007;25:165–81.
Cramton RC, Knowles LP. Professional secrecy and its exceptions: Spaulding v. Zimmerman Revisited. Minn Law Rev. 1998;83:63.
Gentilomo A, Barbieri C. Medicina Legale per I giuristi. Milan: Giuffrè editore; 2021.
Giorgi-Guarnieri D, Janofsky J, Keram E, Lawsky S, Merideth P, Mossman D, Zonona H. AAPL practice guideline for forensic psychiatric evaluation of defendants raising the insanity defense. J Am Acad Psychiatry Law. 2002;30(2 Suppl):S3–40.
Goldner E. Treatment refusal in anorexia nervosa. Int J Eat Disord. 1989;8(3):297–306.
Grisso T, Appelbaum P, Hill-Fotouhi C. The MacCAT-T: a clinical tool to assess patients' capacities to make treatment decisions. Psychiatr Serv. 1997;48:1415–9.
Gutheil TG, Appelbaum PS. Clinical handbook of psychiatry and the law. Philadelphia: Lippincott Williams & Wilkins; 2019.
Helm RK, Ceci SJ, Burd KA. Unpacking insanity defence standards: an experimental study of rationality and control tests in criminal law. Eur J Psychol Appl Legal Context. 2016;8(2):63–8.
Iller J, Schmidt U, Treasure J. Compulsory treatment for anorexia nervosa: compassion or coercion? Br J Psychiatry. 1993;162:679–80.
Luchins DJ, Cooper AE, Hanrahan P, Rasinski K. Psychiatrists' attitudes toward involuntary hospitalization. Psychiatr Serv. 2004;55(9):1058–60.
Merzagora I. Colpevoli si nasce? Criminologia, determinismo, neuroscienze. Milan: Raffaello Cortina Editore; 2013.
Merzagora I, Amadasi A, Blandino A, Travaini G. The expert and the foreigner: reflections of forensic transcultural psychopathology on a total of 86 reports by experts on criminal liability. Int J Law Psychiatry. 2018;57:24–30.
Moser DJ, Schultz SK, Arndt S, et al. Capacity to provide informed consent for participation in schizophrenia and HIV research. Am J Psychiatry. 2002;159:1201–7.
Parmigiani G, Mandarelli G, Meynen G, Tarsitani L, Biondi M, Ferracuti S. Free will, neuroscience, and choice: towards a decisional capacity model for insanity defense evaluations. Riv Psichiatr. 2017;52(1):9–15.
Ponti G, Merzagora I. Compendio di criminologia. Milan: Cortina; 2008.

Correction to: Fundamentals of Psychiatry for Health Care Professionals

Roberto Cavallaro and Cristina Colombo

Correction to: R. Cavallaro and C. Colombo (eds.), *Fundamentals of Psychiatry for Health Care Professionals*, https://doi.org/10.1007/978-3-031-07715-9

This book was inadvertently published with the incorrect affiliations of Roberto Cavallaro and Cristina Colombo. This has now been amended in the Frontmatter.

The correct affiliations are given as follows.

Roberto Cavallaro, Vita-Salute San Raffaele University, General Psychiatry Unit, IRCCS San Raffaele Scientific Institute, Milan, Italy

Cristina Colombo, Vita-Salute San Raffaele University, Mood Disorders Unit, IRCCS San Raffaele Scientific Institute, Milan, Italy

---

The updated online version of the book can be found at
https://doi.org/10.1007/978-3-031-07715-9

R. Cavallaro, C. Colombo (eds.), *Fundamentals of Psychiatry for Health Care Professionals*, https://doi.org/10.1007/978-3-031-07715-9_15

MIX
Papier aus verantwortungsvollen Quellen
Paper from responsible sources
FSC® C105338

If you have any concerns about our products,
you can contact us on
ProductSafety@springernature.com

In case Publisher is established outside the EU,
the EU authorized representative is:
Springer Nature Customer Service Center GmbH
Europaplatz 3, 69115 Heidelberg, Germany

Printed by Libri Plureos GmbH
in Hamburg, Germany